The Art and Science of 3D Audio Recording

Edwin Pfanzagl-Cardone

The Art and Science of 3D Audio Recording

Springer

Edwin Pfanzagl-Cardone
Sound-Engineering and Acoustics
Department
Salzburg Festival
Salzburg, Austria

ISBN 978-3-031-23045-5 ISBN 978-3-031-23046-2 (eBook)
https://doi.org/10.1007/978-3-031-23046-2

This Springer imprint is published by the registered company Springer Nature Switzerland AG
The registered company address is: Gewerbestrasse 11, 6330 Cham, Switzerland

Preface

This is—again—a book written by a sound engineer for sound engineers, which takes not only a look at the theory behind '3D' audio, but also at the practical implementations in the form of microphone techniques for 3D, as well as the most common software platforms which can be used for working on 3D audio.

Strictly speaking, using the term '3D audio' in connection to an array of replay speakers arranged along the x, y and z planes is wrong in itself, as of course a fourth dimension—time—is needed for any sound event to be able to 'happen'. So—maybe—it should be termed '4D audio' instead … but this change is not very likely to happen. (Also, the term 4D seems to have been adopted by theme parks for their use in movie theatres, displaying 3D movies, enhanced with tactile stimuli for the viewers.) However, as author of this book I prefer '3D audio' over the use of 'immersive audio', as I would already consider surround recordings to be 'immersive', and therefore no real distinction to 3D audio.

In the world of theatre, having venues with a 3D audio speaker system available (usually even with one—or more—overhead 'Voice of God' speaker(s) integrated in the ceiling) is common since decades, so '3D' audio was an 'everyday reality' already since many years in those venues.

Another milestone in terms of audio-visual 3D sound happened in the cinema theatre already in 1940—by now 80 years ago—with the creation of Walt Disney's 'Fantasia'. This was the first movie ever to employ an extensive multi-channel sound-track system, which—despite several different technical implementations developing from 'Mark I' to 'Mark X' over the years (for more details, see https://en.wikipedia.org/wiki/Fantasound)—was mainly based on three tracks with audio information and one control track, which would steer the replay volume, but also sound-track positioning in the room, on front, rear and ceiling speaker(s) in later versions. As such, this 'Fantasound' system, as it was called by Walt Disney and its developer, chief audio engineer William E. Garity, was not only the first stereophonic, but also the first 'surround' and also first '3D audio' system with object-based moving sound, realized with analogue technology …

In 3D audio systems, of course not only the specific microphone positions for 3D audio recording are important, but equally so the layout of the loudspeakers for 3D audio reproduction. Therefore, it is also necessary to examine the current proposals with respect to 3D audio, namely Dolby® Atmos™, DTS:X®, Auro 3D®, but also such individualistic approaches as that of the 'Isosceles Triangle'-based MMAD system by Michael Williams (see Chap. 3).

'Quadraphony' in the 1970s was a first attempt at trying to bring a more 'immersive' sound aesthetic to the homes, at least in the context of recorded music. Unfortunately, as we know, this attempt has failed. When 'immersive sound' came along the second time in the form of 'surround sound', the speaker arrangement and channel layout was mainly dictated by needs or conventions already found in the cinematic industry, with a strong emphasis on the front hemisphere, by providing (at least) a L, C, R front loudspeaker system, which did not find equal balance at the rear or at the sides.

Had it been primarily been with a seamless reproduction of moving sound objects in mind, proposals for new 'surround' speaker layouts would likely have had to be more of a symmetrical kind, e.g. with one loudspeaker at every 60 degrees of the full circle (for 2D surround).

With respect to 3D audio systems, Michael Williams' 'MMAD 3D' (Multichannel Microphone Array Design) stands out positively in that it does not adhere to the cinema-standard driven emphasis on a L, C, R system which neglects a balanced splay of speakers in the rear and on the sides, but he rather goes for his own design using side and rear loudspeakers, which partly also uses time difference information in the vertical axis to achieve the sensation of sound source elevation and correct localization, based on the results of his own research, which seem to be somewhat in contrast with the findings of Florian Wendt et al., but also those of Hyunkook Lee.

Williams' interest goes towards providing a complete system, consisting of a microphone array, able to capture true 3D sound with accurate localization, as well as an appropriate loudspeaker system, which is capable of correctly reproducing the signals of his multi-microphone arrays. The resulting speaker layouts (as well as microphone array signals) are compatible with the standard speaker layouts as found with Auro 3D, Dolby Atmos and the like, but actually even go beyond those systems in the sense that they are fully symmetrical in order to ensure accurate reproduction of moving sound sources in all directions.

The only other system which is as complete, by providing both a recording solution for 3D audio, as well as a defined loudspeaker replay system layout, scalable to need, is Ambisonics. Unfortunately, there is a discrepancy between the theoretical versatility and the practically achievable sound quality, which—at least so far—has severe limits, as became evident through various subjective listening tests (see Chap. 5 and Chap. 10 in this book for more details).

Apart from Ambisonics, which was probably the first attempt at capturing 3D audio already in the early 1970s, interest in 3D audio took up momentum mainly from the year 2000 onwards, when more and more sound engineers and scientists started 'to jump on the band wagon'.

This is why today—roughly 20 years after the first system proposals in respect to 3D audio started to appear in the world of audio engineering—we find a relevant number of 3D microphones arrays from a wide variety of authors and with different psycho-acoustic approaches, which will be examined in this book.

During the last 15 years, the awareness on the importance of also capturing the diffuse components of a soundfield with an appropriate microphone technique has significantly increased, as is evident due to the existence of various papers.

In reality, already during the era of pure 2-channel stereo most microphone techniques were not complying to the following two important criteria:

(A) direct sound needs to be captured in a well-defined manner which ensures correct (i.e. ideally non-distorted) localization along the stereophonic reproduction base (between the two loudspeakers) and will result in a correlated signal (with a cross-correlation coefficient of 1, i.e. 'mono') in the L and R channel for on-axis sound sources, and
(B) at the same time the diffuse sound components (i.e. reverb) need to be picked up in a completely de-correlated manner (down to the lowest frequencies, in order to ensure a high degree of spatial impression) by the microphone system.

Unfortunately, most well-known stereo-microphone techniques fail with the second condition: it is mainly the so-called Blumlein Pair of crossed figure-of-eights, which complies to both criteria (A and B), and 'large AB' spaced omni-capsules which are separated at least by the critical distance (or reverberation radius) of the recording venue, complying to criterion B; two other systems which comply well to the criterion of de-correlation are: MS-technique with an effective recording angle of 132°, as well as hyper-cardioids at an included angle of 133° (for more details, see Sect. 2.5 in my first book with Springer publishing *The Art and Science of Surround and Stereo Recording*, see link: The Art and Science of Surround and Stereo Recording | SpringerLink).

Unfortunately, the inadequacies of many of the already well-established stereo-microphone techniques have not been taken into consideration when 'expanding' or 'extrapolating' them into the realm of surround sound. Hence, it is no surprise that for many of them, drawbacks in terms of spatial reproduction have remained. (An in-depth analysis of this can be found in Chaps. 7, 8 and 9 of my first book with Springer.)

With the added complexity in 3D microphone systems, the task of 'natural' reproduction with a convincing spatial impression has opened up new possibilities in respect to achieving higher quality, but at the same time has not necessarily become an easier one.

Due to practical considerations in terms of mechanical set-up of 3D audio microphone systems, with many of these proposals the microphones applied for capturing sound—'from above'—for the so-called 'height layer' are mounted in close vicinity to the microphones of the middle (or main-) layer, often using directional patterns to ensure sufficient channel separation (or signal de-correlation) in order to arrive at a—in psychoacoustic terms—'meaningful' sonic result.

However, spacings between the microphones of this height layer are very often only in the order of 1 to 2 m, so the resulting signal de-correlation will depend very much also on appropriate capsule orientation or the application of time delay schemes to arrive at a high degree of 'spatial impression', which should undoubtedly be the main goal in 3D audio.

Many of the 3D microphone array proposals found today seem mainly concerned with just capturing 'diffuse sound from above' (which may be sufficient, if the direct sound components are radiated only from a frontal stage). However, these systems are not designed to reproduce proper localization of moving sound objects, for example, in the upper hemisphere. As Michael Williams expressed in his AES-papers from May 2022 '… Many pseudo height microphone array recording systems have been created recently, specifically for use in the context of the cinema or home-cinema industry. These pseudo height systems are simply a combination of two surround sound systems placed one above the other—essentially two layers of surround sound reproduction. It must be understood that these systems have absolutely nothing to do with creating any real impression of height. Although the result in reproduction is far from satisfactory for the discerning listener of 3D, it seems to have been accepted, in the context of cinema or home-cinema reproduction, in giving an impression to the public of something happening above their heads—the listener gets the impression of what is now being called 'an immersive' soundfield …'

Setting up the quantity of microphones needed for 3D audio recording can be a quite labour—and time-intensive process, and if the main aim of the capsules applied for capturing information for the height layer is to pick up diffuse sound, one may ask if it was really worth the effort and maybe not more efficient to just use a high-quality digital hardware device or software plug-in, capable of delivering appropriate 3D audio reverb.

However, especially with the advent of the 'bottom' layer (as counterpart to the 'height layer') in 3D audio systems, the engaged sound engineer or Tonmeister will probably strive in not only capturing the diffuse sound in a hall, but also meaningful first reflections from the ceiling, side (and rear-) walls as well as the floor of a concert hall, thereby capturing very important acoustic details which are characteristic for each specific venue. Unfortunately, it seems that many of the microphone proposals for 3D audio out there do not yet cover this important aspect …

In respect to 3D audio terminology used on a practical level by the different manufacturers, there is still a bit of 'variety' to be found out there, not only what concerns abbreviations for the individual loudspeakers:

The loudspeaker 'main' layer, as employed for 2-channel stereo as well as pure 2D surround setups is sometimes also referred to as the 'surround' layer, 'base', or 'front' layer;

- the 'height' layer is sometimes called 'upper' layer;
- the 'bottom' layer is also known as 'lower' layer or 'floor' layer.

Not to forget about the 'top' layer which holds one or more (ceiling) loudspeakers above the head of the listener …

Also the syntax, indicating which speaker belongs to which layer, has not yet been unified on a practical level. A loudspeaker arrangement defined as 'AURO 3D 13.1' is quite different from a 'SONY360 13.0' speaker layout: while the Auro 3D '13.1' consists of seven loudspeakers in the main layer, five loudspeakers in the height layer and one speaker in the top layer ('Voice of God' speaker, right above the sweet spot) and one LFE channel (see Chap. 3), the Sony 360 '13.0' layout foresees five loudspeakers in the main layer, no loudspeakers for the LFE channel, five loudspeakers for the height layer and three loudspeakers for the bottom layer (in Sony360 syntax written as 5.0.5+3B, see Chap. 8).

I hope to have covered a wide range of interesting aspects in respect to 3D audio, relevant to practically working sound engineers, while also shedding new light on some theoretical considerations in relation to 3D sound. The main aim of this book is to present the ideas-of the various authors-concerning '3D audio' in a 'democratic fashion' next to each other, even if they may-at times – be opposed to each other. I believe this is what can help to fuel a 'healthy' academic discourse ... It is almost exclusively in the first chapter that I do allow myself a critical evaluation of other author's ideas, proposals or publications – along with a number of case studies, which I hope will find your interest. Readers who happen to know my first book *The Art and Science of Surround and Stereo Recording* may have noticed that my special interest goes towards the analysis of frequency dependent signal cross-correlation (FCC) from which a number of qualitative conclusions can be drawn, not only for stereo and surround recordings, but certainly also for 3D audio.

A short series of educational videoclips concerning microphone technique analysis (mainly for 2-channel stereo microphone techniques, but also including 3-channel techniques like DECCA and BPT), based on FCC measurements using the '2BCmultiCORR' frequency-dependent signal cross-correlation plug-in can be found on my Youtube channel 'futuresonic100', when searching for 'mic tech analysis' or 'Nevaton BPT'.

I would like to thank all the authors of the plethora of papers which are used for this book and would also like to thank in particular the following people, which have contributed to the making of this book in various ways (listed in alphabetical order, without academic title or affiliation):

Wilfried Van Baelen, Enda Bates, Dennis Baxter, David Bowles, Hans Riekehof-Böhmer, Doug Clarke, David Chesky, Edgar Choueiri, Etienne Corteel, Maximilian Dietel, Gary Elko, Robert Ellis-Geiger, Nino Guillermo, Akira Fukada, William Howie, Dale Johnson, John Johnson, Nigel Jopson, Toru Kamekawa, Gavin Kearney, Sungyoung Kim, Hyunkook Lee, Morten Lindberg, Paul Geluso, Harald Gericke, Christopher Gribben, David Griesinger, Kimio Hamasaki, Simon Hildenbrand, William Howie, Robert Höldrich, Jeff Levison, Juha Merimaa, Sven Mevissen, Olaf Mielke, Masataka Nakahara, Rozenn Nicol, Lasse Nipkow, Markus Noisternig, Iker Olabe, Makoto Otani, Ronald Prent, Darcy Proper, Ville Pulkki, Hashim Riaz, Francis Rumsey, Alex Ryaboy, Mirek Stiles, Lukas Thalhammer, Günther Theile, Floyd Toole, Jeff Turner, Rory Wallis, Michael Williams, Helmut Wittek, Boris Wood, Kathleen Zhang and all the others, which I may have forgotten (my apologies, in case ...).

As a closing remark, I would like to point out that there is, consciously, a minimal amount of occasional redundancy between chapters: as readers of the e-book version have the possibility to also buy only single chapters for download directly at the 'Springer Link' website, at times it seemed necessary to include all the important information in the same chapter, in order to give the reader a chance to follow the 'red thread', without loosing out on details.

As the book consists of many citations from other authors, I have reserved the use of [ ] for inserting occasional comments of mine.

I wish you interesting hours with this book.

Salzburg, Austria
September 2022

Edwin Pfanzagl-Cardone

Contents

About the Author

Edwin Pfanzagl-Cardone is head-of-sound at the acoustics department of the "Salzburg Festival" of classical music in Austria. After completing his degree in electronics engineering and information technology at TGM, Vienna, in 1988, he graduated in 1991 at the 'University of Music and Performing Arts' in Vienna and received a *Tonmeister* (Sound master) degree. During 1994–1999, he was lecturer for theory of sound engineering at the 'Institute of Electro-Acoustics' of the same university. In 2000, he completed a M.A. in Audio Production at the University of Westminster, London. In 2011, he received a Ph.D. in musical acoustics and psychoacoustics from KUG—University of Music and Performing Arts, Graz, Austria. Since the early 1990s, he has been working as a sound engineer for music recording and live sound reinforcement, and sound for film and TV, mainly in Europe, yet also in Japan and in the USA. As arranger and composer, he released with BMG and SONY in the field of pop music, and at the same time, he delivered content to international production music labels, and for commercials in radio and TV. Author of AES-and VDT-convention preprints, Ing. Dr. Pfanzagl-Cardone has published more than 60 articles in magazines for sound engineers, such as Pro Sound News Europe, Studio Sound, Sound-On-Sound, Production Partner, Prospect, among others. He has been a member of the Audio Engineering Society (AES) since 1991, and a member of the 'Austrian Sound Engineer's and Music Producer's Association' ÖTMV. He has been active as an international lecturer and audio consultant and was invited by the AES to present the results of his research in the OEAW (Austrian Academy of Sciences) in 2017.

In the area of classical music, he worked with well-known conductors such as Abbado, Barenboim, Boulez, Dudamel, Gatti, Gergiev and Harnoncourt. He also worked with Jansons, Maazel, Mehta, Metzmacher, Minkowski, Muti, Sir Rattle, Orozco-Estrada, Salonen, Savall and many others. He recorded the Vienna and Berlin Philharmonic Orchestra, The Mozarteum Orchestra of Salzburg and a large number of other European and foreign orchestras. For the Salzburg Summer Festival 2018, he realized the 3D audio sound effects for Mozart's 'Magic Flute' conducted by Constantinos Carydis.

From 2010 to 2021, he has been teaching sound reinforcement technology at the faculty of design, media and arts of the University of Applied Sciences, in Salzburg, Austria. As a composer, he released four international CDs. Besides several hundred archival recordings for the Salzburg Festival, his discography as sound engineer consists of about thirty CD- and three LP-releases with music labels such as Deutsche Grammophon and Orfeo, among others. Finally, he is the inventor of three microphone techniques such as the AB Polycardioid Centerfill (AB-PC), the ORTF-Triple (ORTF-T) and the Blumlein-Pfanzagl-Triple (BPT), and holds a patent in surround microphone technology.

Abbreviations

3D	3-dimensional
3DWA	3D Audio Workstation
AAC	Advanced Audio Coding
AAX	Avid Audio Extension (plug-in format)
AB-PC	AB Polycardioid Centerfill
ADM BWF	Audio Definition Model Broadcast Wave File
ADR	Automated dialogue replacement
AES	Audio Engineering Society
AIDT	Average interaural time delay
AKG	Akustische und Kinogeräte GmbH
AMS	Advanced music systems
APL	Applied Psychoacoustics Lab (Huddersfield University)
ASI	Index of Acoustic Spatial Impression
ASW	Apparent source width
AU	Audio Unit (plug-in format)
AV	Audio-visual
AVC	Advanced Video Codec
AVR	Audio/Video Receiver
BD	Blue-ray disc
BQI_{rep}	Binaural Quality Index of Reproduced Music
BS (ITU-R)	Broadcasting service
C	Cardioid
CAF	Core Audio Format
CBA	Channel-based audio
CD	Compact disc
CEDIA (convention)	Custom Electronic Design and Installation Association
CES	Consumer Electronics Show
CFF	Common file format
CHAB	Centre hemicardioid AB
CLI	Command line interface
CPU	Central processing unit

D	Deutlichkeit
DAMF	Dolby Atmos Master File
DAMP	Dolby Atmos Music Panner
DAPS	Dolby Atmos Production Suite
DAR	Dolby Atmos Renderer
DAW	Digital Audio Workstation
dB	Decibel
DCI	Digital Cinema Initiatives
DCP	Digital Cinema Package
DD	Dolby Digital
DD+JOC	Dolby Digital Plus Joint Object Coding
DF	Directivity factor
DFC 3D	Digital Film Console 3D (from AMS-Neve)
DFC	Diffuse field correlation
DFI	Diffuse field image predictor
DFT	Discrete Fourier transform
DI	Directivity Index
DJ	Disc jockey
DMS-Z	Double MS – Z
DPA	Danish Pro Audio
DRC	Dynamic range compression
DRR	Direct-to-reverberant ratio
DTF	Diffusefield transfer function
DTS	Digital theatre systems
DVD	Digital versatile disc
DVE	Digital video effects
EMI	Electric and Musical Industries Ltd
EQ	Equalizer
ER	Early reflection
ERB	Equivalent rectangular critical band
ERD	Equivalent rectangular distortion
ESMA	Equal segment microphone array
ETO	Electronic time offset
FCC	Frequency-dependent cross-correlation
FIACC	Frequency-dependent interaural cross-correlation
FOA	First-order Ambisonics
GEMS	Geneva Emotional Music Scale
HD	High definition
HDMI	High-Definition Multimedia Interface
HEVC	High-Efficiency Video Coding
HF	High frequency
Hi-fi	High fidelity
Hi-Res	High Resolution
HOA	Higher-order Ambisonics
HRTF	Head-related transfer function

HTC	High Tech Computer Corporation
IAB	Immersive audio bitstream
IACC	Interaural cross-correlation
ICCC	Interchannel cross-correlation
ICLD	Inter-channel level difference
ICTA (convention)	International Cinema Technology Association
ICTD	Inter-channel time difference
IEM (KUG Graz)	Institute of Electronic Music and Acoustics
ILD	Interaural level difference
IMB	Integrated Media Block
IMF IAB	Interoperable Master Format Immersive Audio Bitstream
IMS	Immersive
IP	Internet Protocol
IPD	Interaural phase difference
IRT	Institut für Rundfunktechnik
ISO	International Organization for Standardization
ITD	Interaural time difference
ITU-R	International Telecommunication Union Recommendation
JBL	James B. Lansing
k	Wavenumber
KDM	Key Delivery Message
KFM	Kugelflächenmikrofon ('sphere microphone')
KUG	University of Music and Performing Arts (Graz, Austria)
LEDT	Lateral early decay time
LEV	Listener envelopment
LF	Lateral energy fraction
LF	Lateral fraction
LF	Low frequency
LFE	Low-frequency extension
LG	Lateral gain
LR	Late reflection
LSB	Least significant bit
LTC	Longitudinal time code
MADI	Multichannel Audio Digital Interface
MAGIC	Microphone Array Generating Interformat Compatibility
Mbit/s	Megabits per second
MDA	Multi-dimensional audio
MDS	Multi-dimensional scaling
MF	Mid-frequency
mic	Microphone
MMAD	Multichannel microphone array design
MOV	MOVie data file format
mp3	MPEG-1 Audio Layer 3
mp4	Short for MPEG-4 Part 14 (multimedia data file format)
MPEG	Moving Picture Experts Group

MRIR	Multichannel room impulse response
MS	Mid-side (microphone technique)
MSC	Magnitude-squared coherence
MUSHRA	Multiple Stimuli with Hidden Reference and Anchor
MXF	Material exchange format
MZ	Middle-Z
NDF	Non-drop frame
NGA	Next-Generation Audio
NHK	Nippon Hōsō Kyōkai (Japan Broadcasting Corporation)
NOS	Nederlandse Omroep Stichting (Dutch Broadcasting Corporation)
NTT	Nippon Telegraph and Telephone
OBA	Object-based audio
OCT	Optimized cardioid triangle
OLE	Overall listening experience
ORTF	Office de Radiodiffusion Télévision Française
OS	Operating system
OSC	Open sound control
OSIS	Optimal sound image space
OTQ	Overall tonal quality
OTT	Over-the-top (content)
PA	Public address (loudspeaker system)
PC	Personal computer
PCM	Pulse code modulation
PCMA	Perspective control microphone array
PHRTF	Personalized head-related transfer function
PICC	Perceptual interaural cross-correlation coefficient
PLF	Premium large format
PLP	Preferred listening position
POV	Point of view
PPA	Phantom power adaptor
PZM	Pressure zone microphone
QC	Quality control
R	Measure of spatial impression
REE	Random energy efficiency
RMU	Rendering and Mastering Unit (Dolby)
SBA	Scene-based audio
SBD	Scene-based delivery
SC	Super cardioid
SCA	Segment coverage angle
SCC	Secure content creator
SDK	Software Development Kit
SFX	Sound effects
SMPTE	Society of Motion Picture and Television Engineers
SPL	Sound pressure level

SRA	Stereophonic recording angle
SSL	Solid-state logic
TOA	Third-order Ambisonics
TV	Television
UHD	Ultra-high definition
VBAP	Vector-base amplitude panning
VDT	Verband Deutscher Tonmeister
VOG	Voice of God (speaker position)
VR	Virtual reality
VST	Virtual Studio Technology
WC	Wide cardioid
XTC	Cross-talk cancellation
τ	Time constant

Chapter 1
Introductory Critical Analysis and Case Studies

Abstract After a look back on early 3D audio developments in the form of "Quadraphonic Sound" and Ambisonics in the 1970s, an analysis of 3D mic array comparisons is made. The psychoacoustic value of signal decorrelation in vertically spaced microphones of the base and height layer in 3D mic arrays is briefly discussed. The DFC (Diffuse Field Correlation) of a few common main microphone techniques is documented and the importance of a low degree of diffuse sound signal correlation especially at low frequencies in order to achieve good spatial impression is pointed out. A qualitative evaluation of first order and higher order Ambisonic microphones is made, based on the outcome of the studies of several researchers. Frequency dependent ICCC (InterChannel Cross-Correlation) data are presented as part of a comparison of eight different 3D mic arrays. Frequency dependent ICCC of an OCT-based mic array is analyzed as a case study and conclusions drawn for 3D audio. Another case study analyzes the frequency dependent intercapsule signal correlation in the Hamasaki Square, which is also used in 3D mic arrays in the extended form of the Hamaski-Cube. The use of FIACC (Frequency dependent Inter Aural Cross Correlation) measurement and the introduction of an artificial head (Neumann KU81) as human reference is motivated. As conclusion, the BQI_{rep} (Binaural Quality Index of reproduced music) as a qualitative psychoacoustic measure is proposed also for 3D audio.

Keywords 3D audio · Surround sound · Ambisonics · Frequency dependent Inter-Aural Cross-Correlation (FIACC) · Diffuse Field Correlation (DFC) · Binaural Quality Index of reproduced music (BQI_{rep})

"When I sit in an acoustically perfect hall (full of people), in the best seat for hearing, and listen to an orchestra, I hear such and such sounds. I want to hear precisely this effect from a recording, if that be possible …" An anonymous music critic (K. K.) writing in the 1928 issue of The Gramophone (Torick 1988).

This statement from almost 100 years ago clearly shows that—independent of the available technology—listener expectation towards the quality of reproduced music was always very high. Today, with stereophony being almost 140 years old (see Eargle 1986) and the advent of 3D audio it seems we are closer to this goal than ever before.

E. Pfanzagl-Cardone, *The Art and Science of 3D Audio Recording*,
https://doi.org/10.1007/978-3-031-23046-2_1

In accordance with the statement above I am convinced that the 'right approach' to capturing an event of acoustic music in a concert hall is the aesthetic of "I am there" instead of "They are here" (in the living- or control-room), which is an old debate among Tonmeisters. For this reason I have developed my "Natural Perspective" recording approach, the fundamentals of which are summarized towards the end of Chap. 11 in Pfanzagl-Cardone 2020.

Surround sound in music reproduction, from its early inception in the form of "Quadraphonic sound" in the 1970s to its ripening in the 1990s as "5.1 Surround" has come a long way. Since the early 1930s a lot of experimentation and research has happened in the field of stereo (see Blumlein 1931; Keller 1981) and—later on—surround recording. I have presented a selection of more that 20 stereophonic and about 30 different surround techniques in a previous volume to this publication (see Pfanzagl-Cardone 2020).

Unfortunately, many of the current surround microphone techniques are derived from stereo counterparts which do not provide enough signal decorrelation at low frequencies. In terms of localization accuracy, their output may be convincing—or at least satisfactory—to the listener due to the high-frequency content involved, but on the low-frequency side they fail to capture the sound source in an adequate manner. This is true not only for many of the most well-known stereo microphone techniques—such as 'small AB', for example—but even for some of the most commonly used surround microphone systems (for more details on this see the preface and Chaps. 7 and 8 of Pfanzagl-Cardone 2020).

If we take a look at the currently proposed 3D techniques, many of which are presented in more detail in the following chapters of this book, we realize that a lot of 3D mic-arrays have their roots in related 2D surround techniques. This is understandable and similar to what happened in the technical transition from stereo to surround techniques, but also in the case of progressing from 2D surround to 3D audio the question needs to be asked whether the new techniques are sufficiently de-correlated at low frequencies so as to provide convincing spatial impression, which should be a primary goal, especially with 3D audio.

Michael Gerzon's invention of the "Ambisonics" principle is most likely the best known among the first efforts to capture sound in 3D (see Gerzon 1973).

After the commercial failure and fading out of "Quadraphonic Sound" during the 1970s, it took considerable time for surround sound (in respect to music consumption) to reemerge in the form of "5.1" from the mid 1990s onwards.

Despite the fact that a lot of research has happened since the early efforts of an introduction of a 3D audio format in the early 2000s (see Dabringhaus 2000), up until now there are still only a few studies which are evaluating various 3D audio techniques, either through an analysis of their acoustic attributes in the form of subjective listening tests or by acoustic signal analysis, but usually not covering both aspects, which I consider a relevant lack.

In 'Multichannel 3D Microphone Arrays: A Review' Hyunkook Lee has categorized existing 3D microphone arrays according to their physical configurations, design philosophies, and purposes, followed by an overview of each array. In his AES-paper (see Lee 2021) he examines studies that have subjectively or objectively evaluated different microphone arrays. In this, different approaches in the configuration of the upper microphone layer are discussed, aiming to provide theoretical and practical insights into how they can contribute to creating an immersive auditory experience. Finally, Lee identifies limitations of previous studies and future research topics in 3D sound recording.

Before the publication of this paper, Lee has conducted extensive 3D microphone array recordings, involving seven 3D arrays and a total of 71 microphones, which he reports on in (Lee 2019). The 3D mic arrays involved were: PCMA-3D, OCT-3D, 2L Cube, Decca Cuboid, Hamasaki Square+Height, 32-channel spherical mic array (Eigenmike EM32) and first-order Ambisonics (in the form of a Sennheiser Ambeo VR) Additionally, microphones for the NHK 22.2 format (side and side height, overhead (i.e. so-called "The Voice of God" channel), floor as well as the ORTF stereo pair, KU100 dummy head and spot microphones for individual instruments were used in the recording which took place at the St. Paul's hall, a church-converted concert venue with a high ceiling and an average reverb time of 2.1 s. In respect to the latter aspect the acoustics of this church are close to the most renowned high quality concert halls like Amsterdam Concertgebouw, Boston Symphony Hall and Vienna Musikverein.

A similarly extensive comparative 3D audio microphone array test like the one by Lee has taken place at Abbey Road Studios in 2017, led by Hashim Riaz and Mirek Stiles, which has involved a total of 11 different microphone systems: mh acoustics EM32 'Eigenmike®'(HOA), SoundField ST450 MKII, ESMA (Equal Segment Microphone Array), ORTF-3D Surround, Sennheiser AMBEO (FOA) , OCT-9 Surround, PCMA 'Perspective Control Microphone Array', Stereo XY-pair, IRT Cross, Hamasaki Cube, and a Neumann KU100 dummy head (as reference), which is reported on in Riaz et al. (2017) and Stiles (2018), as well as in Chap. 10 of this book. However, so far no results on a subjective evaluation of these recordings are published.

In my previous publication "The Art and Science of Surround and Stereo Recordings" five surround microphone techniques were examined in much detail by analyzing them through subjective listening tests with 50 participants, as well as through measuring signal correlation over frequency ("FCC"—Frequency dependent Cross-Correlation). The outcome of the subjective listening test was then compared with the objective physical factors, which had been measured by means of the correlation function.

In order to make the FCC more meaningful, an artificial human head (Neumann KU81i), as well as a software plug-in using the HRTFs of the KEMAR dummy head, had been introduced to serve as a 'human reference' when measuring the FIACC (Frequency-dependent Inter-Aural Cross-Correlation).

By measuring the signal cross-correlation over frequency between channels for the various microphone techniques (or the binaural equivalent, FIACC of the final result of replaying the recordings of such microphone arrays via a standard 2-channel stereo or 5.1 surround loudspeaker setup), it was shown that every stereo and surround microphone technique is characterized by an individual 'sonic fingerprint.'.

At the current state a similar test for 3D microphone arrays, which uses extensive evaluation based both on a subjective listening comparison as well as an objective (acoustic) evaluation, seems to be still missing.

Almost 20 years ago, in my first paper on the matter, which was based on previous research by David Griesinger, presented at a convention of the VDT (Verband Deutscher Tonmeister) in 2002, with the title 'Über die Wichtigkeit ausreichender Dekorrelation bei 5.1 Surround-Mikrofonsignalen zur Erzielung besserer Räumlichkeit,' I tried to point out the importance of sufficient signal decorrelation in 5.1 surround recordings in order to achieve a better spatial impression. At that time, I think there was hardly any awareness among colleagues on the importance of how not only direct sound, but also diffuse sound gets picked up by a specific microphone technique in use. Furthermore, certain authorities in the field actually reinforced an opposite opinion of a minimum signal coherence necessary for a ' … natural reproduction of space and envelopment …'. Fortunately, during the years to follow more papers were published by various authors which were in the same vein as mine, and about ten years later, Helmut Wittek, an important expert in the field concluded that ' … the diffuse sound field plays an enormously important role for spatial perception, as well as for sound color. Therefore it must be ensured that it is reproduced with low correlation.

Many good or bad properties of a stereophonic [microphone] arrangement do not have anything to do with their localization properties, but only with their ability to reproduce a nice, open-sounding spaciousness. XY cardioids are a good example: good localization properties, but bad spatial reproduction. In the future, much more importance should be paid to diffuse field correlation. Very often only the effective recording angle is considered when choosing a specific arrangement. Diffuse field correlation is at least as important' (from Wittek 2012).

At the same convention—as part of another paper presentation—the authors noted that ' … Diffuse sound (i.e. reverb or background noise) needs to be reproduced diffusely. This can be achieved using Auro-3D if appropriate signals are fed to the extra speakers. Diffuse signals must be sufficiently different on each speaker, that is, they need to be decorrelated over the entire frequency range.

A sufficient degree of independence is necessary, in particular, in the low-frequency range as it is the basis of envelopment perception. However, increasing the number of channels that need to be independent makes recording more complex. It is a tough job to generate decorrelated signals using first-order microphones—for example, a coincident array such as a double-MS array or a SoundField microphone allows for generating a maximum of four channels providing a sufficient degree of independence. Therefore, the microphone array needs to be enlarged to ensure decorrelation. It is worth noting here that measuring diffuse field correlation is not trivial. There are two reasons for this: First, measuring the correlation requires the diffuse sound level to be much higher than direct and reflection levels, so the distance from the source needs to be sufficiently long. Secondly, considering the degree of correlation is not sufficient; this does not account for the fact that low-frequency (de)correlation is particularly important …' (from Theile and Wittek 2012).

1.1 Diffuse Field Image Predictor (DFI) and Signal Correlation in Stereo and Surround Systems

In Wittek (2012), he has revisited the question how high a value of DFC ("Diffuse Feld Correlation") should be allowed in order not to degrade spatial impression.

In previous research (Riekehof-Böhmer et al. 2010), listening tests were performed in relation to diffuse sound, for which coincident, equivalence and runtime (i.e. AB) microphone techniques were used. Artificial reverb was reproduced, applying these techniques and test listeners had to rate its 'Apparent Width' (or Stereophonic Width), the results of which are displayed in Fig. 1.1.

Translated and partly summarized from Wittek (2012):

"…It can be assumed that a large apparent width is optimal for diffuse sound/reverb. For AB-based (runtime) microphone techniques the DFC is not uniform over frequency (see Fig. 1.2) but it can be assumed that it is preferable if the area below the function is small, especially for low frequencies (see also Griesinger 1998). This is why in Riekehof-Böhmer et al. (2010) an asymptotic value for the DFC was determined, which—in essence—is the integral of the squared, weighted coherence function (low frequencies being more important).With this—the DFI ('Diffuse-Field Image') Predictor—a value is obtained, by which different mic techniques can be compared and the 'apparent width' can be determined (Fig. 1.1).

If we deem a minimum quantitation value of +2 sufficient, the following conclusions can be drawn:

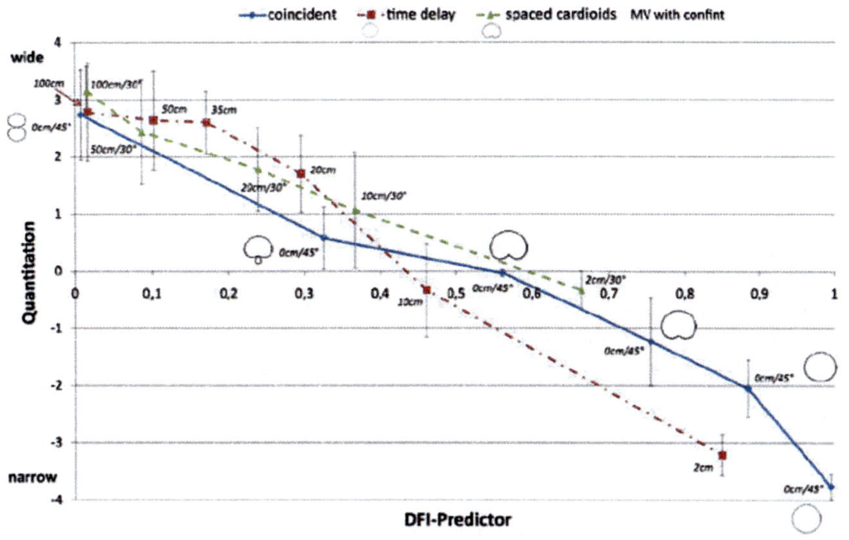

Fig. 1.1 Listener evaluation of apparent width of diffuse sound (reverb), recorded by various microphone techniques (from Riekehof-Böhmer et al. 2010); mean value with confidence intervals

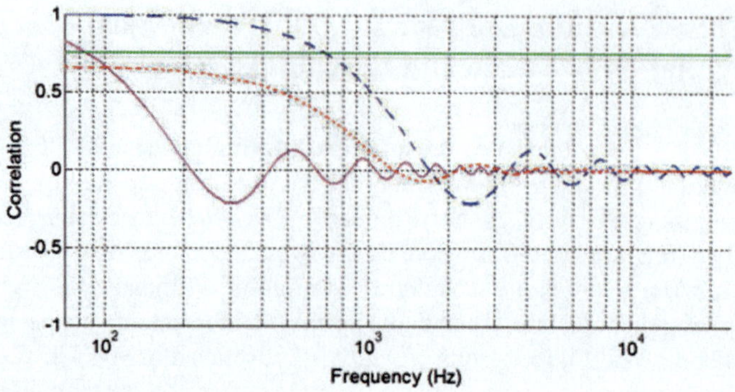

Fig. 1.2 Diffusefield correlation (DFC) for various stereo microphone techniques: (listed from top to bottom) blue dashed line: small AB omnis $d = 10$ cm; green straight line: XY-Cardioids 90°; red dotted line: ORTF Cardioids, 17 cm, 110°; violet: small AB, omnis 80 cm (Fig. 7.33 from Pfanzagl-Cardone 2020; modified from Wittek 2012)

- The Blumlein-Pair was the only 'good' coincident technique, according to the listening test. Its DFC is 0.
- With equivalence and pure runtime techniques the DFI-predictor needs to be smaller than 0.5.

  For pure runtime techniques, this means the capsule spacing has to be ≥35 cm. For equivalence techniques, the spacing can be smaller, depending on the polar pattern and orientation of the microphones involved.
- An optimum, 'wide' reproduction of diffuse sound can be achieved both via coincident, as well as spaced techniques.
- As pure runtime techniques need to have a capsule spacing of at least 35 cm in order to achieve optimum results, multichannel setups using baffles with a smaller effective diameter will not work to satisfaction, as the baffle-effect vanishes with lowering frequency (due to wavelength λ). …"

A short overview for a few selected stereo 2-channel microphone techniques on how the combination of different microphone patterns and capsule spacings result in terms of diffuse field correlation (DFC) can be found in Fig. 1.2.

A much more detailed analysis on frequency dependent cross-correlation (FCC) for a plenitude of stereo and surround microphone systems can be found in Chaps. 7 and 8 of (Pfanzagl-Cardone 2020).

In relation to the question of an optimal inter aural correlation for concert hall listeners it is worth to look back on the research by Gottlob (1973), who has found this to be at a value of 0.23 (more details can be found in Sect. 2.23 of Pfanzagl-Cardone 2020).

Also this finding seems to indicate that microphone systems with high signal correlation should most likely better be abandoned for the sake of systems that are largely decorrelated. As was found in research by Nakahara (2005) and can be seen in Fig. 1.3, low correlation of the playback signals in a 5.1 surround setup (L/R vs. C

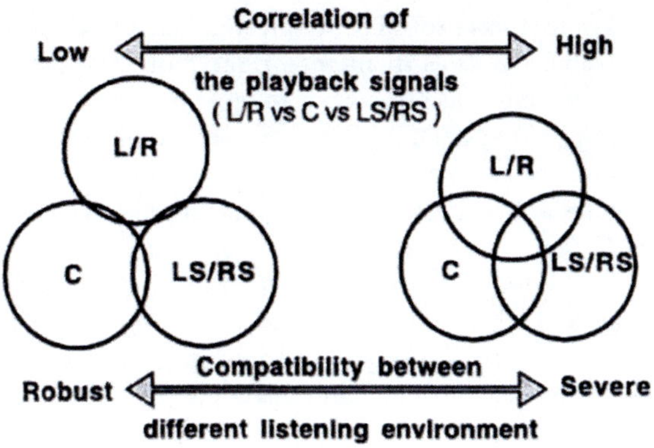

Fig. 1.3 Amount of signal cross-correlation in relation to compatibility between different 5.1 surround listening environments (from Nakahara 2005)

vs. LS/RS) ensures a better compatibility between different listening environments and also leads to an enlarged sweet spot (see also Prokofieva 2007).

What was found to be true for 2D audio in the form of '5.1 Surround', is certainly also applicable for 3D audio, therefore: microphone techniques should be used which are able to provide low signal correlation over the entire frequency range, if possible. Among the four surround microphone systems (OCT, DECCA, KFM, ABPC) which were evaluated via subjective listening tests, as well as objective measurements (FCC and FIACC) in Pfanzagl-Cardone (2020), the low-correlation AB-PC system (using a 'large AB' main microphone technique in combination with an appropriate centerfill-microphone ORTF-T) turned out most successful in terms of listener preference. It was also reassuring to see that -in addition- AB-PC was the microphone system, which managed to physically approximate the original sound field of the concert hall better than any of the other systems under test (see Chap. 2 and the results in Chap. 10 in Pfanzagl-Cardone 2020).

1.2 Qualitative Considerations Concerning Ambisonics

In 3D-Audio,'Ambisonic' (FOA—First Order Ambisonic or HOA—Higher Order Ambisonic) is probably the most important format in respect to 360°VR gaming applications, but also as a format for multichannel 3D music, broadcast (see Baxter 2016) and film-sound related reproduction.

Given the fact that the 'SoundField' Microphone (with its underlying first-order Ambisonics principle, as invented by Michael Gerzon in the early 1970s; see Gerzon 1973) has consistently scored very low in Surround-Microphone comparisons (see -among others- Hildebrandt and Braun 2000; Camerer et al 2001) it is surprising to see that the Ambisonic format has become the 'de facto' standard also for many 3D audio applications, even though already in 5.1 surround other, dedicated microphone arrays proved sonically superior.

While it has proved to sound significantly 'drier' than its competitor systems in the subjective tests mentioned right above and therefore much less able to create a convincing sense of spatial impression which would be desirable for a surround microphone system, it has proved a certain flexibility in respect of 2-channel stereo recordings, as processing of the 4 hemi-cardioid capsule signals allows for decoding any desire first-order pressure gradient polar pattern. (for more details see Chap. 4 in Pfanzagl-Cardone 2020).

I would assume one of the main reason for the widespread appreciation of the Ambisonics principle also for 3D audio might lie in the great flexibility this system offers in terms of post-production signal processing, which –with HOA-systems— allows for beamforming and rendering the content also to a binaural format. Since the average prospective listener is likely to not have a multi-channel 3D-audio loudspeaker system at his/her disposal, replaying this information via headphone (or earbuds) —with, or without head-tracking- is the next best option.

An analysis and technique of 3D binaural sound reproduction using a real-time virtual Ambisonic approach can be found –among others- in Noisternig et al. (2003).

In this context it is interesting to note that some people in the recording industry have arrived at that conclusion already more than 25 years ago: New York based Chesky Records (www.chesky.com) have adopted recording many of their artists (but also orchestras) with an artificial human head -using also additional spot-microphones- a long time ago (see examples of recording sessions in Fig. 1.4). It seems history is repeating itself, or –at least– 'the circle is closing'.

The relatively widespread use of the Ambisonic technique is most likely rooted primarily in the relative computational ease with which the multi-channel signals of a first order, or higher order Ambisonic microphone can be manipulated and rendered to suit various replay speaker layouts or a binaural reproduction via headphones, which is of course also of high interest for the gaming and VR community. More info on this can be found in (Oliveri et al. 2019) "Scene-based Audio and Higher Order Ambisonics: A technology Overview and Application to Next-Generation Audio, VR and 360° Video".

Another -also mathematically very detailed- analysis regarding the Ambisonic technique can be found in the book by (Zotter and Frank 2019), which seems to be a rather extensive reflection on the matter, but unfortunately pretty void of any thorough analysis of the inherent limits or drawbacks of the system.

A critical analysis concerning the resolution -and therefore also the limits in terms of reproduction accuracy- of the Ambisonic principle can be found in (Nicol 2018), where the author writes: " '…The tetrahedral microphone arrangement of the 'SoundField' Ambisonic microphone allows the capturing of a full 3D sound field. However, since it is composed only of four capsules, its spatial resolution is low, which means that the discrimination between the sound components is not accurate…" (Fig. 1.5). Further analysis on the limits of HOA based microphones can be found in Chap. 4 of Pfanzagl-Cardone 2020, based on research by (Nicol 2018) and (Ward and Abhayapala 2001).

According to the subjective evaluations as documented in (Hildebrandt and Braun 2000) and (Camerer et al 2001) signals of the 'SoundField' microphone seem to be

Fig. 1.4 Photos from various Chesky Records recording session, using an artificial human head. Top left: ensemble recording, bottom: jazz duo including vocal recording (reproduced with kind permission of Chesky Records)

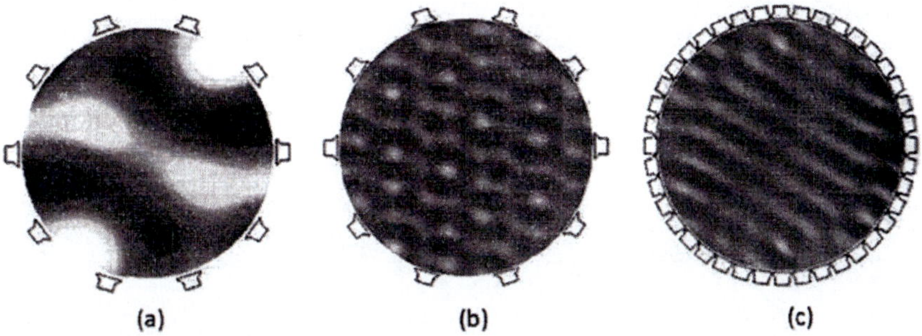

Fig. 1.5 Benefits of adding Higher-Order Ambisonics components: **a** the 250 Hz plane wave is well reconstructed over a fourth-order system; **b** increasing the frequency to 1 kHz yields a poorer reconstruction using the same fourth-order system; **c** to reconstruct the 1 kHz plane wave with the same accuracy as in (**a**), the system must be increased to a 19th order (Fig. 4.5 from Pfanzagl-Cardone 2020; originally from Nicol 2018)

characterized by a noticeably drier sound than those of the competitor surround microphone systems. An explanation for this might be that—due to the use of semi-cardioid capsules, as well as the following signal-processing- a large part of the captured diffuse sound is effectively eliminated.

In fact, it is visible in Fig. 1.6 that there is almost no influence on the interchannel signal correlation caused by diffuse sound (which would usually be of decorrelated nature, depending of course on how it is picked up):

Signal correlation for the first-order Ambisonics 'SoundField' microphone is usually very high (with values above 0.6) between all channels except for high frequencies above 3 kHz in the rear-channel pairing of LS, RS and above 600 Hz for the lateral channel-pairing of L, LS.

A signal correlation that high is quite unusual and definitely to the disadvantage for good spatial reproduction, which requires signal decorrelation at frequencies below 500 Hz.

It seems that the unusually high signal correlation values found with the Sound-Field microphone are really an inherent characteristic of the Ambisonic principle, at least with current decoding algorithms: in Fig. 1.7 the results of early time segment

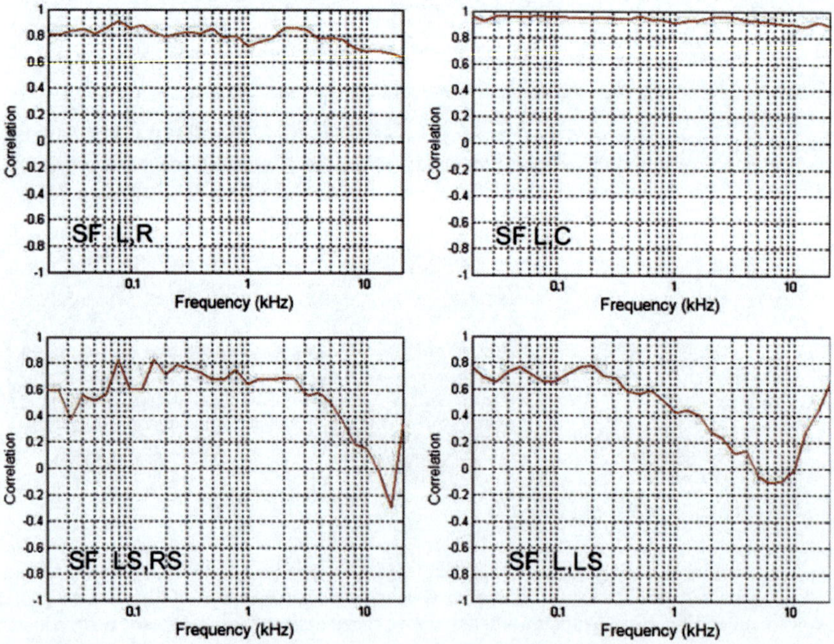

Fig. 1.6 Paired channel correlation over frequency (i.e. FCC) for the first-order Ambisonic Sound-Field MK-V surround microphone system signals (grouped to 31 frequency bands, 1/3rd octave; center frequencies according to ISO; music, 60 s) decoded via SoundField MK-V Surround Processor (Fig. 7.22 from Pfanzagl-Cardone 2020)

(i.e. 0–80 ms) interchannel correlation measurements for the LF-, MF- and HF-band of the comparative 3D mic-array recording by (Lee and Johnson 2020) are displayed.

What sticks out visually are the bars representing the 1st order and 4th order decodings of the Eigenmike® HOA-microphone with 32 transducer elements, which both display very high correlation values, independent of the frequency band. While

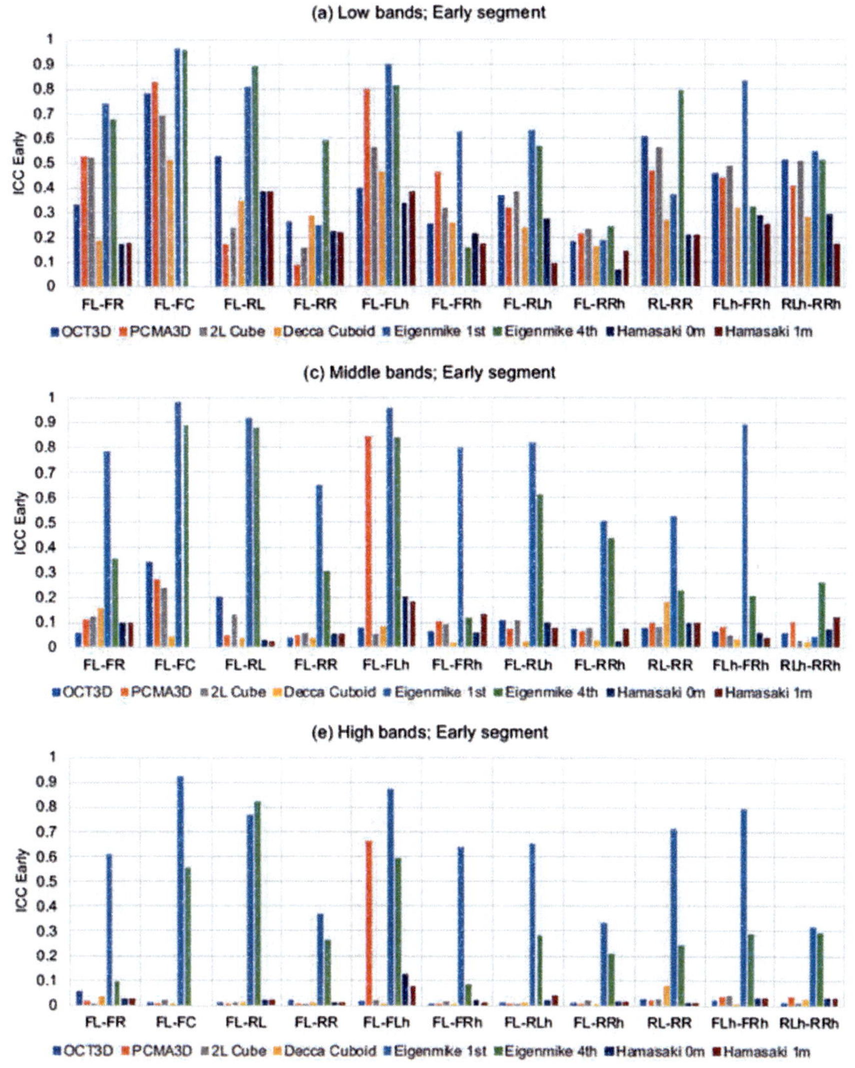

Fig. 1.7 Interchannel correlation coefficients for different pairs of microphone signals of the base and height layer (left part of Fig. 7 from Lee and Johnson 2020); L = Left, R = Right, C = Center, F = Front, R = Rear, h = height

the other 3D mic array techniques are characterized by moderate low correlation values of usually below 0.5 also in the LF-band, not only the 1st order, but also the 4th order HOA-decoding has significantly higher correlation values than all other mic-arrays also for the MF- and HF-band. In respect to the base layer channel pairs, the values measured in the study be Lee and Johnson coincide very well with my own findings from previous research in (Pfanzagl-Cardone and Höldrich 2008), as displayed in Fig. 1.6.

In Fig. 1.8. some of the 3D-audio microphone arrays, which were used in the MARCo (3D-'Multichannel Array Recording Comparison', see Lee and Johnson 2020) recording are displayed, including their physical position and capsule directivity. Figure 1.9 shows the placement of all the microphone array systems used in relation to the recording venue, as well as the sound sources (loudspeakers) which were used for the test.

The above mentioned, comparatively dry sound of the Ambisonic recording principle can also be evidenced in Fig. 1.10, in which the Direct-to-Reverberant Ratio (DRR) is documented for all microphone channels for the objective measurements which used a loudspeaker at +45° left of the microphone arrays main axis' as sound source. While the DRR is usually in favor of the diffuse sound component for all microphones that are facing away from the sound source, as was to be expected, this is not necessarily the case for the 1st order and 4th order Eigenmike® signals.

The fact that the signals of both decoded "Left height" microphone channels (for Front and Rear; i.e. FLh and RLh) have positive values in the favor of direct sound

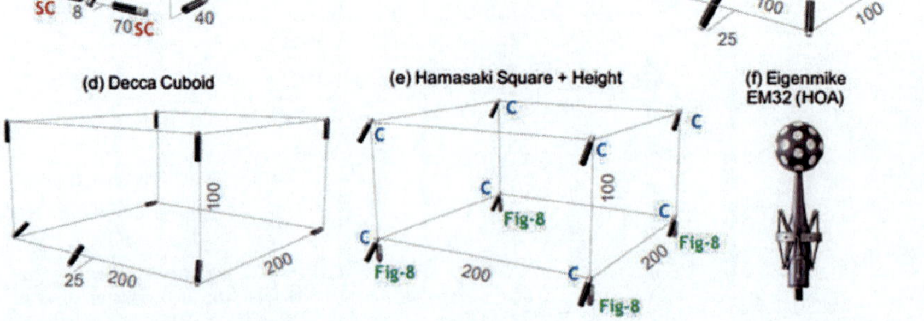

Fig. 1.8 Some of the 3D-Audio mic arrays as used in the MARCo recording. All microphones except for the Eigenmike® and the Hamasaki-Square (Schoeps CCM8) were from the DPA d:dicate series. Capsules were omnis, except if noted otherwise (C = cardioid, SC = super-cardioid, Fig. 8 = Figure of eight) (Fig. 1.8 was adapted from Fig. 1 from Lee and Johnson 2020)

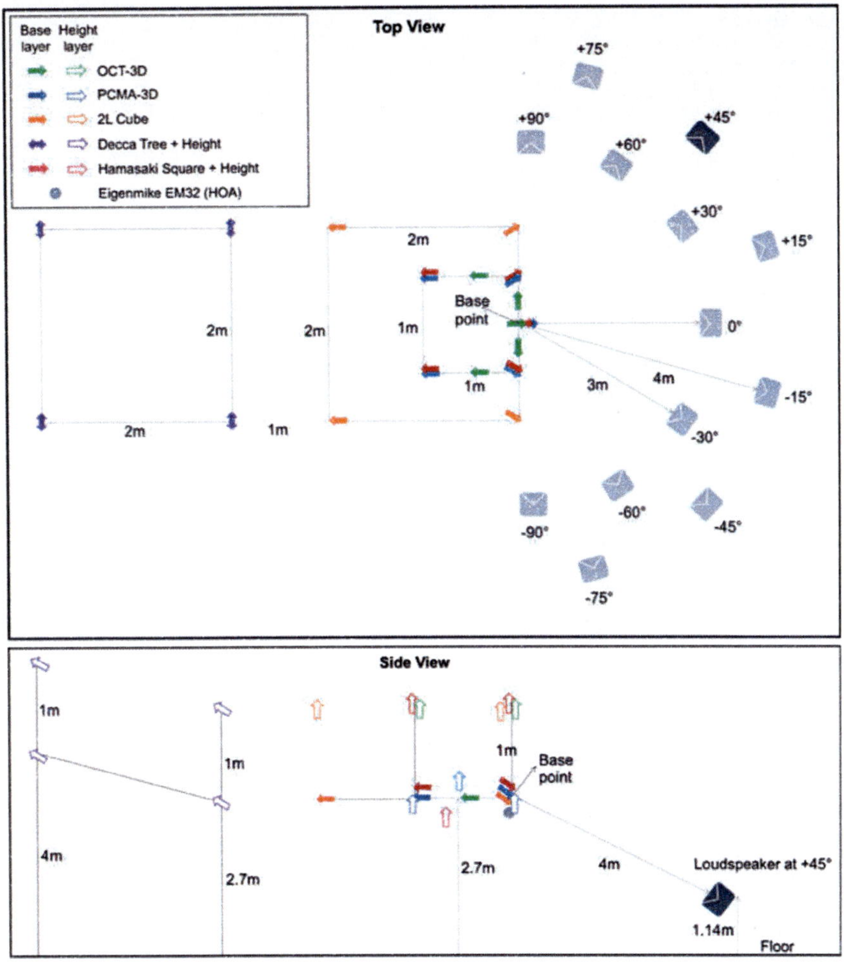

Fig. 1.9 layout schematic of the microphones and loudspeakers used for capturing the multichannel room impulse responses (MRIRs) in 3D-MARCo. For the objective measurements, presented in Figs. 1.7, 1.10 and 1.11, the MRIRs for the source at +45° were used (graphic equivalent to Fig. 2 from Lee and Johnson 2020)

components clearly shows that the decodings of the Eigenmike® EM32 result in a much lower (sic) directivity than for the other microphone techniques which used dedicated single microphone patterns.

Hence it should be no surprise if localization resolution and—for sure—spatial impression proved worse for the EM32 1st and 4th order recordings -in relation to the other techniques- in subjective listening tests.

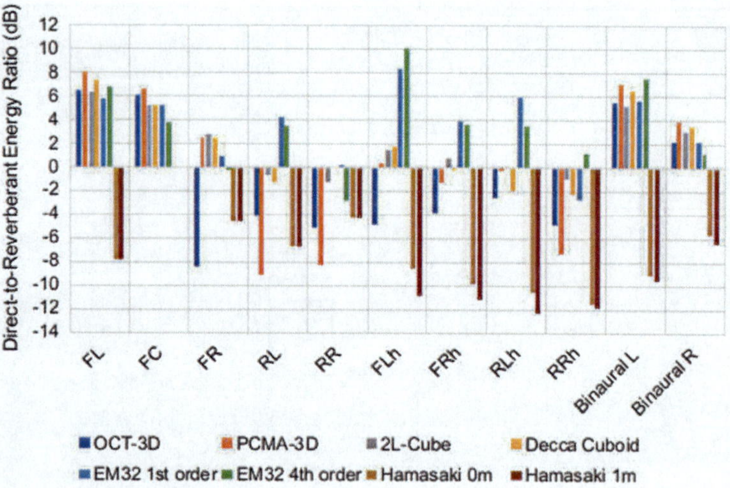

Fig. 1.10 Direct-to-Reverberant Ratio (DRR) for each microphone and ear-input signal (Fig. 9 from Lee and Johnson 2020)

The overall significantly higher signal correlation in both EM32 Ambisonics recordings shows very clearly also in the results of the inter aural cross correlation measurements, which were derived from a binaurally synthesized 9-channel 3D loudspeaker system and are displayed in Fig. 1.11.

While the IACCs (Inter Aural Cross-Correlation Coefficients) for the various 3D microphone arrays usually stay below a value of 0.4 for the base layer, as well as the height layer (with the exception of the base-layer signals of the PCMA-3D and OCT-3D techniques), for the EM32 1st order and EM32 4th order these values are usually above 0.6, i.e. highly correlated. In broad terms: the Ambisonic techniques sound more 'narrow' in comparison to the other techniques.

A very interesting and up-to-date comparison between several first-order Ambisonics, as well as higher-order Ambisonics commercially available microphones including the Sennheiser 'Ambeo,'Core Sound 'TetraMic'; SoundField 'MKV,' mh Acoustics 'Eigenmike®' (with 32 transducer elements) and Zoom 'H2n' was carried out by Enda Bates et al., about which they report in Bates et al. (2016, 2017), the outcome of which is also presented in more detail in Chap. 5 of this book on Higher Order Ambisonics.

As it turns out, it seems dubious whether HOA is really able to provide overall superior sonic results in comparison to FOA, as the increase of the number of capsules also seems to have its trade-offs: In (Bates et al. 2017) it is reported that "… As might be expected, the more elaborate designs of the Eigenmike® produced the best results overall in terms of directional accuracy, but with decreased performance in terms of timbre. As with part 1 of this study, this again suggests a trade-off between timbre quality and directionality, particularly when the number of individual capsules in the microphone is significantly increased."

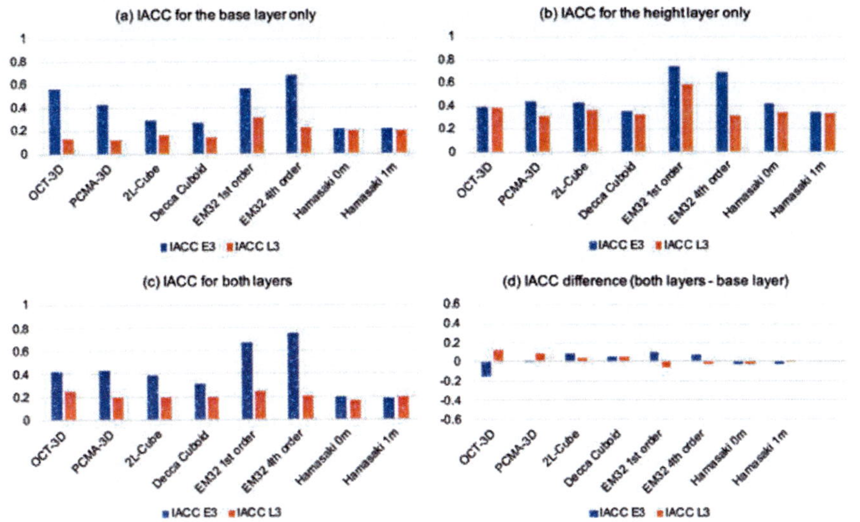

Fig. 1.11 Interaural cross-correlation coefficients (IACCs) for ear-input signals resulting from different microphone signals reproduced from a binaurally synthesized 9-channel 3D loudspeaker system (Fig. 8 from Lee and Johnson 2020)

It is therefore interesting to note that the FOA SoundField MKV microphone, despite being the oldest version of the Ambisonic microphones tested in (Bates et al 2016, 2017) still seems to be—'overall quality' wise—at par with the HOA 'Eigenmike'® with the latter being superior in terms of localization accuracy, but less favorable in terms of 'timbre' or sound-color …

Also in the results from the objective measurements on 3D microphone arrays by (Lee and Johnson 2020) there seems to be evidence that going to a higher order in Ambisonics does not necessary resolve all problems "… this might suggest that, in the current 9-channel loudspeaker reproduction, the well-known limitation of Ambisonic loudspeaker reproduction regarding 'phasiness' during head movement would still exist even at higher order." (from Lee and Johnson 2020)

However, this tendency that the overall sound quality is compromised for microphone techniques which are based on generating their output by combining signal content from various capsules seems to have found evidence also in the results of the comparative listening test by Kamekawa et al 2007, in which the DMS ("Double MS") technique repeatedly has scored as lowest (see Figs. 9.48, 9.49 and 9.50 in Chap. 9 of Pfanzagl-Cardone 2020).

This is why I have proposed the BPT ("Blumlein-Pfanzagl-Triple") microphone technique (see Pfanzagl-Cardone 2005), which uses signals, that are directly derived from three single capsules, which can be switched to a high directivity (i.e. figure-8), as this arrangement –while including also the possibility for signal matrixing or decording- in its direct form provides discreet, highly de-correlated signals for the (front) channels of a surround system (see Pfanzagl-Cardone and Höldrich 2008).

More information on this system, which takes advantage of its DFC being zero for the L, R capsule pairing (see Fig. 1.1. in Sect. 1.1), can also be found in Chaps. 3, 4, 6, 7, and 9 of Pfanzagl-Cardone 2020. A subsequent development of this technique for application in 3D audio is presented in Chap. 8 of this book (see section on the AB-BPT-3D technique).

1.3 Surround Microphone Case Study: OCT-Surround

With 5.1 surround applications, each of the full-range speakers involved accounts roughly for about one fifth (or 20%) of the overall sound energy which finally arrives at the ears of the listener (neglecting the fact that the front channel signals will likely be louder than the rear channel signals in case of a 'standard' sound source positioning in front), hence it makes very much sense to go into a detailed analysis of inter-channel phase- and signal-content relations in the form of a frequency dependent cross-correlation analysis ('FCC'), as was performed in my studies (see Pfanzagl-Cardone and Höldrich 2008; Pfanzagl-Cardone 2020).

Figure 1.12, as an example, shows the set-up details of the OCT-Surround ("Optimal Cardioid Triangle") system as proposed by Theile (see Theile 2000, 2001) (Fig. 1.13).

Despite the use of directional capsules it can be noted that signal cross-correlation is relatively high, both in the front system (for the capsule pairings L, C [and also R, C which is not shown]), as well as in the rear system LS, RS, but also between the front and the rear system (see the L, LS pairing). If we consider a cross-correlation value of +0.6 to be 'highly correlated' then this is the case for signal content in the front system below 400 Hz, as well as signal content below about 150 Hz in the rear system (see LS, RS pairing).

However, even from the detailed, frequency dependent signal cross-correlation measurements as documented in Fig. 1.11 it would not be easy to make deductions

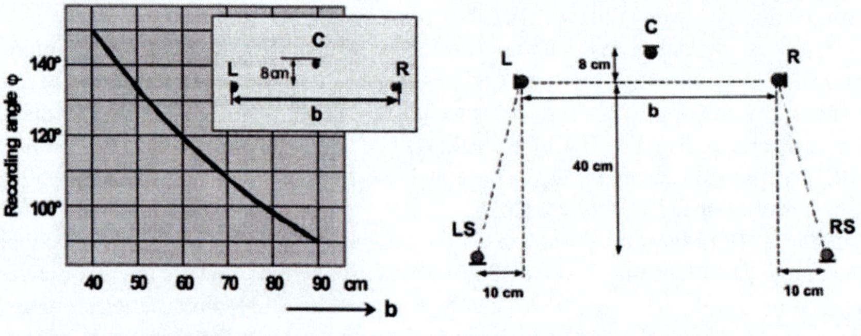

Fig. 1.12 OCT surround: Left—relation between recording angle and capsule spacing; Right—schematic capsule layout; L + R front mics are supercardioids, center and rear mics are cardioids (Fig. 4.29 from Pfanzagl-Cardone 2020; composed from Theile 2000, 2001)

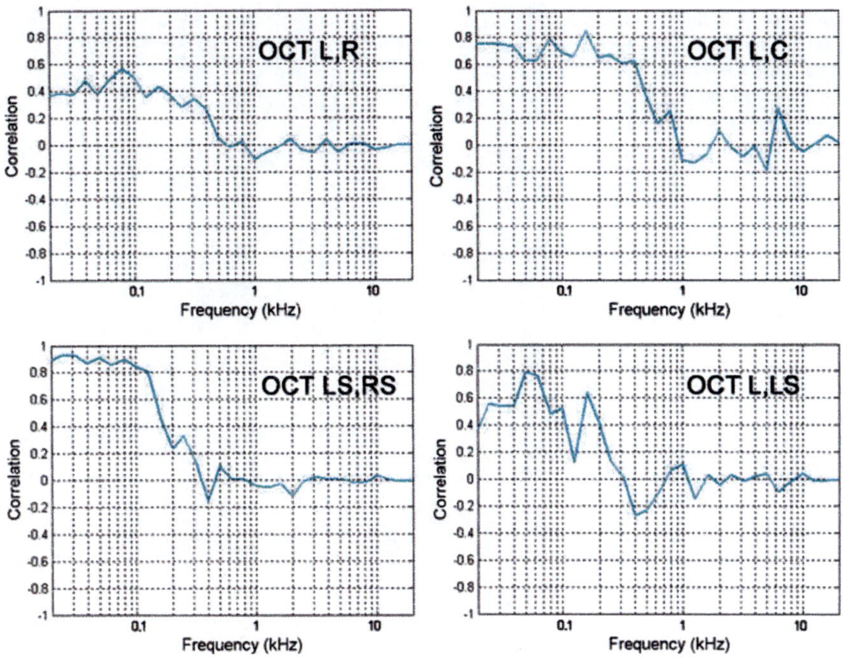

Fig. 1.13 Paired channel correlation over frequency for OCT surround microphone system signals (2048-point DFT (Discrete Fourier Transform) grouped to 31 frequency bands, 1/3rd octave; center frequencies according to ISO; orchestral music, 60 s, concert hall); capsule spacing for L/C and C/R is 31 cm, capsule spacing for L/LS, R/RS is 41 cm, capsule spacing for LS/RS about 80 cm (Fig. 7.13 from Pfanzagl-Cardone 2020)

for the resulting overall sound field and sound pressure levels, which will arrive at the ear membranes of a listener.

Therefore it was decided to introduce an artificial human head (Neumann KU81i) as a kind of 'human reference' in order to be able to break down the complex sound field in the listening room to a much simpler, 2-channel binaural result, by re-recording the replayed output of the OCT 5.1 surround recording to the same dummy head, which had been used in the concert hall (Figs. 1.14 and 1.15).

The resulting sound fields, or respective binaural FIACCs (Frequency-dependent InterAural Cross-correlation Coefficients) from the re-recording setup in Fig. 1.11 are displayed in Fig. 1.16 for four surround microphone systems (OCT, DECCA, KFM - Kugelflächenmikrofon and AB-PC—AB-Polycardioid Centerfill).

In Fig. 1.16, the FIACC for the OCT microphone array shows high correlation in the LF-band for frequencies below 800 Hz and in this respect is similar to both the DECCA as well as the KFM mic-array techniques, but relevantly different to the AB-PC system where this frequency lies about one octave below, at 400 Hz. This difference in terms of 'border-frequency' of just one octave (800 Hz vs. 400 Hz) may seem negligible, but in reality is of relevant importance in relation to the spatial

Fig. 1.14 Neumann KU81i artificial human head set up in the sixth row at the Salzburg Festival hall for a binaural recording (Fig. 6.3 from Pfanzagl-Cardone 2020)

impression, which can be achieved for human listeners, as will be shown in the two paragraphs below.

As can be seen in, the FIACC of the AB-PC system was the closest to the original FIACC of the Neumann KU81i as captured in the concert hall (dotted line). The other three systems displayed much higher correlation values for high frequencies and only the AB-PC system followed closely also for frequencies below 600 Hz. This is of major relevance for the spatial quality of a recording, as previous research has shown that a high degree of signal decorrelation at frequencies below 500 Hz is key to creating a convincing "spatial impression" in human listeners. (see—among others—Baron and Marshall 1981; Griesinger 1986; Hidaka et al 1995; Hidaka et al 1997; Beranek 2004).

The OCT system seems to be characterized by good localization properties, but the overall spatial impression is degraded due to high signal-correlation at low frequencies, as evident in Fig. 1.13, which has also been noticed and commented by practicing

Fig. 1.15 Re-recording of the surround microphone signals by means of a Neumann KU81i artificial human head in the sweet spot of the control room of the IEM—Institute of Electronic Music and Acoustics at the KUG University of Music and Performing Arts, Graz, Austria (rear-channel loudspeakers outside of the picture; rem: clothes placed on and around the mixing desk to avoid unwanted reflections) (Fig. 6.6 from Pfanzagl-Cardone 2020)

sound engineers (see Appendix B, pg. 395 in Pfanzagl-Cardone 2020, downloadable here as "backmatter" PDF: https://link.springer.com/content/pdf/bbm%3A978-3-7091-4891-4%2F1.pdf).

Further evidence of this relatively high cross-correlation value found with the OCT-technique, due to the rather small capsule distance between the backward facing cardioid microphones and the front microphones of only 40 cm, also shows up in the measurements of (Lee and Johnson 2020), as displayed in Fig. 1.11 (see OCT-3D 'IACCs for the base layer only', E3 early segment in blue).

1.4 Naturalness and Related Aspects in the Perception of Reproduced Music

I have conducted a more details analysis on "Naturalness and Related Aspects in the Perception of Reproduced Music" (Pfanzagl-Cardone 2012) based on the data of the comparative mic-array listening test concerning the AB-PC, DECCA, KFM and OCT

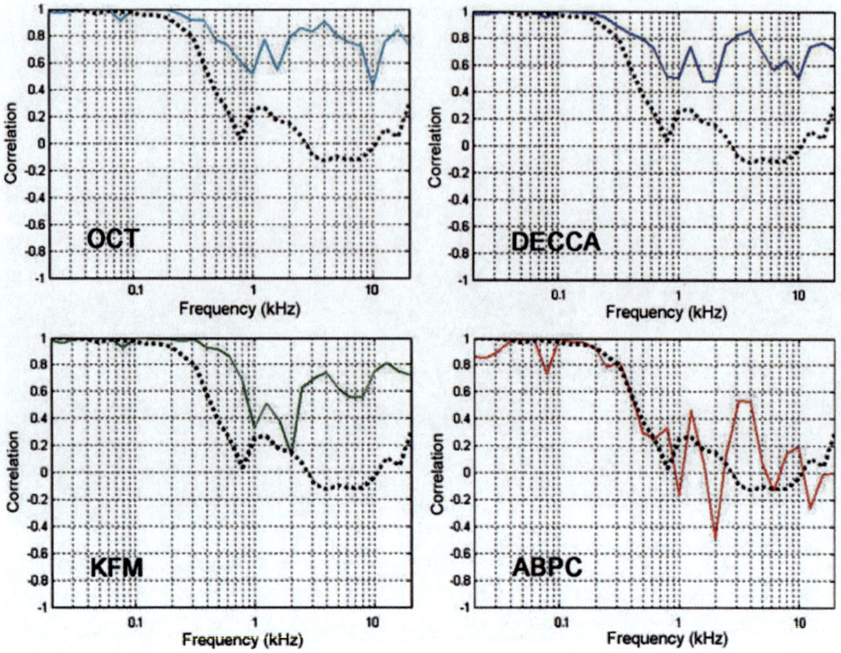

Fig. 1.16 FIACC of various surround-microphone techniques (OCT, DECCA, KFM, ABPC) as re-recorded through a Neumann KU81i artificial head (solid line) compared with the original FIACC, recorded through the same artificial head in a "best seat" position (see Fig. 1.10) in the concert hall (dotted line); (orchestral sample of 60 s duration, soft to loud dynamics) (from Fig. 7.4 in Pfanzagl-Cardone 2020)

surround microphone techniques. The results of the 'ORCH5.1' listening test (see Pfanzagl-Cardone and Höldrich 2008) were subject to correlation analysis, based on the Pearson product moment correlation coefficient, using MATLAB software.

It seems of primary interest to note which attributes have the highest correlation to listener preference: These were naturalness ($r = 0.84$), balance ($r = 0.69$), spatial impression ($r = 0.68$), sound color ($r = 0.67$) and width ($r = 0.65$).

The attributes, which have the highest correlation to naturalness are sound color (0.75) and spaciousness (0.64). Stability, balance and localization follow with correlation values of around 0.6. Apart from preference and naturalness, sound color is most strongly correlated with spaciousness (0.57), followed by localization and width, both with values around 0.51.

Apart from preference and naturalness, spaciousness is strongly correlated with width (0.65). Not surprisingly, it is also thoroughly correlated with reverberation (wet/dry balance) ($r = 0.51$).

A listing of the six strongest correlations, as found in the test can be found in (Table 1.1).

Table 1.1 Highest attribute correlations found for the listening test ORCH 5.1 (Table 6.3 from Pfanzagl-Cardone 2020), originally from (Pfanzagl-Cardone and Höldrich 2008)

Attribute combination	Corr.-coefficient
Preference—naturalness	0.84
Naturalness—sound color	0.75
Preference—balance	0.69
Preference—spaciousness	0.68
Preference—sound color	0.67
Spaciousness—width	0.65

1.5 Inter Aural Cross-Correlation (IACC) and the Binaural Quality Index of Reproduced Music (BQI_{rep})

As the proposed AB-PC microphone technique had received the highest listener preference ratings (in both loudspeaker, as well as binaural listening tests; for more details see Chap. 6 of Pfanzagl-Cardone 2020) among the mic-techniques under test, it seemed worthwhile to further investigate possible underlying reasons. Given the high correlation between listener preference and spatial impression of $r = 0.68$, as pointed out in the previous paragraph, research was undertaken, making use of a newly defined "Binaural Quality Index of Reproduced Music" (BQI_{rep}), which had been derived from the "Binaural Quality Index", as defined in (de Keet 1968) for the listener evaluation of concert hall acoustics:

$$BQI = (1 - IACC_{E3}) \tag{1.1}$$

While IACC stands for the Inter Aural Cross-correlation Coefficient, subindex E3 denotes the early sound energy in the time window from 0 to 80 ms in the octave bands with center frequencies at 500 Hz, 1 kHz and 2 kHz. In order to achieve a high Binaural Quality Index rating - ideally close to the value of 1 -, the $IACC_{E3}$ measured at the eardrums of the listeners in the concert hall should be as low as possible, meaning that the signals presented to both ears should be as different as possible in the three octave bands around 500, 1000 and 2000 Hz. The importance of the BQI as a valid measure was reconfirmed in the research of Beranek, which has found the BQI to be " … one of the most effective indicators of the acoustic quality of concert halls" (see Beranek 2004).

In analogy to the BQI, the BQI_{rep} is defined as:

$$BQI_{rep} = (1 - IACC_3) \tag{1.2}$$

with $IACC_3$ being the mean value of the cross-correlation coefficients of the octave bands 500 Hz, 1 kHz and 2 kHz measured between the L and R binaural signals (no time-window applied).

(Rem.: MATLAB-code for calculating the BQI_{rep} from binaural samples can be found in Appendix C, pg. 404 of Pfanzagl-Cardone 2020, free for download here: https://link.springer.com/content/pdf/bbm%3A978-3-7091-4891-4%2F1.pdf).

In connection to this, the acoustics of the listening room, as well as the actual radiation characteristic of the replay-loudspeakers involved is of course also of major importance. Standing wave phenomena and room modes will largely influence the sonic outcome at the 'sweet spot' (or 'reproduction sphere', in the case of 3D audio) as was pointed out—among others—in (Griesinger 1997) as well as (Toole 2008).

Taking a look at the horizontal and vertical radiation characteristics of the d&b Y7, a professional grade loudspeaker, as frequently used in current sound reinforcement systems, we notice that—due to the laws of acoustics and the increase of wavelength λ—with lowering frequency the dispersion characteristic of the loudspeaker becomes more and more omnidirectional. In fact, for this model, below 500 Hz the loudspeaker radiates at an angle larger than 180° in the horizontal plane, while maintaining a quite steady radiation characteristic of about 80° for frequencies above 1.5 kHz, due to the use of a 'constant directivity' horn.

In the vertical plane, directivity also drops off below 500 Hz, but at a slower pace (see Fig. 1.17).

As a consequence, there is the obvious danger that in 3D multichannel loudspeaker systems, IACC at the "reproduction sphere" listening position will be rather high for low frequencies, as the loudspeakers radiate with very low directivity in this frequency range.

In addition, due to sound being able to bend around the human head at low frequencies, (for which the head diameter is very small in relation to the wavelength λ) the level difference between the left and right ear will be rather small, which makes localization and also spatial impression more difficult to achieve for the human listener.

Already in 1907 Lord Rayleigh published his findings that sound with a wavelength smaller than the diameter of the human head will effectively be shaded off for the ear on the other side, which results in an Interaural Level Difference (ILD) between the two ears. In addition to that, sound takes a different amount of time to

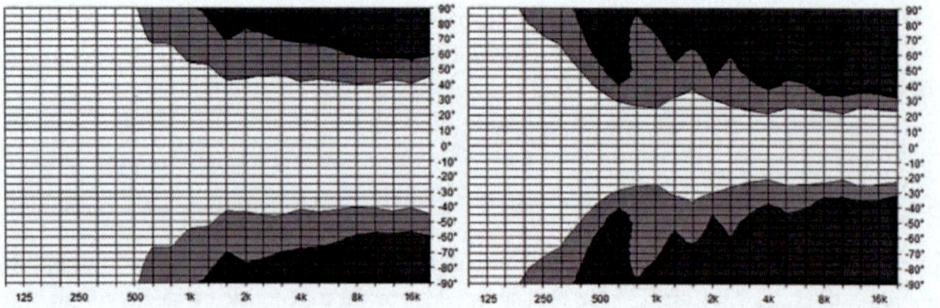

Fig. 1.17 Dispersion angle over frequency of d&b model 'Y7' loudspeaker, plotted using lines of equal pressure (isobars) at −6 dB (grey) and −12db (black); left: horizontal dispersion, right: vertical dispersion (loudspeaker vertically oriented) (reproduced with kind permission of d&b Germany)

arrive at both ears, which results in Interaural Time Differences (ITDs). (Rayleigh 1907).

More detailed research into this was undertaken by Steinberg and Snow (see Fig. 1.18 from Steinberg and Snow 1934).

The perceived change in terms of spaciousness toward low frequencies for a sound source presented to the human listener can be checked by listening selectively to isolated frequency bands to see (or better: hear) which spatial impression a microphone technique is able to provide in different frequency ranges. In this respect, the entire frequency band below approx. 800 Hz is especially important, as the human head is not yet effective as a baffle and sound signals bend around it.

Above 800 Hz, the shadowing effect of the human head becomes more and more evident and thus human hearing is based mainly on ILD (Interaural Level Difference), while at low frequencies it is mainly based on an analysis of phase and time differences (More details on the fundamentals of spatial hearing can be found in Chap. 1 of Pfanzagl-Cardone 2020).

In this context, research by Yost et al. (1971) can be mentioned, which has shown that the low-frequency components of transient binaural signals are of highest importance for localization: the high-pass filtering of clicks (similar to Dirac impulses) with a cutoff frequency at 1500 Hz leads to a clear deterioration of localization, while low-pass filtering with the same frequency resulted only in a minimal change (deterioration) of localization.

The risk of a deterioration of spatial impression (simply due to law of acoustics when applied to a standard loudspeaker layout) becomes evident already when only 2 loudspeakers are used, which was pointed out in (Hirata 1983).

In his research Hirata has dealt with the phenomenon of localization distortion of the low-frequency components of a stereo signal upon loudspeaker playback. He proposes PICC, a "Perceptual Inter Aural Cross-correlation Coefficient":

$$\mathrm{PICC} = \mathrm{DR_0} + (1 - \mathrm{D})\mathrm{R_E} \tag{1.3}$$

with D (Deutlichkeit), as defined by R. Thiele (see Thiele 1953)

$$\mathrm{D} = \frac{\int_0^{50\ \mathrm{ms}} p^2(t)\mathrm{d}t}{\int_{0\ \mathrm{ms}}^{\infty} p^2(t)\mathrm{d}t} \tag{1.4}$$

with:

R_0 the interaural cross-correlation coefficient of direct sound (which is 1 for sound from 0°),

R_E the interaural cross-correlation coefficient of diffuse sound, which is defined as:

$$\mathrm{R_E} = \frac{\sin \mathrm{kr(f)}}{\mathrm{kr(f)}} \tag{1.5}$$

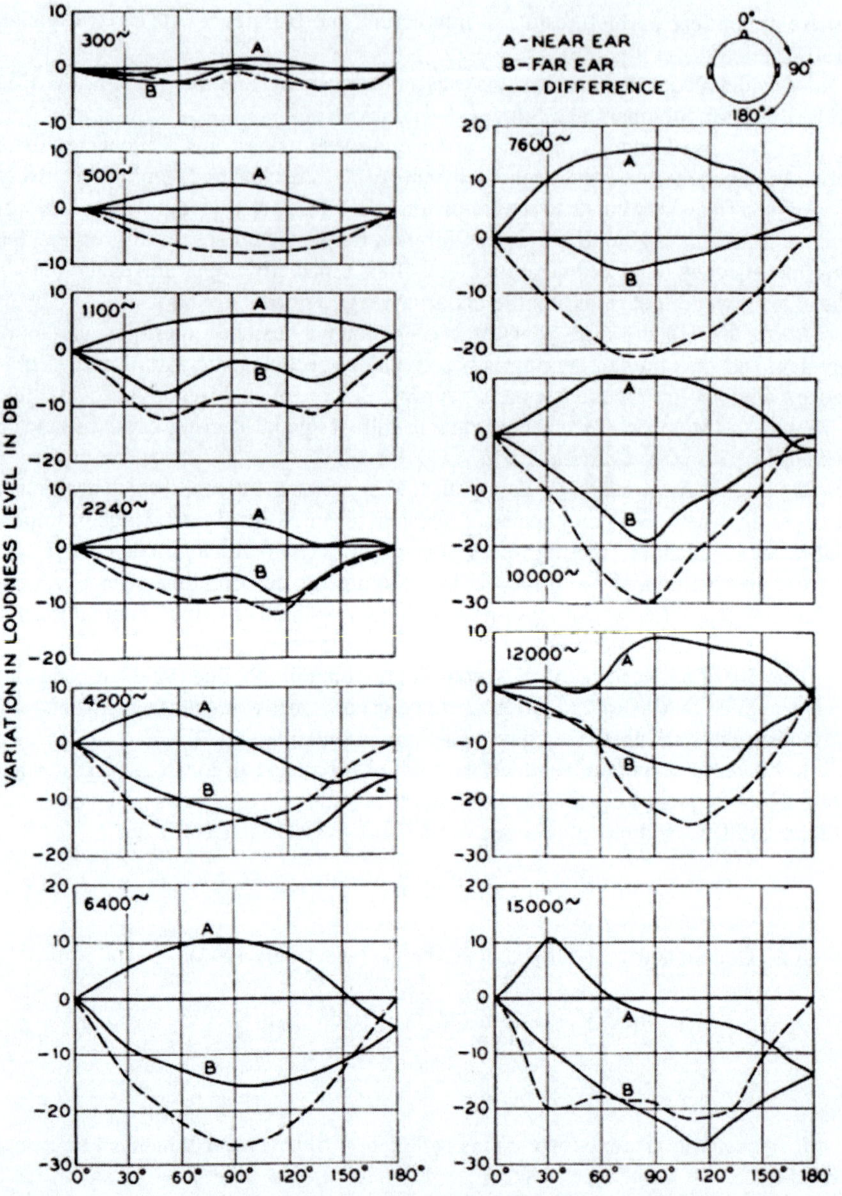

Fig. 1.18 Interaural Level Difference over source angle for different frequencies (graphic from Steinberg and Snow 1934)

and the wavenumber k as: $k = 2\pi f/c$ with c, the speed of sound and $r(f)$ the effective acoustic distance between the human ears, which is 30 cm [see (Yanagawa et al. 1976), as well as (Suzuki and Tohyama 1981)].

In addition he defines ASI, an "Index of Acoustic Spatial Impression", as follows:

$$\mathrm{ASI} = 100(1 - \mathrm{D})\ (\%) \tag{1.6}$$

Total spatial impression signifies ASI = 100%, while the complete absence of spatial impression is ASI = 0%.

Figure 1.19 shows that in a standard listening room ($RT_{60} = 0.3$ s), ASI is small for frequencies below 800 Hz, but high for frequencies above 800 Hz in comparison with ASI at a seat in the concert hall, for which ASI equals 60%.

Research by Griesinger concerning listener envelopment (Griesinger 1999) has shown that for rising frequency the ideal loudspeaker position moves toward the median plane. For frequencies below 700 Hz instead, a speaker layout is ideal which enables maximal lateral separation between the transducers, i.e. at a position left and right of the listener at ±90° (see Fig. 1.20).

Given the evidence of what was discussed above, I believe the importance of looking for 3D microphone arrays which offer high channel signal separation and are also characterized by sufficient signal decorrelation especially in the low frequency range below 500 Hz-in order to be able to convey convincing spatial impression via an appropriately laid out 3D multichannel loudspeaker system- cannot be over emphasized.

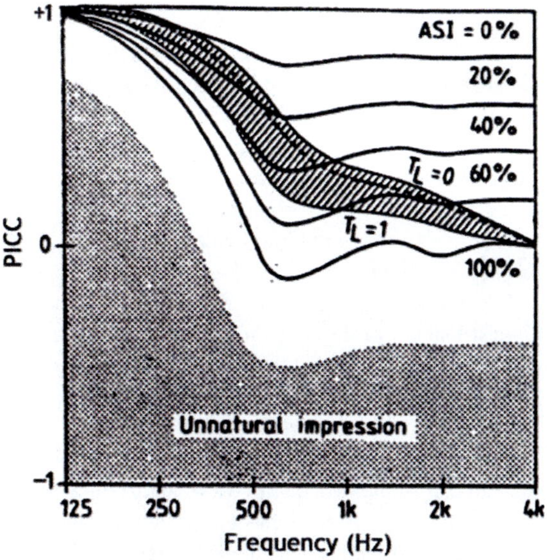

Fig. 1.19 PICC curves for stereo reproduction in a listening room with a reverb time T_L (0–1 s) show small ASI values at low frequencies compared to an ASI = 60% for the seats in the middle section of a concert hall. The dashed curve stands for $T_L = 0.3$ s (from Pfanzagl-Cardone 2020, originally from Hirata 1983)

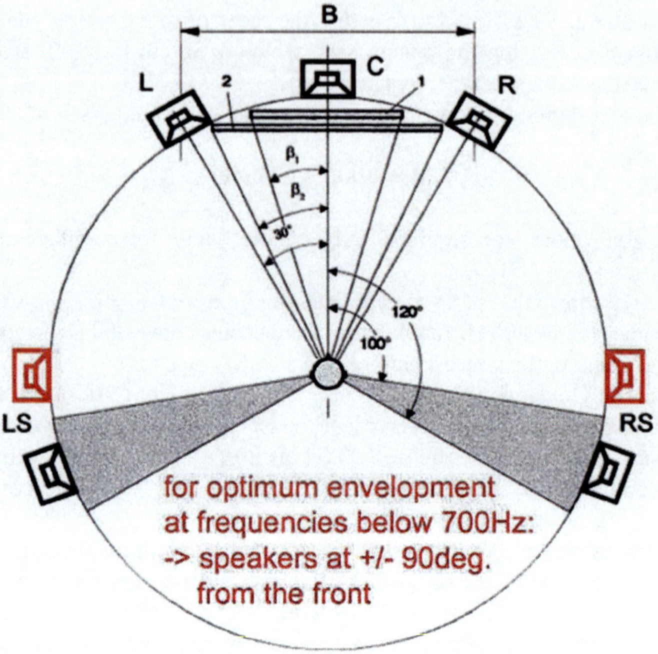

Fig. 1.20 A '5.2' arrangement with subwoofers at the sides at ±90°,optimized for the spatial reproduction of low frequencies, according to the recommendation by Griesinger (1999) (from Pfanzagl-Cardone 2020)

Since this applies also to stereo and surround recording and replay systems, I have tried to point out these fundamentals already almost 20 years ago in an AES—as well as a VDT-convention preprint (see Pfanzagl-Cardone 2002; Pfanzagl 2002).

Returning to the above mentioned BQI_{rep} and the results of my own research into dedicated 5.1 surround microphone arrays, it was very interesting to see the actual importance of their performance in terms of low-frequency signal separation (or decorrelation) as put into evidence by the BQI_{rep} measurements, documented in Table 1.2.

It turned out that AB-PC, the microphone technique, which had received the highest listener preference ratings in the subjective listening tests, was also the one which was closest in terms of the resulting rho (i.e. signal cross-correlation coefficient) and BQI_{rep} values in relation to the reference values of the KU81 dummy head recording from the concert hall (see Table 1.2).

As I have already pointed out above, with 5.1 surround, each full-range loudspeaker contributes to a maximum of approximately 20% of the overall signal content; in 3D audio instead, which has a much higher count of replay channels, the percentage is reduced accordingly. Due to the higher number of loudspeakers involved, the inter-channel signal relationships tend to become more complex, and at the same time the importance of the contribution of each single loudspeaker signal is diminished.

Table 1.2 BQI_{rep} and binaural signal correlation values 'rho' measured for octave bands (with center frequencies at 500, 1000, 2000 Hz), based on acoustic measurements of the re-recordings of a sample of orchestral music (60 s duration) as captured through the surround microphone techniques OCT, DECCA, KFM and AB-PC by means of the Neumann KU81 dummy head; also the original value of the KU81 dummy head recording from the concert hall is displayed (i.e. Reference, **bold**); *italics* denote the rho and BQI_{rep} values that are closest to the reference

MicArray	rho500	rho1000	rho2000	BQIrep
OCT	0.83	0.64	0.66	0.29
DECCA	0.77	0.61	0.57	0.35
KFM	0.90	0.61	0.39	0.37
AB-PC	*0.43*	*0.26*	*0.11*	*0.73*
KU81	**0.41**	**0.19**	**0.20**	**0.73**

1.6 Case Study: Intercapsule Signal Correlation in a Hamasaki Square

I would like to take a look at a specific mic-array, devised for the capturing of diffuse sound, which is commonly used in both surround, as well as 3D audio microphone systems: the "Hamasaki Square" (for 2D surround) and its successor, the "Hamasaki Square + Height" mic-array system (3D audio). For this example we want to reduce our analysis to the Hamasaki square, which forms also the basis of the Hamasaki Cube, to which a top-layer of upward facing super-cardioids is added for capturing height information (see Fig. 1.21). (see also the Hamasaki-Cube in Chap. 10, Fig. 10.16)

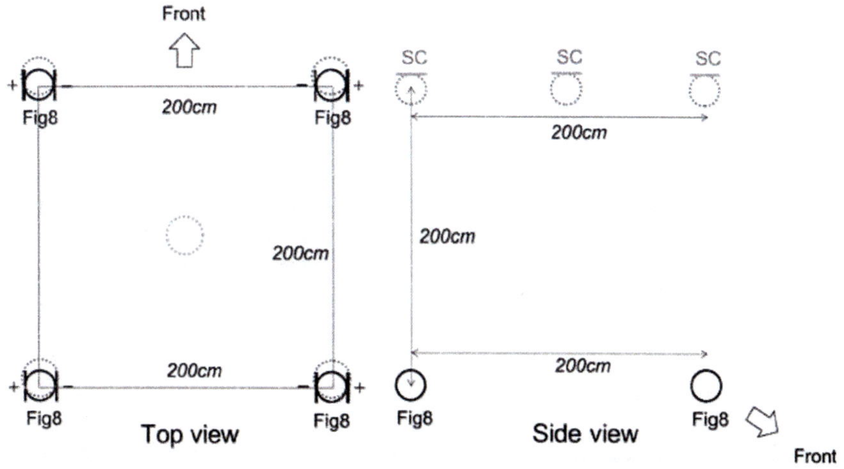

Fig. 1.21 Top and side views of 'Hamasaki Square + Height'. The solid black and dotted grey circles represent the middle and upper layer microphones, respectively. SC = super-cardioid, figure-8 = figure-of-eight. (Fig. 8 from Lee 2021)

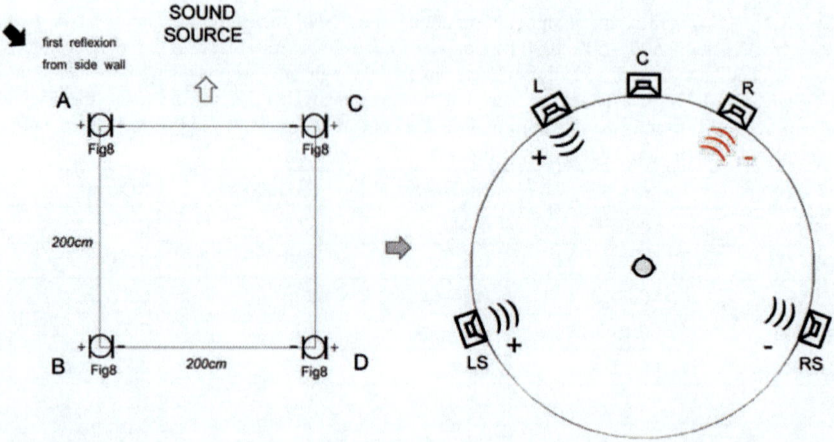

Fig. 1.22 Analysis of phase relationship of Hamasaki-square signals replayed through a 5.1 surround loudspeaker system; LFE-subwoofer not pictured (using graphical elements from Lee 2021, Fig. 8)

The Hamasaki Square was designed with the aim of optimal attenuation of direct sound (from the front) in mind, while focusing on diffuse-sound pickup, hence the use of the laterally oriented Fig8 microphones (see Hamasaki 2003).

Now we want to consider the case of a first reflection on the left sidewall of a concert venue, which impinges on the membranes of the four figure-of-8 microphones, in order to examine the resulting phase relationships upon playback for the mapping to a standard 5.1 surround loudspeaker configuration (Fig. 1.22). As is common practice, in this case the signals of the front Fig. 8s will be assigned by the sound engineer or Tonmeister to the front speakers L, R with appropriate level (to which also signals from potential front main microphone systems will be routed or mixed), and the signals of the rear fig-8s will be assigned the surround speaker-channels LS and RS.

While the first reflection from the left side wall of a concert hall will be picked up 'in-phase' by the A and B fig-8s of the Hamasaki square in Fig. 1.22, the same signal will be picked up slightly later -and with inverted polarity- by the figure-8 microphones C and D, which form the right half of the Hamasaki square.

When being put through the 5.1 loudspeaker setup in Fig. 1.22, the signal information replayed via the LS and RS speaker will complement each other in an constructive manner, as the two speaker-membranes will move in an almost'coherent' fashion: when the LS-membrane moves out to reproduce a positive λ half-wave, the RS-membrane will move in, which means it moves in a similar direction, as the LS an RS loudspeakers are at opposing positions and orientations to each other (in relation to the listener). Therefore the overall soundwave movement for the positive half-wave will be 'from left to right'. Unfortunately this is not the case for the signals of the A and C figure-8 microphones which are replayed through the L and R front-speakers: here, the listener will experience two very similarly sounding signals, which are emitted with opposed polarity from these two front speakers, which is problematic—to say the least- and under worst conditions could lead to the impression of 'phased sound'.

A similarly problematic situation occurs, when replaying the Hamasaki-square signals through the side and rear speakers of a 7.1 surround loudspeaker setup (Fig. 1.23).

In that situation, the A and B signals will usually be mapped to the LS side speaker and LR left rear speaker, while the C and CD signals will be mapped to RS and RR.

In respect to the first reflection, the membranes of LS and RS will move coherently, but the signals radiated from the LR and RR speakers will be of inverted signal polarity, which may be detrimental to the overall sonic impression (Fig. 1.23).

Now in connection to all that was said above, something else deserves consideration: the fact that the particular way in which the four figure-of-eight capsules are arranged in the Hamasaki-square will lead to a characteristic signal cross-correlation, the fundamentals of which are analyzed in (Elko 2001). While the A and B (as well as the C and D) capsule each form a pair -similar to a Faulkner "Phased Array", albeit with a much larger capsule spacing, as the traditional Phased Array spacing of Faulkner's two 'parallel figure-8s' is only 20 cm-, the combination of the A and C, as well as the B and D capsule can be considered to be two 'serial figure-8s'. (with the C and D capsule of 'inverted polarity', to be precise).

The resulting signal coherence (in respect to the diffuse sound of a spherical isotropic sound field) from two serially arranged figure-of-eight microphones can be seen in Fig. 1.24 (dotted line). We notice that signal coherence converges towards 1 for kr dropping below 2 and that it reaches a first maximum at kr = 4 with a relatively high value of about 0.55, as well as a rather low coherence of 0.15 around a value of kr = 7.3.

To explain the constants and variables used in Fig. 1.24:

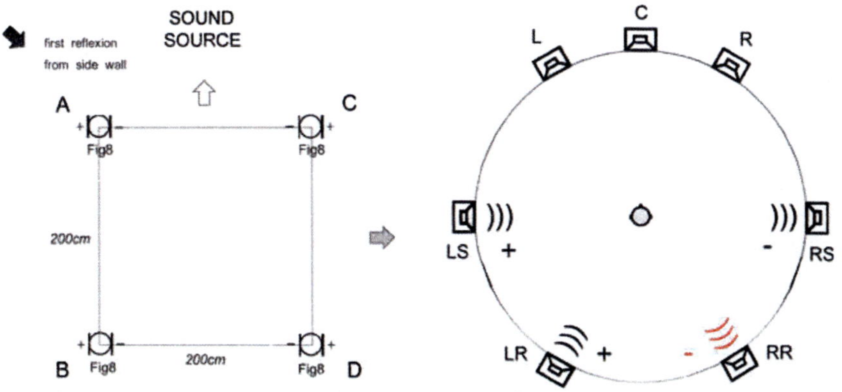

Fig. 1.23 Analysis of phase relationship of Hamasaki-square signals replayed through a 7.1 surround loudspeaker system; LFE-subwoofer not pictured (using graphical elements from Lee 2021, Fig. 8)

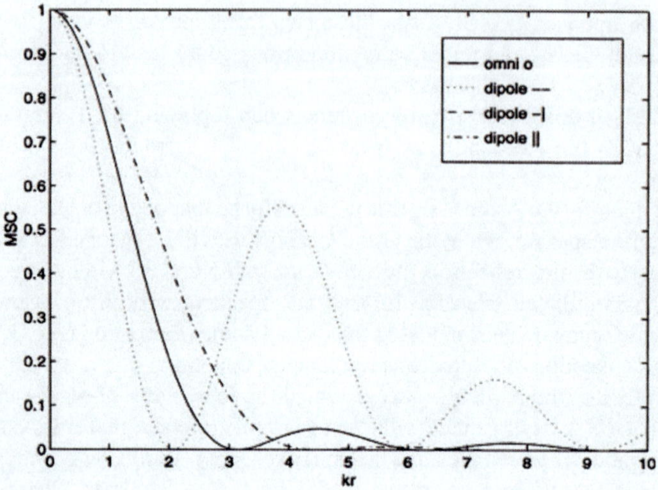

Fig. 1.24 Magnitude-squared coherence (MSC) for omni-directional and bidirectional (i.e. Figure 8) microphones in a spherical isotropic sound field. Note that the coherence function for the orthogonal fig-8s coincides with the *x*-axis (from Elko 2001)

$$R = (\sin kr)/kr \tag{1.7}$$

with: $k = 2\ \pi/\ \lambda$ (k = wave-number; λ = wavelength), therefore

$$kr = 2\pi r/\lambda \tag{1.8}$$

or $k = \omega/c$, since $\lambda = c/f$ (with c … speed of sound, i.e. approx. 340 m/s)

The complete deduction can be found in (Elko 2001).

Figure 1.24 contains interesting information for some basic 2-channel main microphone techniques (from top to bottom):

- AB (with two omni capsules spaced by r),
- two serial figure-of-eight capsules, spaced by r (dotted line; "Half Hamasaki-Square").
- two crossed figure-of-eight capsules, spaced by r (dashed line, not visible as it coincides with the x-axis; the "Blumlein-Pair" coincident figure-of -eights is a common application)
- two parallel figure-of-eight microphones, spaced by r (dot-dash line; in the style of Faulkner's "Phased Array").

Taken from (Elko 2001), in Chap. 2 of (Pfanzagl-Cardone 2020) we find a similar figure, which contains the coherence values for various combinations of omni with dipole (i.e. figure-of-eight) and omni with cardioid microphones, the first of which is essentially an MS-technique with various orientations of the figure-of-eight microphone (rem.: MS stands for Mid-Side technique, which combines an omni-capsule with a side-facing figure-of-eight capsule).

Now we come back to our analysis of the signal coherence, which applies to the figure-8 microphones in the 'half Hamasaki square' with 200 cm spacing. Considering the first coherence maximum of value 0.55, which is reached for $kr = 4$: through substituting k with $2\,\pi/\lambda$, we obtain the corresponding frequency of 108 Hz. For the second, smaller coherence maximum, which is reached for $kr = 7.3$, we obtain a corresponding frequency of 197.5 Hz.

In the low frequency region, below $kr = 2$ (which means below 54 Hz in our current example with $r = 200$ cm) the cross-correlation of the "half Hamasaki-Square" fig-8 microphone signals is somewhat similar to spaced omni capsules, which would exhibit the same tendency of an increase towards a value of 1 with frequency lowering towards zero. However, the fig-8s cross-correlation differs significantly for the fact that at $kr = 4$ they exhibit this highly coherent 'hump', which applies not only for diffuse sound, but also for direct sound.

It is interesting to note that—in comparison—the signal output of a coincident 'Blumlein-pair' combination of crossed figure-of-eight microphones is ideally decorrelated over the entire frequency range, which is also displayed in its DFC value of 0 (remember Fig. 1.1).

Now for the 200 cm spacing and the moderately high cross-correlation maximum of 0.55 at a frequency of 108 Hz, the psychoacoustic effect—when replaying through a 5.1 or 7.1 surround loudspeaker system—should not be too detrimental in terms of achieving a good sense of spatial impression, which—ideally—calls for complete decorrelation at frequencies in the low frequency region below 500 Hz.

What is not evident in Fig. 1.24 is the fact that the signal content of the two channels, which are analyzed, at the first coherency maximum at $kr = 4$, is actually of inverted polarity, as the Magnitude-squared Coherence estimate does not take into account the signal phase relationship. However, a comparison with Fig. 2.4 from Pfanzagl-Cardone 2020 shows that the same is true for the coherence function of two omni capsules, which clearly show a first "out-of-phase maximum" around $kr = 4.5$.

With our above analysis of the cross-correlation characteristics of the "Half Hamaski Square" we see that inter-capsule signal cross-correlation characteristics can become quite complex, even with just a few capsules involved, considering also the fact that the Hamaski square alone already accounts for the direct combination of two parallel (A/B, C/D) and two serial (A/C, B/D) figure-of-eight capsules, not counting the relationships of the diagonally opposed figure-of-eight pairings A/D and B/C, which are likely to exhibit a 'crossover' characteristic half-way between Faulkner 'Phased Array' and 'Half Hamasaki Square'.

However: all these signal correlations matter at the end of the day, when the sound from each capsule -after being radiated trough the assigned loudspeaker in a multichannel 2D or 3D system- gets superimposed on the signals from the other microphone-capsules involved, when summing up at the entrance to the ear canal of the human listener in the sweet spot. If we take a look at a rather elaborate

3D microphone system layout in Fig. 1.25, which is comprised of a plenitude of microphones arranged in an upper, middle and bottom layer, we quickly understand that the interaction of all these signals is quite complex.

And while -due to the sheer number of signals involved- each single one may have much less importance for the final sonic outcome than in the comparatively 'Spartan' situation of a 5.1 surround setup with just 5 full-range signals, their signal cross-correlation of course matters.

1.7 Some Thoughts on Psychoacoustic Signal Interaction in Multichannel Microphone Array Systems

As is described also from a practical point of view in relation to the Auro-3D system by Van Baelen in (AURO-Technologies NV 2015): " … A large part of what we hear in a natural sound field is the reflections of sound around their sources. The goal of Auro-3D is to use a minimum amount of channels to achieve a good 'sound spread' in the hemisphere. *Every additional channel can become a possible cause for problems with phase, workflow, distribution, bandwidth* etc.… Therefore the goal of the Auro-3D format is to deliver the most immersive listening experience with the minimum amount of channels/speakers."

Research conducted on the acoustical characteristics at typical microphone positions in a music studio (see Yamamoto and Nagata 1970) has shown that the definition D of a typical 'industry standard' recorded source for symphonic music is 0.5. This is the result of a certain 'accepted' direct-/diffuse sound ratio for the final (2-channel stereo) mix of such type of music. Assuming that the taste of the average listener, regarding the direct-/diffuse sound ratio perceived at the sweet spot, has not changed with the technological shift from stereo via surround sound to 3D audio, it is clear that the direct-/diffuse sound ratio for the various channel signals must change drastically, depending on the respective reproduction format used: while in 2-channel stereo direct, as well as diffuse sound can be radiated only through the 2 front speakers, in surround we have—for the first time- the possibility to separate the directions of incidence of direct and diffuse sound for the listener.

While the radiation of direct sound will normally be reserved for the front speakers (L, C, R), diffuse sound should be radiated both through the rear (and side) speakers in a surround setup, as well as –normally to a lesser degree- through the front speakers, if a good sense of 'acoustic envelopment' is desired. This—apart from psychoacoustic considerations- will depend on the mixing practices and taste of the sound engineer, which is reflected also in a statement by Jürg Jecklin, Tonmeister of Swiss National Radio, when defining the principle characteristics for his OSIS (Optimal Sound Image Space) surround sound microphone system: "No image in space, no space in image" (in which 'image' refers to the representation of the sound source in the front channels, while the rear channel signals are reserved for reproduction of 'space', i.e. diffuse sound only) (see Jecklin 2002).

The experiment by Lee and Johnson, about which we have already reported above (see Lee and Johnson 2020, as well as Lee 2021), is currently among the most extensive ones, using a total of 71 microphones simultaneously, from which 17 different 3D-audio microphone setups can be derived (see also Fig. 1.8 in this context). Conducted by Prof. Hyunkook Lee, head of the APL—Applied Psychoacoustics Lab at Huddersfield University, the resulting analysis conducted so far is mainly about the inter-channel signal-correlations between the microphones. As sound source for the correlation measurements, a loudspeaker array, slightly above floor level and aimed at the microphone arrays, was used in the recording venue, St. Paul's church. Later-on other recordings were undertaken at the same venue, using real acoustic instruments in the form of a small ensemble, as well as single instruments. However, at the time of this writing (i.e. spring 2022) no formal subjective listener evaluation has taken place, but Prof. Lee has generously made the recordings freely available for research purposes, downloadable from this source: https://zenodo.org/record/3477602#.YgUh-6btxjs.

The main microphone arrays consisted of PCMA-3D (a layout based on proposals found in Lee and Gribben 2014, which is horizontally spaced and vertically coincident), OCT-3D (based on Theile and Wittek 2012), 2L-Cube (after Lindberg 2014), Decca Cuboid, First-order Ambisonics (FOA - Sennheiser Ambeo), Higher-order Ambisonics (HOA – Eigenmike EM32) and Hamasaki Square with height. In addition, ORTF, side/height, 'Voice of God' and floor channel microphones as well as a dummy head and spot microphones were included. The sound sources recorded were string quartet, piano trio, piano solo, organ, a cappella group, various single sources and room impulse responses of a virtual ensemble with 13 source positions captured by all of the microphones.

Looking at the rather elaborate multi-channel multi-layer microphone setups in Figs. 1.8 and 1.25 we notice that various microphone pattern characteristics are used, ranging from omni via wide-cardioid (WC) to cardioid (C) and super-cardioid (SC), and even Figure-of-Eight in the case of the Hamaski-square.

In connection to the resulting signal correlations, as documented in in Figs. 1.6, 1.7, 1.8, 1.9, 1.10 and 1.11 and the accompanying text, of course the question arises, whether there is an overall 'ideal' mic-pattern, which could be used in 3D-audio microphone setups, or—at least—if it was possible to relate a particular 'overall sound characteristic' to microphone systems which use essentially only one specific type of mic pattern. In this respect an interesting experiment, including a subjective listening test, applying repertory grid technique and MDS (multi-dimensional-scaling) methodology, was conducted at Tokyo University of the Arts. The results are reported in Kamekawa and Marui (2020), on which more information can be found in Chap. 10 at the end of this book, which deals with comparison and subjective listener evaluation of 3D-audio microphone systems. For details of the experiment see Figs. 1.26, 1.27, 1.28, 1.29 and 1.30.

In their study, the researchers compared three microphone techniques for 22.2 multichannel sound, namely a spaced microphone array (using 17 omni-microphones and 3 cardioids), a near-coincident microphone array using 24 shortgun-microphones, and a coincident microphone array (First Order Ambisonics FOA)

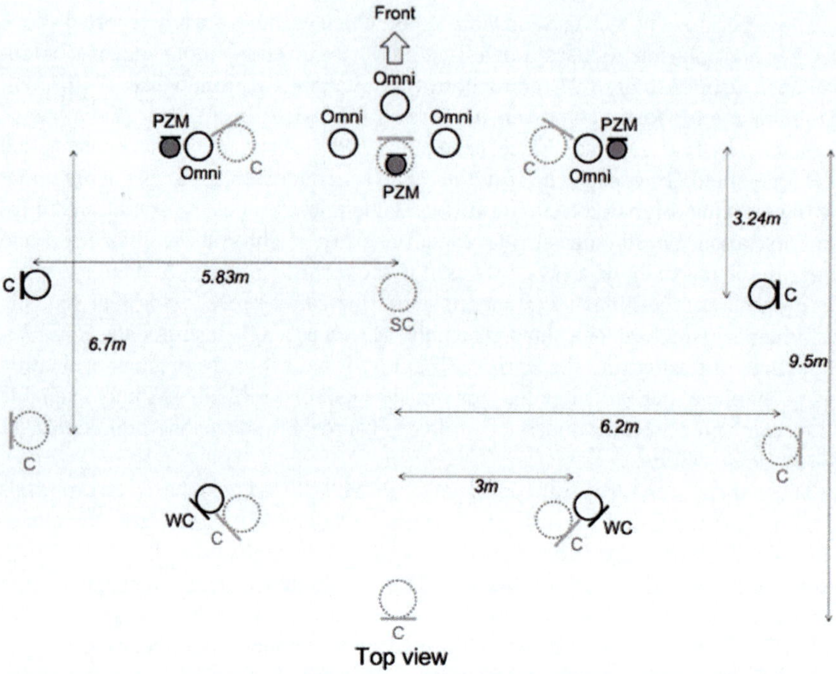

Fig. 1.25 Top view of Howie et al.'s 9 + 10 + 3 microphone arrangement (see Howie et al 2017) used for recording a large orchestra. The solid black and dotted grey circles represent the middle and upper layer microphones, respectively. The filled grey circles represent bottom layer microphones. This layout uses the Decca Tree as the main microphone array; middle layer height = 3 m, upper layer height = 5.5 m. PZM (Pressure Zone Microphones) at floor level (partial reproduction of Fig. 9 from Lee 2021)

First, the evaluation attributes were extracted by referring to the repertory grid technique. Through this, the following ten attributes where selected for listener evaluation: 'rich', 'bright', 'hard', 'wide sound image' (width), 'near', 'clear sound image' (clear), 'listener envelopment' (LEV), 'wide space' (broad), 'more reverberation' (rev), 'clear localization' (loc), 'the sense of being there' (presence), and 'preference' (pref). Using these attributes, participants had to compare the differences between these microphone techniques, including the difference in the listening position through two experiments. From the results it was observed that the difference, depending on the listening position, was the smallest in the spaced array. Also it was found that FOA gave the impression of 'hard'-sounding, the near-coincident array resulted as 'rich' and 'wide', while the spaced array gave the impressions 'clear' and 'presence'. Furthermore, 'presence' was evaluated from the viewpoints of clarity and richness of reverberation, with a negative correlation with the spectral centroid and a positive correlation with the reflection from lateral and vertical sides. [Rem.: the 'Spectral Centroid' is a measure used to characterize a frequency spectrum. It indicates where the 'Center of mass' of the spectrum is. Perceptually, it has a robust

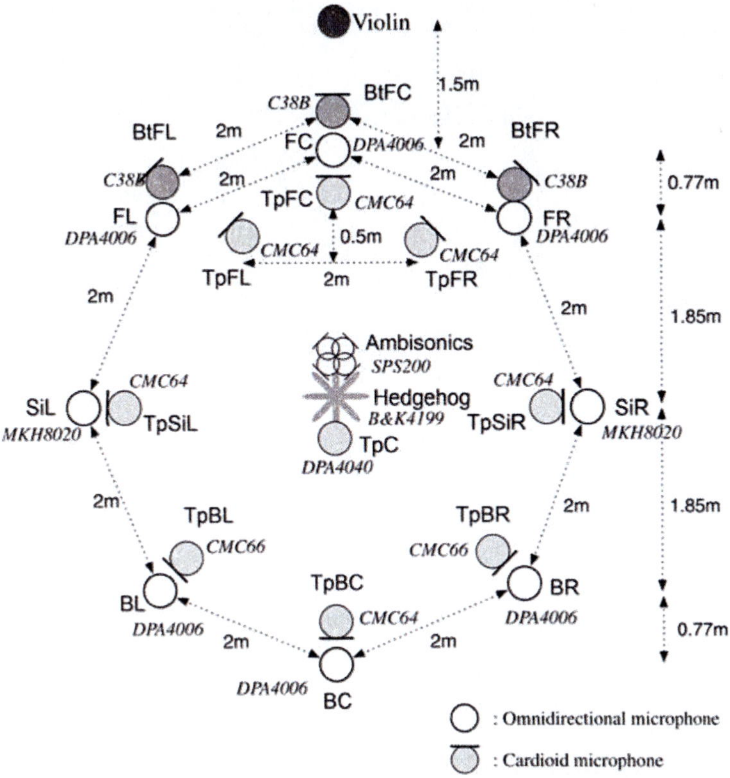

Fig. 1.26 Top view of the microphone layout as used in the experiment of (Kamekawa and Marui 2020) Italics indicate the model names of microphone (graphic is the upper part of Fig. 1 from Kamekawa and Marui 2020)

connection with the impression of 'brightness' of a sound (see Grey and Gordon 1978). It is calculated as the weighted mean of the frequencies present in the signal, determined using a Fourier transform, with their magnitudes as weights (see IRCAM 2003; Schubert et al. 2004)].

1.8 A Few Thoughts on Microphone Pattern Choice and Capsule Orientation in 3D Audio

Sound-engineers and Tonmeisters of course need to make an informed choice of microphone patterns for the use in their 3D-audio main microphone system, which will usually be strongly related to the chosen overall layout of the 3D-audio main microphone system, as well as the individual room acoustic situation ('wet' or 'dry'

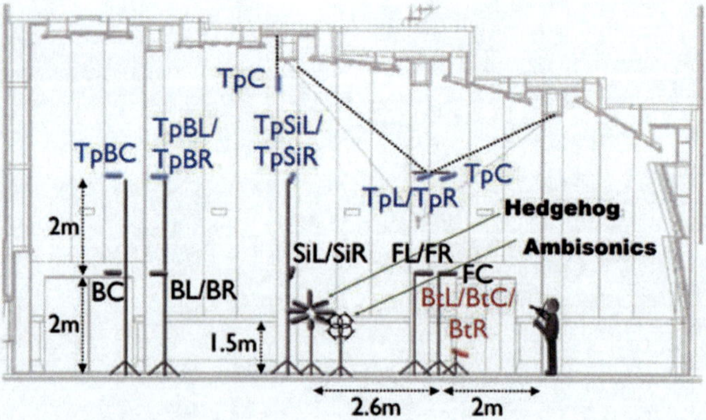

Fig. 1.27 Sectional view of the microphone layout as used in Kamekawa and Marui 2020. Symbols such as FL (Front Left) correspond to the channels of the 22.2 multi-channel audio system. 'Tp' indicates the top layer and 'Bt' indicates the bottom channel (graphic is the lower part of Fig. 1 from Kamekawa and Marui 2020)

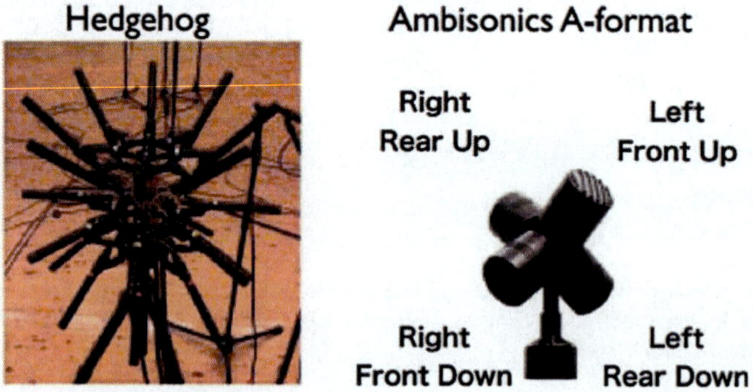

Fig. 1.28 The 'Hedghog' (near coincident) and First-Order Ambisonic (coincident) microphone systems as used in the experiment of Kamekawa and Marui 2020 (graphic is Fig. 2 from Kamekawa and Marui 2020)

venue acoustics) which will also determine how far the main microphone system can be placed from the sound source(s).

In this respect the results from the above mentioned studies by Lee (2021), Lee and Johnson (2020), as well as Kamekawa and Marui (2020) can be indicative.

In previous research it was shown that in order to avoid unwanted upwards shifting of source image in 3D reproduction, a direct sound captured or reproduced from a height channel (i.e. vertical interchannel crosstalk) should be at least 7 dB attenuated compared to the same sound captured or reproduced from the main channel (Lee 2011; Wallis and Lee 2016, 2017). This implies however that also a directional microphone serving a height channel has to be sufficiently spaced and/or angled upwards to reduce the amount of direct sound.

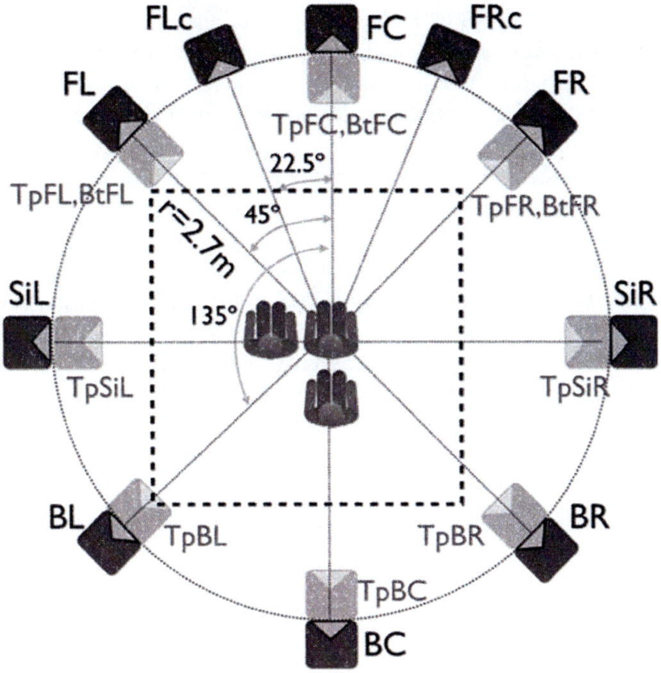

Fig. 1.29 Top view of the loudspeaker setup of the experiment's listening part by Kamekawa and Marui 2020. The participants moved to each of these three listening positions as shown in the above figure (graphic is upper part of Fig. 3 from Kamekawa and Marui 2020)

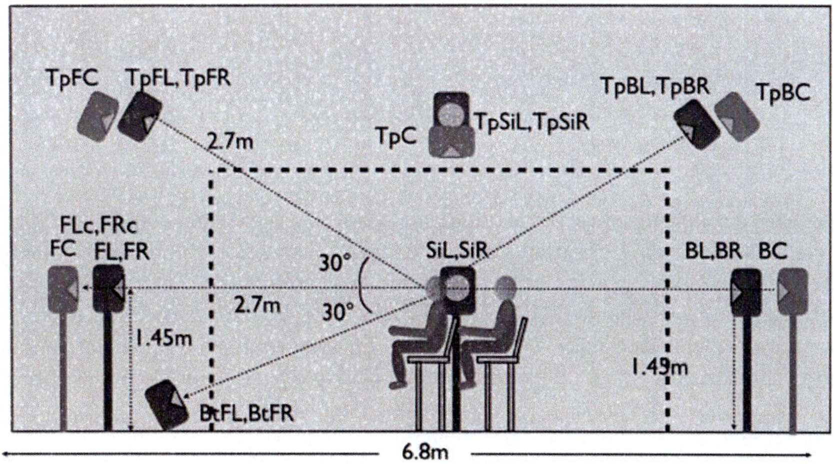

Fig. 1.30 Sectional view of the loudspeaker setup in the listening part of the experiment by (Kamekawa and Marui 2020). The dotted line indicates acoustically transparent black screens used to conceal the loudspeakers. The participants moved to each of these three listening positions as shown in the figure. (graphic is the lower part of Fig. 3 from Kamekawa and Marui 2020)

In addition to the choice of main microphone, also a decision on the microphone pattern or directivity for potential spot microphones needs to be made. Some sound-engineers have reported that, deviating from their usual practices which they are applying for stereo and surround recordings, they find themselves preferring more directional microphones (with super- or hyper-cardioid patterns) for their spot-microphones when working in 3D-audio, due to the more precise localization and 'overall image' which they are able to achieve in this way.

In this context it is interesting to give a look to the paper—which is by now 50 years old—by Nakayama et al. (1971) "Subjective Assessment of Multichannel Reproduction", in which the subjective effects of one- through eight- channel sound recording and reproduction were studied in relation to fullness, clearness, and depth of the image sources. Of these, fullness had the greatest weight in determining preferences among the reproduced sounds. It was also found that there was a close relationship between the perception of 'fullness' and a small value for the cross-correlation coefficient between sounds at the listener's ears (i.e. incoherence). The source material for the experiment were two popular musical selections which were played at the Beethoven Hall of Musashino Music College. The one- through eight-channel reproductions were performed in an anechoic chamber.

The multi-channel microphone system consisted of eight Neumann U-87 microphones, switched to cardioid directivity, set-up as described in Fig. 1.31. It is interesting to note the directional orientation of the cardioid microphones in respect to the sweet-spot listening position, which seems to have been chosen as an absolute 'reference point' for the desired incident (reproduction) angle for both direct (front microphones), as well as reflected/diffuse sound (side- and rear microphones). This is an approach that seems to have its merits but apparently is not reflected in many of today's 3D-audio microphone system proposals.

1.9 Some Thoughts on Time-Alignment in 3D Microphone Array Systems

Recently a colleague expressed his concerns in respect to exaggerated front- and rear-microphone spacing. He is an advocate of 'compact' 3D microphone arrays, claiming that setting up the rear microphones too far away from the front microphones (and therefore closer to the rear wall) would "bring the rear wall closer to the listener".

It is of course true that—being closer to the rear wall—the rear microphones will capture a different soundfield, in comparison to a position in the middle of the concert hall. But as long as correct time-alignment within the 3D microphone array is respected, this should not cause any acoustic problems.

In case there is a strong rear-wall reflection, which arrives at the rear microphones with a delay time larger than 30 ms (in relation to the original direct sound that is picked up by the front microphones), then an audible echo might occur for the listener.

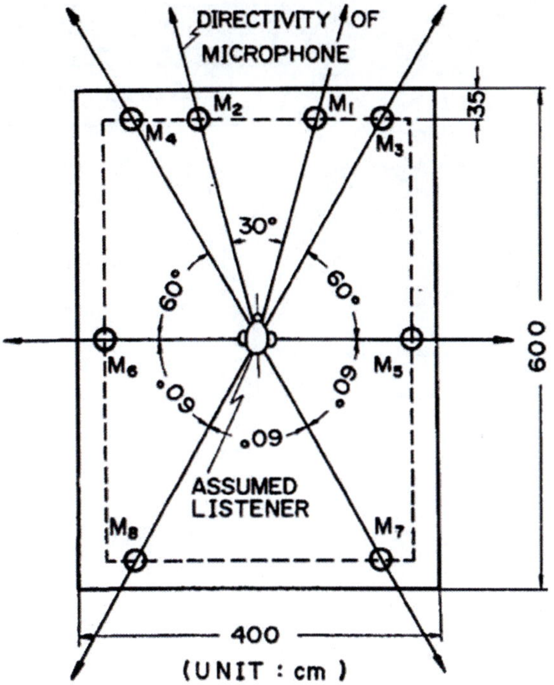

Fig. 1.31 Recording microphone set-up with Neumann U-87 cardioids (Fig. 2 from Nakayama et al. 1971)

So—here are my suggestions for 3D microphone arrays in which the front-rear capsule spacings are larger than 4 m or exceeding the reverberation radius of the respective venue:

(Rem: in strict theoretical terms time alignment applied to the signals of 3D microphone arrays makes sense once the capsules are physically separated and the technique deviates from a pure 'one-point' set-up. However, for smaller capsule spacings time alignment may not be necessary, as the sonic difference may be negligible—but this will also depend on the room-acoustics, of course).

If the front- and rear-microphone signals in Fig. 1.32 are not treated time-wise (i.e. no time alignment) then a larger physical spacing will likely result in a more spacious feeling for the listener at playback, than if the microphones are closer to each other. This is of course also a creative 'tool', which can be used by the sound engineer to evoke a more 'spectacular' spatial impression in the listener.

However, if the sound-engineer's concern is sonic accuracy, then he/she may want to apply a 'one-point' microphone technique and, for example, place a (higher order) Ambisonics microphone in an appropriate position. (Rem.: unfortunately both first order Ambisonics, as well as higher order Ambisonics currently provide only sonically inferior results in comparison to competitor techniques; see Chaps. 5 and 10 for details. If the use of a coincident microphone system is however preferred, a 'decorrelated' system based on crossed figure-of-eight patterns like the AB-BPT-3D configuration of Sect. 9.6 may be considered).

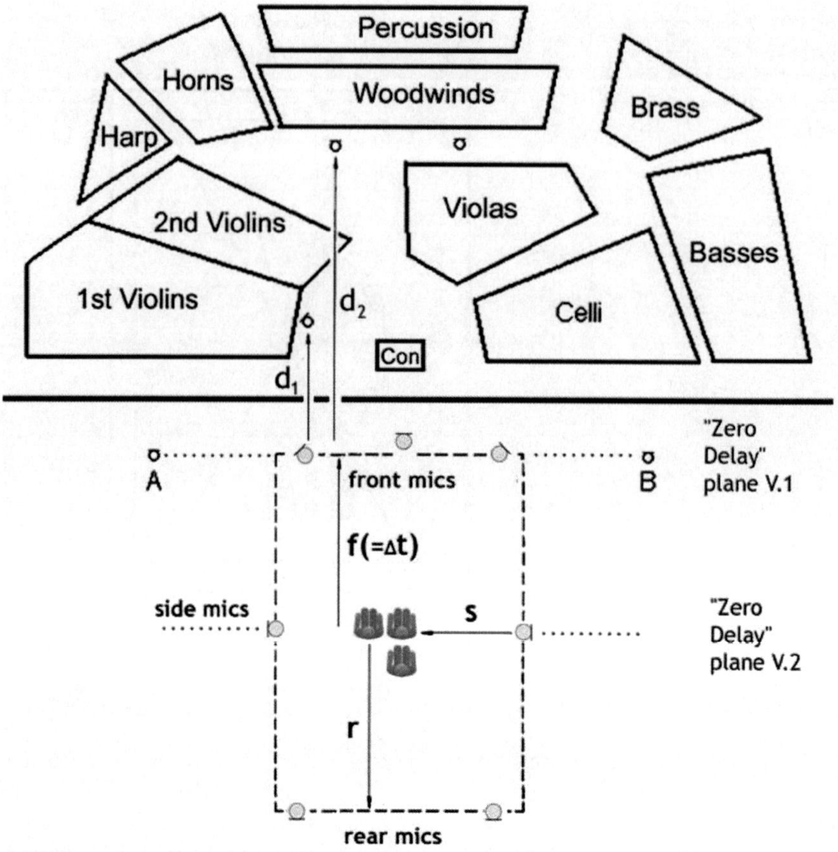

Fig. 1.32 Schematic view of a generic 3D microphone array and spot microphones for an orchestral recording, including virtual listeners at sweet spot/sweet area position (using also graphical elements of Fig. 3 from Kamekawa and Marui 2020)

However, when using a spacing between the front and rear microphones, there are some simple guidelines to consider in order to achieve 'correct' (as opposed to 'artistic') time alignment: as depicted in Fig. 1.32, spot microphones in the orchestra are usually time-aligned by the mix engineer, according to their distance d_1, d_2 … d_n to the main microphone. In case a one-point technique was used for the main microphone its physical position will become the reference point for time alignment. However, in the case of more elaborate main microphones, that are utilizing physically spaced capsules, the 'Zero Delay Plane' concept should be applied. In this case the distance d of the respective spot-microphone has to be measured perpendicular to the 'Zero Delay' plane V.1, as indicated by the according line in Fig. 1.32.

Now—in our case of the 3D audio recording—the plane as defined by the positions of the 3D microphone capsules involved, becomes the reference.

Sound travelling from the stage to the rear wall will be reflected and picked up by the rear microphones on its way back. In relation to the 'Zero Delay plane V.1' this sound will be picked up 'too early' by the rear microphones. If we keep the 'one-point' recording approach in mind, we have to delay the rear-mic signals by an amount corresponding to the sum of the distances r + f, where each meter of distance translates into roughly 3 ms of time added. (To be precise: for sound travelling at 343 m/s at 20 °C the correct amount of time would be 2.915 ms per meter).

In the way that we see the spot microphones on stage as a support to the main microphone (in order to arrive at the desired direct/diffuse sound ratio for the signals which shall be mixed for replay via the front speakers of the main layer), we can also regard the rear and side microphones as support mics for the 3D microphone array as a whole in respect to creating the right 'overall amount' of diffuse sound, in order to create the sensation of musical envelopment for the listener in the 'sweet spot' or 'sweet area'.

If we go for the 'spaced microphone' (in respect to the distance between front mics and rear mics) aesthetic approach, then the 'Zero Delay' plane V.2 becomes valid, which uses the (prospective) listener in the sweet spot as reference position.

In this case, the signals of the front-microphones should be time-aligned (i.e. delayed) according to the distance f and those of the rear microphones according to distance r. (Rem.: the signals of the spot microphones will have to be time aligned according to $f + d_n$, individually).

Of course it is also possible to use all microphone signals of a 3D microphone array without any time alignment in the mix, if the desired spatial impression can be achieved without. The spot microphones on stage, however, should always be time aligned to the front microphones in order to ensure optimized conditions also in respect to individual sound source localisation.

If we look at Fig. 1.33, which is a side view of the microphone setup from Fig. 1.32, we can distinguish the microphones of the main layer and the height layer.

Assuming the case of the 'Zero Delay' plane V.2: if we also want to apply the time alignment concept to the microphones of the height layer, the question arises, which distance to the listener needs to be considered. Should it be x or f ?

I believe there are strong arguments to use f, if we want to follow a 'wave-front' approach which should provide optimized listening conditions not only for one listener in the sweet spot, but preferably for several listeners in the sweet zone (or the whole listening area).

By applying a delay time according to distance f to the height front microphones, we try to create a 'coherent' soundfield, by radiating correctly time-aligned signals from the main and height front speakers.

The same applies of course for the soundfield radiated by the rear speakers (main and height). So what we are essentially trying to do with this approach is to re-create the soundfields (which would have been reaching a listener in the sweet spot/best seat in the house position in the concert hall) from the front wall, rear wall and side walls (and also the ceiling and floor, in case of the use of 'top-layer' and 'bottom-layer' speakers).

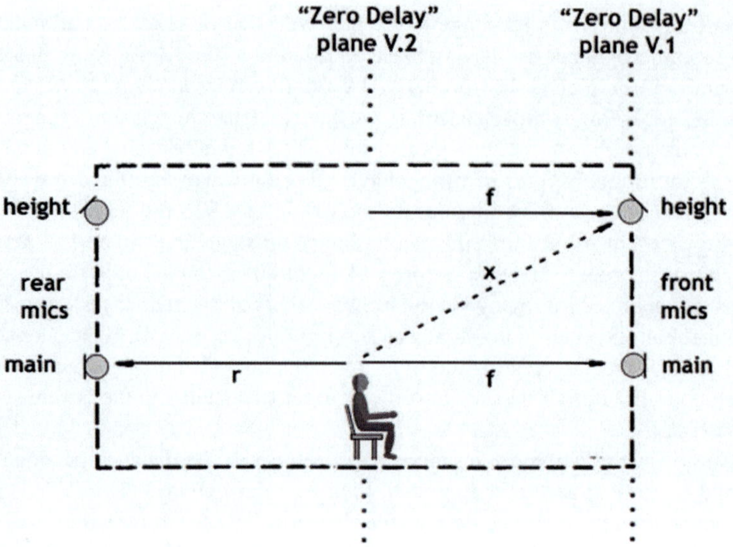

Fig. 1.33 Schematic side view of 3D microphone array, including virtual listener at the sweet spot position (using graphical elements of Fig. 3 from Kamekawa and Marui 2020)

It is understood that with the relatively small number of loudspeakers involved, this is of course far from the well known 'wavefield' approach (which needs a much higher number of speakers), but the above approach has some roots in the 'wavefront' principle, as applied by Arthur Keller of Bell Laboratories already back in 1934 (Keller 1981).

Going back the Fig. 1.32, the question remains why to also apply a time alignment for the signals of the side microphones ?

With the same argument as above, that we do not want 'to bring the side walls (sonically) closer', we can apply an appropriate time delay according to s. However, assuming that correctly chosen (and correctly positioned) side microphones will mainly capture diffuse sound, the perceived difference may be negligible.

In respect to time alignment in 3D microphone arrays it is also interesting to analyze the approach used by Tonmeister Lasse Nipkow, as described in (Nipkow 2012) as well as in Sect. 9.2 of this book.

1.10 Distribution of Direct- and Diffuse-Sound Reproduction in Multichannel Loudspeaker Systems

In 3D audio, for a traditional European-style ensemble setup, the musical sound source will remain in a frontal position on stage, therefore the reproduction of direct sound will have to remain with the main layer L, C, R speakers, while the recording of first ceiling (and floor) reflections may be reproduced through the height (and base) layer.

Therefore, in a 5.1 surround speaker setup the direct sound will be reproduced mainly through 3 front speakers, while (at least half of) the diffuse sound will be reproduced via only 2 rear/side speakers, hence a speaker number ratio of 3:2 = 1.5, or 60% of all speakers are mainly dedicated to direct sound reproduction..

In an NHK 22.2 3D audio speaker layout, according to (Hamasaki et al. 2007) the reproduction of direct sound in classical music mixing is likely reserved to the 5 front speakers of the base layer, while first reflections and diffuse sound may be reproduced through the rest of a total of 17 full range speakers, arranged in the upper, middle and lower layer (not including the two LFE speakers). Hence a speaker number ratio of 5:17 = 0.29, meaning that less than 23% of all speakers are dedicated to direct sound radiation (see Fig. 1.34).

Assuming the same volume balance ratio between direct- and diffuse sound as in traditional stereo, it seems clear that in any multichannel 3D audio setup the sound energy radiated through the speakers assigned to 'diffuse sound' has to be much less for each individual speaker in comparison to a small scale 5.1 (2D) surround setup, otherwise the total amount of radiated diffuse sound energy would certainly exceed the amount of radiated direct sound energy.

I believe this to be also a common trap, frequently encountered already with 2D surround but also 3D audio recordings, of sound engineers establishing too high levels of diffuse sound for the ‚surround' channels: to my taste, sitting in the sweet spot, if one is able to discern sound information coming from one individual reproduction channel (i.e. loudspeaker), which is originally dedicated for diffuse sound playback, then the goal of a 'natural reproduction' has not been achieved, as in 'natural hearing' in a concert hall one would not have the impression of diffuse sound coming from one particular, single direction. (Rem.: for a 'first reflection', this may sometimes

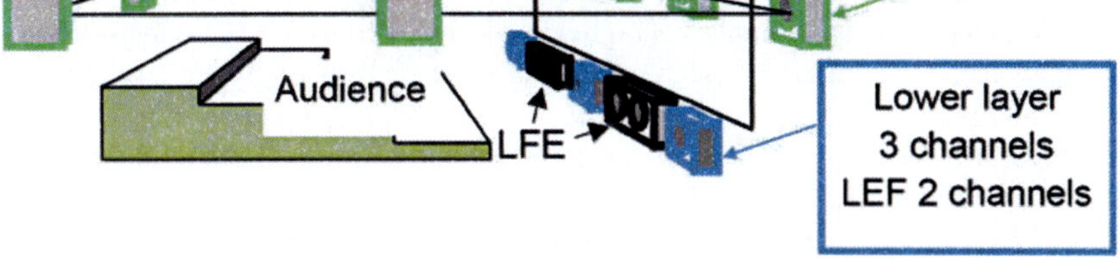

Fig. 1.34 22.2 multichannel sound system with upper, middle and lower loudspeaker layer (reproduction of Fig. 1 from Hamasaki et al. 2007)

be the case, depending on your listening capabilities and the specific room acoustics / architectural conditions involved).

On the event of a purely ‘academic’ recording for the purpose of a subjective listening test, the adjustment for the signal level of diffuse sound may be that of ‘unity level’ (in respect to the front channels, which carry mainly the ‘direct sound’ signal) and therefore it is a consequence of adhering to ‘scientific correctness’, if the actual diffuse sound level of the original recording venue is carried through to the reproduction room.

What seems ‘correct’ in a scientific sense, may—due to the limitations found even in extensive multichannel loudspeaker systems—turn out wrong on the basis of a psychoacoustic evaluation concerning ‘spatial impression’, the direct/diffuse sound ratio in music reproduction or the overall sense of ‘spatial envelopment’, if the aim is to re-created the original ‘feeling’ a listener would have had, when positioned in the ‘best seat in the house’ of a high quality concert hall. This is the reason why classical music mixing engineers will usually have to re-balance the levels between direct sound and diffuse sound radiating loudspeaker channels.

However, there is an interesting comment by Morten Lindberg in this context, explaining the work practice with his ‘2L-Cube’ (see Chap. 9 for details): “ … All microphones go through the same type of mic pre and I know the sensitivity specs of each microphone model. My unity gain at recording is set to equal acoustic sensitivity for all microphones.” (from Inglis 2022).

However, in the evaluation of the informal 3D mic array listening comparison session in (Gericke and Mielke 2022) it was found that the levels of the omni surround-microphones (LS/RS) of the 2L-Cube were “ ... significantly louder than, for example, the rear-facing cardioids of the PCMA-3D, - in the course of the listening tests— [so] we decided to lower their level a little …” (see also Chap. 10, Sect. 10.7).

1.11 Some Thoughts on Optimized Loudspeaker Directivity

This has to do with the fact that (unless we have a very extensive wavefield-style loudspeaker system at our disposal) the reduction of sound reproduction to a limited number of actual, physical reproduction sources in the form of a multi channel (usually direct radiator point source) loudspeaker system has its limits in various respects: considerable debate has taken place between experts in the field concerning the advantages and disadvantages of higher or lesser directivity for reproduction loudspeakers (see Holman 2000) and a critical detailed analysis on this matter can also be found in Chap. 2 of my book (Pfanzagl-Cardone 2020), which touches on many related topics, such as optimized signal correlation in surround microphone systems (and, accordingly, in the related surround loudspeaker replay systems), the question of an ‘optimized inter-aural correlation coefficient’ (in concert hall acoustics, but- consequently- also for loudspeaker listening at home), as well as evaluations concerning the interaction between loudspeaker directivity and listener envelopment.

Various researchers seem to have divergent answers to the question of optimized loudspeaker directivity, which is related to the ease with which a listener will be able 'to discern the loudspeaker itself': (Kates 1960) favors loudspeakers with higher directionality, as he thinks that it is of advantage to avoid unnecessary room-reflections. Also (Zacharov 1998) shares this opinion, but encounters strong criticism by important leaders within the audio-engineering community [see—among others—(Holman 2000)]. To underline his arguments, Holman points out the results of his own research (Holman 1991), in which he was able to show that a higher directivity index of loudspeakers is helpful for better localization, but a smaller directivity index is in favor of better 'envelopment' for the listener (especially with a surround-replay setup).

Likewise (see (Toole 2008), pg. 137)—based on the analysis of his previous listening tests (Toole 1985, 1986)—arrives at the conclusion that the majority of listeners apparently prefers loudspeakers with a broader dispersion angle (i.e. smaller directivity index). In his case this also had to do with the fact that he was using listening rooms with acoustically untreated side walls which caused relevant reflections due to the wide dispersion angle of the loudspeakers. In turn these reflections were responsible for a lowering of the IACC at the listeners which leads to an increase in perceived ASW. This resulted in a higher listener preference for loudspeakers which have a wider dispersion characteristic.

Chapter 2 on spatial hearing covers some related topics such as the influence of loudspeakers and listening room acoustics on listener preference and psychoacoustic effects concerning localization and spatial impression with loudspeaker reproduction. As part of this, frequency-dependent localization distortion in the horizontal, as well as the vertical plane is examined, and also effects concerning the reproduction of spaciousness.

1.12 Conclusion

Calculating the 'overall resulting' IACC for a human listener in the 'reproduction sphere' (in analogy to the previous 'sweet spot' in a 2-channel stereo playback environment, or 'sweet area' of a 2D surround-sound playback environment) of a 3D-audio loudspeaker system, based on the microphone setup from Fig. 1.25 for example, would require HRTFs which do not only cover the horizontal plane (as would suffice for 2D surround) but are highly accurate (and appropriately spaced also in terms of degree resolution) in the vertical plane, which—given the high number of input signals—is not exactly an easy computational exercise.

With this in mind, it might seem much more efficient to come back to my idea of introducing the above mentioned 'human reference' in the form of an artificial human head, which can be placed in the sweet sphere of a 3D-audio system and measure the FIACC and related BQI_{rep} in order to arrive at an acoustic analysis and qualitative evaluation of any given 3D-microphone setup / loudspeaker-system combination (see preface of Pfanzagl-Cardone 2020): no matter how many reproduction loudspeaker channels are involved as part of a 3D replay setup—finally, the sound vibrations have to enter the two ear canals of the human head, so the resulting FIACC, which is effective between the eardrums, is what counts in the end.

A short series of educational videoclips concerning microphone technique analysis (mainly for 2-channel stereo microphone techniques, but also including 3-channel techniques like DECCA and BPT), based on FCC measurements using the "2BCmultiCORR" frequency dependent signal cross-correlation plug-in can be found on the author's Youtube channel "futuresonic100", when searching for 'mic tech analysis' or 'Nevaton BPT'.

I am convinced that—as I have already been able to point out for 5.1 surround microphone systems in Pfanzagl-Cardone 2020—a high degree of signal decorrelation, especially at frequencies below 500 Hz, is key to achieving good spatial impression. I would hope that colleagues might take this into account as I have the impression that there is quite a number of 3D microphone systems out in the field, which do not comply to this criterion: at times it seems their mechanical design is determined by an—understandable—wish for ease in terms of setup, when top-layer microphones (admittedly, with directional patterns) are rigged in close proximity to the main-layer microphones, while a sufficient spacing also in the vertical domain might be necessary to ensure sufficient signal-decorrelation for low frequencies, which is essential for the reproduction of 'spatial impression' or to achieve a truly convincing feeling of 'spaciousness'.

References

AURO-Technologies NV (2015) AURO-3D home theater setup—installation guidelines. Rev 6. http://www.auro-3D.com. Accessed 28 Oct 2015

Bates E, Gorzel M, Ferguson L, O'Dwyer H, Boland FM (2016) Comparing ambisonics microphones: part 1. Paper presented at the conference on sound field control, Audio Engineering Society, Gilford, 18–20 July 2016

Barron M, Marshall AH (1981) Spatial impression due to early lateral reflections in concert halls: the derivation of a physical measure. J Sound Vib 77:211–232

Bates E, Doonery S, Gorzel M, O'Dwyer H, Ferguson L, Boland FM (2017) Comparing ambisonics microphones—part 2. Paper presented at the 142nd audio engineering society convention, Berlin, 20–23 May 2017

Baxter D (2016) Dimensional sound. Resolution 15(5):41–48

Beranek L (2004) Concert halls and opera houses: music, acoustics and architecture, 2nd edn. Springer, New York

Blumlein AD (1931) Improvements in and relating to Sound-transmission, sound-recording and sound-reproducing systems. British Patent 394,325, 14 Dec 1931 (reprinted in: Anthology of Stereophonic Techniques. Audio Eng Soc, 1986, pp 32–40

Camerer F, Sodl C, Wittek H (2001) Results from the Vienna listening test. http://www.hauptmikrofon.de/ORF/ORF_und_FHD.htm. Accessed 1 Dec 2016

Dabringhaus W (2000) 2+2+2 - kompatible Nutzung des 5.1 Übertragungsweges für ein System dreidimensionaler Klangwiedergabe klassischer Musik mit drei stereophonen Kanälen. In: Proceedings to the 21. Tonmeistertagung des VDT de Keet VW (1968) The influence of early lateral reflections on spatial impression. In: 6th Int Congress on Acoustics, Tokyo

Eargle J (ed) (1986) Stereophonic techniques—an anthology of reprinted articles on stereophonic techniques. Audio Eng Soc

Elko GW (2001) Spatial coherence functions for differential microphones in isotropic noise fields. In: Brandstein M, Ward D (eds) Microphone arrays. Springer, Heidelberg p 61

Gerzon M (1973) Periphony: with-height sound reproduction. J Audio Eng Soc 21(1)

Gottlob D (1973) Vergleich objektiver akustischer Parameter mit Ergebnissen subjektiver Untersuchungen an Konzertsälen. Dissertation, Universität Göttingen. Hidaka T, Beranek L, Okano T (1995) Interaural cross-correlation, lateral

Grey JM, Gordon JW (1978) Perceptual effects of spectral modifications on musical timbres. J Acoust Soc Am 63(5):1493–1500

Griesinger D (1986) Spaciousness and localization in listening rooms and their effects on the recording technique. J Audio Eng Soc 34(4):255–268

Griesinger D (1997) Spatial impression and envelopment in small rooms. Paper 4638 presented at the 103rd Audio Eng Soc Convention

Griesinger D (1998) General overview of spatial impression, envelopment, localization and externalization. In: Proceedings to the audio engineering society 15th international conference on small room acoustics, Denmark, Oct/Nov 1998, pp 136–149

Griesinger D (1999) Objective measures of spaciousness and envelopment. Paper 16-003 presented at the Audio Eng Soc 16th int conference on spatial sound reproduction

Hamasaki K, Nishiguchi T, Okumura T, Nakayama Y, Ando A (2007) 22.2 multichannel sound system for ultra high-definition TV, paper presented at SMPTE Technical Conference

Hamasaki K (2003) Multichannel recording techniques for reproducing adequate spatial impression. In: Proceedings to the Audio Eng Soc 24th international conference on multichannel audio—the new reality, Banff, Canada

Hidaka T, Beranek L, Okano T (1995) Interaural cross-correlation, lateral fraction, and low- and high-frequency sound levels as measures of acoustical quality in concert halls. J Acoust Soc Am 98(2)

Hidaka T, Beranek L, Okano T (1997) Some considerations of interaural cross correlation and lateral fraction as measures of spaciousness in concert halls. In: Ando Y, Noson D (eds) Music and concert hall acoustics. Academic Press, London

Hildebrandt A, Braun D (2000) Untersuchungen zum Centerkanal im 3/2 Stereo-Format. In: Proceedings to the 21. Tonmeistertagung des VDT, p 455

Hirata Y (1983) Improving stereo at L.F. Wireless World, pp 60

Holman T (1991) New factors in sound for cinema and television. J Audio Eng Soc 39:529–539

Holman T (2000) Comments on the 'subjective appraisal of loudspeaker directivity for multi-channel reproduction.' J Audio Eng Soc 48(4):314–317

Howie W, King R, Martin D, Grond F (2017) Subjective evaluation of orchestral music recording techniques for three-dimensional audio. Paper 9797 presented at the 142nd Audio Eng Soc Convention, Berlin, 20–23 May 2017

Inglis S (2022) Mixing atmos: Morten Lindberg. Sound-On-Sound, July 2020, pp 112–117

IRCAM (2003) A large set of audio features for sound description. Section 6.1.1, Technical Report IRCAM

Jecklin J (2002) Surround-Aufnahmetechnik OSIS 321. In: Proceedings to the 21. Tonmeistertagung des VDT, Hannover, Nov 2002

Kamekawa T, Marui A, Irimajiri H (2007) Correspondence relationship between physical factors and psychological impressions of microphone arrays for orchestra recording. Paper 7233 presented at the 123rd Audio Eng Soc Convention, New York, Oct 2007

Kamekawa T, Marui A (2020) Evaluation of recording techniques for three-dimensional audio recordings: comparison of listening impressions based on difference between listening positions and three recording techniques. J Acoust. Sci & Tech. 41:1

Kates JM (1960) Optimum loudspeaker directional patterns. J Audio Eng Soc 28:787–794

Keller AC (1981) Early Hi-Fi and stereo recording at bell laboratories (1931–1932). J Audio Eng Soc 29:274–280

Lee H (2011) The relationship between interchannel time and level differences in vertical sound localization and masking. Paper 8556 presented at the 131st Audio Eng Soc Convention, New York

Lee H (2019) 3D microphone arrays—a 3D recording comparison. Resolution Mag 18(7):52–55

Lee H (2021) Multichannel 3D microphone arrays: a review. J Audio Eng Soc 69(1/2):5–26. https://doi.org/10.17743/jaes.2020.0069

Lee H, Gribben C (2014) Effect of vertical microphone layer spacing for a 3D microphone array. J Audio Eng Soc 62(12):870–884

Lee H, Johnson D (2020) 3D Microphone Array Recording Comparison (3D-MARCo): objective measurements. https://doi.org/10.5281/zenodo.4018009

Lindberg M (2014) 3D Recording with the '2L-cube'. http://www.2l.no/artikler/2L-VDT.pdf. Accessed 10 Feb 2022

Nakahara M (2005) Multichannel monitoring tutorial booklet (M2TB) rev. 3.5.2. Yamaha Corp 2005, SONA Corp, p 41

Nakayama T, Miura T, Kosaka O, Okamoto M, Shiga T (1971) Subjective assessment of multichannel reproduction. J Audio Eng Soc 19(9):744–751

Nicol R (2018) Sound field. In: Geluso P, Roginska A (eds) Immersive sound. Focal Press, Routledge

Nipkow L (2012) Eigenschaften von mikrofonierten Raumsignalen bei 3D Audio/Auro 3D. In: Proceedings to the 27. Tonmeistertagung des VDT, Cologne, Nov 2012

Noisternig M, Musil T, Sontacchi A, Höldrich R (2003) 3D binaural sound reproduction using a virtual ambisonic approach, paper presented at VECIMS 2003. In: International symposium, (Virtual environments, human–computer interfaces, and measurement systems), Lugano, Switzerland, 27–29 July 2003, pp 174–178. IEEE Xplore. https://doi.org/10.1109/VECIMS.2003.1227050

Oliveri F, Peters N, Sen D (2019) Scene-based audio and higher order ambisonics: a technology overview and application to next-generation audio, VR and 360° Video. https://tech.ebu.ch/publications. Accessed 27 Aug 2020

Pfanzagl E (2002) Über die Wichtigkeit ausreichender Dekorrelation bei 5.1 Surround-Mikrofonsignalen zur Erzielung besserer Räumlichkeit (engl: The Importance of Sufficient Decorrelation of 5.1 Surround Main-Microphone Signals for Better Spatial Reproduction). Proceedings to the 22.Tonmeistertagung des VDT

Pfanzagl-Cardone E (2002) In the light of 5.1 Surround: why AB-PC is Superior for symphony-Orchestra Recording, preprint 5565, 112th AES-Convention Munich, 2002

Pfanzagl-Cardone E (2005) 3.0 microphone for surround-recording. AU patent AU2005100255

Pfanzagl-Cardone E, Höldrich R (2008) Frequency-dependent signal-correlation in surround- and stereo-microphone systems and the Blumlein-Pfanzagl-Triple (BPT). Paper 7476 presented at the 124th Audio Eng Soc Convention, Amsterdam

Pfanzagl-Cardone E (2012) 'Naturalness' and related aspects in the perception of reproduced music. In: Proceedings to the 27. Tonmeistertagung des VTD, Köln

Pfanzagl-Cardone E (2020) The art and science of surround and stereo recording. Springer Nature. https://doi.org/10.1007/978-3-7091-4891-4

Prokofieva E (2007) Relation between correlation characteristics of sound field and width of listening zone. Paper 7089 presented at the 122nd Audio Eng Soc Convention, Vienna

Rayleigh (1907) On our perception of sound direction. Philos Mag 13

Riaz H, Stiles M, Armstrong C, Chadwick A, Lee H, Kearney G (2017) Multichannel microphone array recording for popular music production in virtual reality. E-brief presented at the 134rd audio engineering society convention, New York

Riekehof-Böhmer H, Wittek H, Mores R (2010) Voraussage der wahrgenommenen räumlichen Breite einer beliebigen stereofonen Mikrofonanordnung. In: Proceedings to the 26. Tonmeistertagung des VDT, pp 481–492

Schubert E, Wolfe J, Tarnopolsky A (2004) Spectral centroid and timbre in complex, multiple instrumental textures. In: Proceedings to the 8th international conference on music perception & cognition, North Western University, Illinois
Steinberg JC, Snow WB (1934) Auditory perspective—physical factors. Electr Eng 53(1):12–15
Stiles M (2018) Recording spatial audio. Resolution 17(2):49–51
Suzuki A, Tohyama M (1981) Interaural cross-correlation coefficient of Kemar head and torso simulator. IECE Japan, Tech Rep EA80–78
Theile G (2000) Mikrofon- und Mischungskonzepte für 5.1 Mehrkanal-Musikaufnahmen. In: Proceedings to the 21. Tonmeistertagung des VDT, Hannover, p 348
Theile G (2001) Multichannel natural music recording based on psychoacoustic principles. In: Proceedings to the 19th international conference of the audio engineering society, pp 201–229
Theile G, Wittek H (2012) 3D audio natural recording. In: Proceedings to the 27. Tonmeistertagung des VDT, Cologne, p 731
Thiele R (1953) Richtungsverteilung und Zeitfolge der Schallrückwürfe in Sälen. Acustica 3:291–302
Toole FE (1985) Subjective measurements of loudspeaker quality and listener performance. J Audio Eng Soc 33(1/2):2–32
Toole FE (1986) Loudspeaker measurements and their relationship to listener preferences. J Audio Eng Soc 34:227–235
Toole FE (2008) Sound reproduction—loudspeakers and rooms. Focal Press (Elsevier)
Torick E (1988) Highlights in the history of multichannel sound. JAES 46(1/2):27
Wallis R, Lee H (2016) Vertical stereophonic localization in the presence of interchannel crosstalk: the analysis of frequency-dependent localization thresholds. J Audio Eng Soc 64(10):762–770
Wallis R, Lee H (2017) The reduction of vertical interchannel crosstalk: the analysis of localisation thresholds for natural sound sources. Appl Sci 7(279):2017
Ward DB, Abhayapala TD (2001) Reproduction of a plane wave sound field using an array of loudspeakers. IEEE Trans Speech Audio Process 9(6):697–707
Wittek H (2012) Mikrofontechniken für Atmoaufnahmen in 2.0 und 5.1 und deren Eigenschaften. In: Proceedings to the 27. Tonmeistertagung des VDT, Cologne, p 804
Yamamoto T, Nagata M (1970) Acoustical characteristics at microphone positions in music studios. NHK Tech Rep 22:475–489
Yanagawa H, Higashi H, Mori S (1976) Interaural correlation coefficients of the dummy head and the feeling of wideness. Acoust Soc Jap Tech Rep H-35–1
Yost WA, Wightman FL, Green DM (1971) Lateralisation of filtered clicks. J Acoust Soc Am 50:1526–1531
Zacharov N (1998) Subjective appraisal of loudspeaker directivity for multichannel reproduction. J Audio Eng Soc 46(4):288–303
Zotter F, Frank M (2019) Ambisonics: a practical 3D audio theory for recording, studio production, sound reinforcement, and virtual reality, 1st edn. Springer, New York

Chapter 2
'3D'- or 'Immersive' Audio—The Basics and a Primer on Spatial Hearing

Abstract The commercially most accepted 3D audio formats 'Auro 3D', 'Dolby Atmos' and 'Ambisonics' are mentioned and the three different approaches of Channel-Based Audio (CBA), Scene-Based Audio (SBA) and Object-Based Audio (OBA) are pointed out. A very short primer on the current state of 3D Audio and related psychoacoustic knowledge—from a paper by Lee and Gribben—is presented. The fundamentals of spatial perception in human hearing are outlined. We are taking a look at this phenomenon mainly from the perspectives of physical acoustics and psycho-acoustics as these two disciplines are majorly relevant in respect to the research and fields of interest which will be presented in this book. For the sake of compactness, we are refraining from giving an outline of the historic development of research into human hearing, but are trying to give a short summary of the current state of knowledge: HRTF's, cone of confusion, ILD's and ITD's, ERB, clarity, 'Deutlichkeit', various acoustic measures related to 'spatial impression'—to name a few relevant topics. Special focus is put on frequency-dependent aspects of human hearing concerning localization in the horizontal and vertical plane, distance perception and spatial impression both for 'live hearing', as well as in relation to loudspeaker production. In doing so, this chapter draws on findings by many specialists in the field of psycho-acoustics, as well as scientific research in the realm of audio-engineering, as well as concert-hall acoustics.

Keywords 3D audio · Spatial hearing · Localization perception · Channel-Based-Audio (CBA) · Scene-Based Audio (SBA) · Object-Based Audio (OBA)

Currently there seem to be mainly three widely accepted systems—or industry defined 'standards'—for the representation of 3D audio, at least for music and film reproduction:

- Auro 3D

Content of this chapter is based on—and partly cites—the following papers: Merimaa and Pulkki (2005), with some additions from Rumsey (2001) and Toole (2008), as well as translated text-citations from Fellner and Höldrich (1998a, b). This whole chapter is an extended and updated version of the chapter on spatial hearing previously published in Pfanzagl-Cardone (2020).

E. Pfanzagl-Cardone, *The Art and Science of 3D Audio Recording*,
https://doi.org/10.1007/978-3-031-23046-2_2

- Dolby Atmos, and
- the Ambisonic Format (First Order Ambisonics FOA or Higher Order Ambisonics HOA), as an appropriate means for capturing—or representing—3D audio in the form of a recording/rendering from a 'one point' perspective.

Apart from these standards there is a number of less well known proposals for the recording and reproduction of immersive sound. (e.g. Ellis-Geiger Triangle, or the MMAD—'Multichannel Microphone Array Design' by Michael Williams, based on the Isosceles Triangle, etc.)

In Corteel et al. (2016, p. 2) the following distinction between the three principally differing approaches is made (citings are slightly altered):

> … If we look at the current technical situation, 2D and 3D audio will essentially be implemented using one of the three following principles, in which each of the following 3D audio formats uses multiple sound streams:
>
> – **Channel-Based Audio (CBA)**: each audio stream is assigned to a fixed loudspeaker. This is the case for the traditional and well established delivery formats of 2-channel stereo and 5.1 surround. Sound positions of objects are a consequence of their representation through the use of a main microphone system (with the possibility of the additional use of spot-microphones, included during the mixdown process), or a kind of pre-rendering (through bus allocation and mixing) of the sound-sources during the mixing stage as well as a well-defined loudspeaker setup (number and position of speakers) for replay.
>
> – **Scene-Based Audio (SBA)**: directional encoding of a sound scene, based on a set of Eigen-functions of radiation/directionality (spherical harmonics in the case of Higher Order Ambisonics or 'HOA').
>
> This is a scalable format which offers increasing spatial accuracy with the order (number of HOA components/streams) it uses. It requires some matrixing on the rendering side since each channel reproduction is neither a source nor attached to a direction where loudspeakers could be located.
>
> – **Object-Based Audio (OBA)**: in which each audio stream is described as a sound object with associated metadata (position, spatial extent…) that may evolve over time. Apart from its complete independence from the reproduction system, shared with the scene-based approach, the advantage of the object-based principle is the possibility to manipulate each object independently from the others. Theoretically, the same content can be used on different devices and different platforms. … from Corteel et al. (2016)

The next generation audio formats—according to the aim of the EDISON 3D audio project (see Corteel et al. 2016) will be supporting either all or a subset of the three principles explained above.

The film and music industry has recently introduced various 3D audio systems to the movie theatre (and home theatre) market, with proprietary approaches, which we will describe in the following chapters.

In Lee and Gribben (2014) a short primer on the current state of 3D-Audio and related psychoacoustic knowledge can be found:

> … The multi-channel audio formats such as 22.2 (ITU-R BS.2159-4 2012) and Auro-3D (Van Daele and Van Baelen 2012) use height channels to evoke the acoustic impression of a 'three-dimensional (3D)' space. In the area of cinema sound or pop music production, the height channels could be used for creative 'panning' of the sound sources in the vertical plane as well as for creating an enhanced and more convincing impression of the acoustic environment.

> On the other hand, the use of height channels in acoustic recordings made in a concert hall is likely to focus on extra acoustic ambience since source images would not need to be elevated in most cases (an exception to this might be choir singers standing on a pedestal). Already some years ago some main microphone techniques using height channels were proposed (see Theile and Wittek 2011; Williams 2013; Geluso 2012). For example, Theile and Wittek (2011) proposed a technique called 'OCT-9' [Rem.: now called OCT-3D] that uses four upward-facing cardioid microphones positioned over the front-left, front-right, rear-left, and rear right microphones of the 'OCT-5' main microphone array. A distance of 1m between the main and height microphones is recommended for this arrangement. Williams (2013) also designed a 3D microphone array with four elevation microphones aligned vertically to the main microphone. The suggested distance between the lower and upper layers is also 1 m, but the polar pattern of the height microphones in this case is a figure eight. In contrast, Geluso (2012) proposed a 'coincident' microphone technique as a method for acquiring height information; a vertically aligned S-microphone with a figure-8 characteristic is combined with a forward-facing M-microphone. In the field of horizontal stereophony, it is well known that a 'wide AB pair' creates a more spatial impression in the reproduction (see Lipshitz 1986; Hamasaki 2003; Rumsey and Lewis 2002). This is because a larger distance between the microphones leads to a lower inter-channel correlation between the signals (Hamasaki 2003). [Rem.: more information on this can also be found in Pfanzagl (2002), as well as in several chapters of Pfanzagl-Cardone (2020)]. However, research indicates that the principles of horizontal stereophony do not apply directly to vertical stereo. With regard to localization, it is known that vertical localization relies on spectral cues (HRTF) rather than inter-aural cues (see Roffler and Buttler 1968; Blauert 1997). It has been reported that the generation of a phantom sound source by amplitude-panning is unstable in vertical stereophonic playback (Barbour 2003). It was also found that the precedence effect between vertically arranged loudspeakers does not fully work, regardless of the time difference applied to them (see Lee 2011; Wallis and Lee 2014), and some research suggest that ICTD is ineffective in the vertical plane (Wendt et al. 2014). Research on spatial impression by Lee and Cribben examined the effectiveness of inter-channel decorrelation in controlling the perceived spread of band-passed pink noise, with two loudspeakers positioned vertically in the median-plane, and two loudspeakers in the horizontal plane (Gribben and Lee 2014). As a result, it was found that the signal decorrelation efficiency in the vertical plane was not as strong as that in the horizontal plane, depending on the frequency. However, it must also be stated that the mechanism of perception of vertical spatial impression has not yet been fully researched and further investigations are still required. … [free citation from Lee and Gribben (2014)]

However, the results of the experiments by Michael Williams (see Williams 2016; Williams 2022a) in respect to the importance of time difference between vertically arranged speakers for localization seem to contradict the findings of Wendt and other researchers mentioned above, so more research on this matter may be needed for a conclusive answer. More detailed information on the results obtained by Michael Williams can be found in Chap. 6, as well as Michael William's book (see Williams 2022b).

Further insight in to the complexity of the human hearing mechanism was gained through research by Otani et al. regarding the origin of frequency dependence of interaural time difference (see Otani et al. 2021): "… The interaural time difference (ITD) plays an important role in spatial hearing, particularly in azimuthal localization of sound images. Although the ITD is essentially determined by the geodesic distance between two ears, researchers have reported that the ITD is greater for lower frequencies. However, the origin of this frequency-dependence has not been revealed. This study investigates how the ITD is physically characterized to have a frequency-dependent nature by conducting measurements and numerical simulations. Dummy head measurements show that the ITD varies with frequency because the apparent propagation time to the ipsilateral ear decreases for low frequency. Dummy head simulations confirmed this phenomenon and revealed that the apparent propagation time decreases because of a sound pressure phase shift due to reflections from the head. Circular plate simulations revealed that the circular profile including its lateral surface and edge produces reflections that are relevant to the phase shift, yielding the frequency-dependence of the apparent propagation time. Furthermore, rigid sphere simulations showed that such reflections are produced even by smooth convex surfaces without clear-cut edges. These results strongly suggest that a major factor in the production of the frequency-dependence of ITDs is backscatter diffractions from convex surfaces of the head and the pinna."

In the absence of real 3D audio listening rooms, equipped with loudspeakers arranged in various layers (bottom layer, main layer, height layer, top layer) many 3D audio producers use binaural simulations of 3D audio sound-fields instead, which is also reflected in the fact that basically all commercial 3D audio mixing and encoding solutions (Auro-3D, Dolby Atmos, DTS:X, Sony 360RA—see the respective chapters in this book) offer binaural simulation plug-ins. In this respect Otani et al. have researched on the optimization and application for auralization of binaural Ambisonics (see Otani et al. 2020), as well as on the reproduction accuracy of Higher—Order Ambisonics with Max-rE and/or least norm solution in decoding (see Otani and Shigetani 2019).

2.1 Influence of Listening Environment and Subjective Evaluation of 3D, Surround and Stereo Loudspeaker Reproductions

Hahn from the Centre for Interdisciplinary Research in Music Media and Technology (CIRMMT), McGill University, Montréal has researched on the musical emotions evoked in untrained listeners, musicians and Tonmeisters, comparing stereo, surround and Auro-3D recordings (see Hahn 2018) and found that: "… In this study, listeners were presented with classical musical excerpts in Stereo, 5.1 Surround, and Auro-3D 9.1 playback formats. Participants were asked to report their emotional states on the Geneva Emotional Music Scale (GEMS) while listening to two contrasting excerpts of Arnold Schönberg's string sextet "Verklärte Nacht" op. 5. The results provide evidence that 3D audio can invoke a stronger overall emotional arousal in listeners."

Eaton and Lee have researched the perception of 3D audio by subjective evaluations of three-dimensional, surround and stereo loudspeaker reproductions using classical music recordings (see Eaton and Lee 2022): "… The present study subjectively evaluated loudspeaker reproductions of four different classical recordings in 0

+ 2 + 0 (stereo), 0 + 5 + 0 (surround), 4 + 5 + 0 (surround with four height channels), each of which was downmixed from the original 9 + 10 + 3 (i.e. NHK 22.2), in terms of four attributes: listener envelopment (LEV), presence (i.e. sense of being there), overall tonal quality (OTQ) and overall listening experience (OLE). Prior to the main experiment, the playback levels of the upper and bottom loudspeaker layers relative to the middle layer level were subjectively adjusted for each of the original 9 + 10 + 3 recordings. It was found that the preferred levels of the upper and bottom layers were around 4 and 6 dB lower than that of the middle layer, on average. From multiple comparison listening tests, the perceived degradation from the original 9 + 10 + 3 to 4 + 5 + 0 was found to be significantly dependent on the recording technique used as well as the programme material. It was also found that 0 + 5 + 0 was not significantly different from 4 + 5 + 0 in general. Overall, LEV (Listener Envelopment) was most correlated with OLE (Overall Listening Experience), whilst Presence and OTQ (Overall Tonal Quality) tended to have a strong association."

Kim and Howie have published on the "Influence of the Listening Environment on Recognition of Immersive Reproduction of Orchestral Music Sound Scenes" (see Kim and Howie 2021): "... This study investigates how a listening environment (the combination of a room's acoustics and reproduction loudspeaker) influences a listener's perception of reproduced sound fields. Three distinct listening environments with different reverberation times and clarity indices were compared for their perceptual characteristics. Binaural recordings were made of orchestral music, mixed for 22.2 and 2-channel audio reproduction, within each of the three listening rooms. In a subjective listening test, 48 listeners evaluate these binaural recordings in terms of overall preference and five auditory attributes: perceived width, perceived depth, spatial clarity, impression of being enveloped, and spectral fidelity. Factor analyses of these five attribute ratings show that listener perception of the reproduced sound fields focused on two salient factors, spatial and spectral fidelity, yet the attributes' weightings in those two factors differs depending on a listener's previous experience with audio production and 3D immersive audio listening. For the experienced group, the impression of being enveloped was the most salient attribute, with spectral fidelity being the most important for the non-experienced group." The results reported in the above study concerning the two salient factors coincide very well with the findings in Pfanzagl-Cardone (2011), as well as Pfanzagl-Cardone (2012).

2.2 Basics of Sound Perception in Humans

Sound impinging on the human head is altered mainly by the 'pinnae' (outer ear) as well as the shoulders, chest and—of course—the head, before it reaches the entrance to the ear canal (cavum conchae). After that, the frequency response (and other parameters) of sound are affected by the ear canal (meatus), which leads to the middle ear. The middle ear consists of the tympanic membrane, hammer, anvil and stirrup. The mechanical energy which arrives to the middle ear by means of the

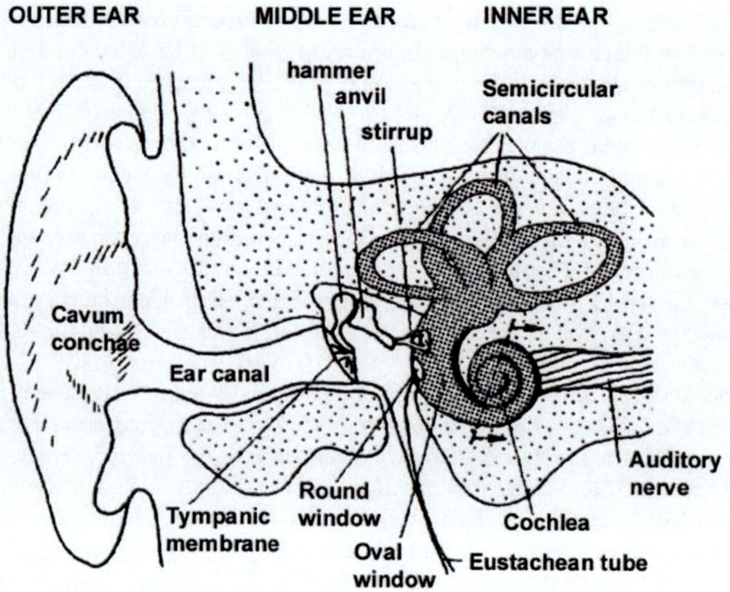

Fig. 2.1 Anatomy of the human inner, middle and outer ear (after Zollner and Zwicker 2003)

sound-wave is submitted to the inner ear via the 'oval window' and leads to a change of pressure of the liquid inside the inner ear (cochlea) (Fig. 2.1).

This pressure leads to frequency dependent excitation patterns on the Basilar membrane, which cause the hair cells to respond and fire nerve impulses, which in turn trigger electrical action-potentials in the neurons of the auditory system. These nerve impulses are dealt with and combined with the information which arrives from the other ear on a higher level of the brain (Fig. 2.2).

Perception of a sound source in space is determined by the following acoustical parameters: localization in the horizontal (azimuth) and vertical plane (elevation), perception of distance, perception of spatial impression (diffuse sound/impulse response of the room). The following sections try to give a short explanation of the underlying psycho-acoustic principles related to these parameters.

2.3 Mechanisms of Localization

The localization of sound sources is mainly based on the following five frequency-dependent information contained in a 'live' sound event (meaning a sound event, which happens in real space, as opposed to an electronically created sound):

1. Interaural Time Difference—ITD, and

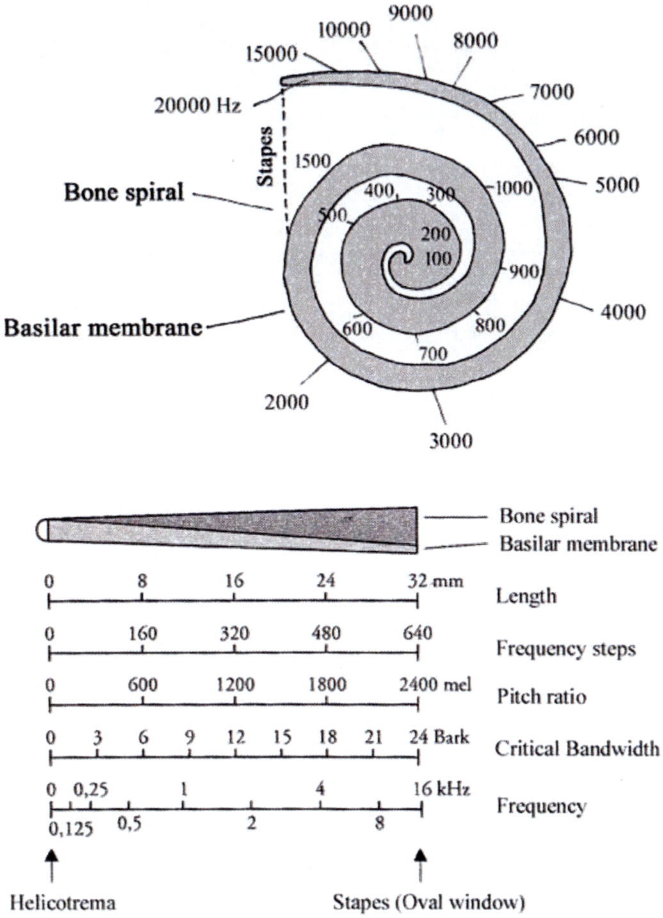

Fig. 2.2 Frequency coding on the Basilar membrane (after Eska 1997, Fig. 34, p. 123). Top: Areas which are most sensitive to indicated frequencies (after K.-H. Plattig). bottom: Schematic of stretched Cochlea (from Zwicker and Fastl 1990)

2. Interaural Level Difference—ILD are the two most important information, based on which human hearing decides in which 'cone of confusion' a sound source is situated (Figs. 2.3 and 2.4).

Lord Rayleigh was among the first to understand that—based on his experiments concerning human hearing towards the end of the nineteenth century—sound with a wavelength smaller than the diameter of the human head will effectively be shaded off for the ear on the other side, which results in an Interaural Level Difference (ILD) between the two ears. In addition to that, sound takes a different amount of time to arrive at both ears, which results in Interaural Time Differences (ITDs) (Fig. 2.3).

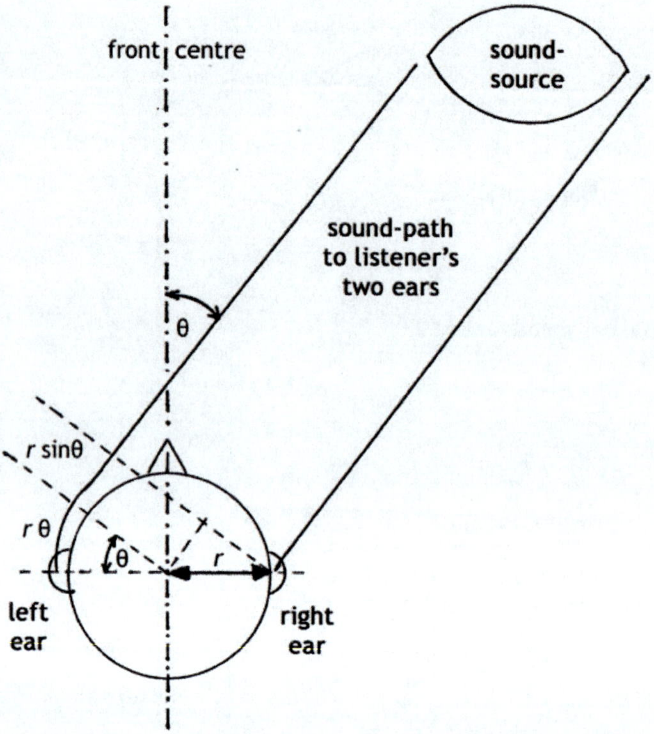

Fig. 2.3 The effective ITD for an individual listener depends on the angle of incidence, as this determines the additional pathway, which the sound needs to travel in order to arrive at the other ear. In this model ITD can be calculated as follows: $\Delta t = r(\theta + \sin\theta)/c$ (with speed of sound $c = 340$ m/s and θ in radians); (after Rumsey 2001, Fig. 2.1, p. 22)

The maximum ITD of natural sound sources is reached for sound impinging from the side of the head at 90°. This amounts to 650 μs (=0.65 ms) for a mean ear spacing of 21 cm (see Rumsey 2001, p. 22; Blauert 1997, p. 143). Other research (Yanagawa et al. 1976) and (Suzuki and Tohyama 1981), (Tohyama and Suzuki 1989) showed that the acoustically effective distance between the ears is—in reality—much larger with about 33 cm (see (Blauert 1997), p. 146–149). Based on the speed-of-sound with 340 m/s this would result in an ITD of 0.96 ms.

Lord Rayleigh was able to show that ITDs are especially important at low frequencies, at which no ILDs at relevant levels occur (Rayleigh 1907). His conclusion was that at low frequencies localization is based on ITDs, while at high-frequencies it is determined by ILDs. The crossover frequency range between these two psycho-acoustic mechanisms lies approximately at 1.5 kHz.

If one does not take into account the filter-effect caused by the pinnae, for an abstract sphere-shaped head, symmetrical lateral sound-source positions slightly in front (a) or behind (b) the dummy head in Fig. 2.6 will have the same ITDs and

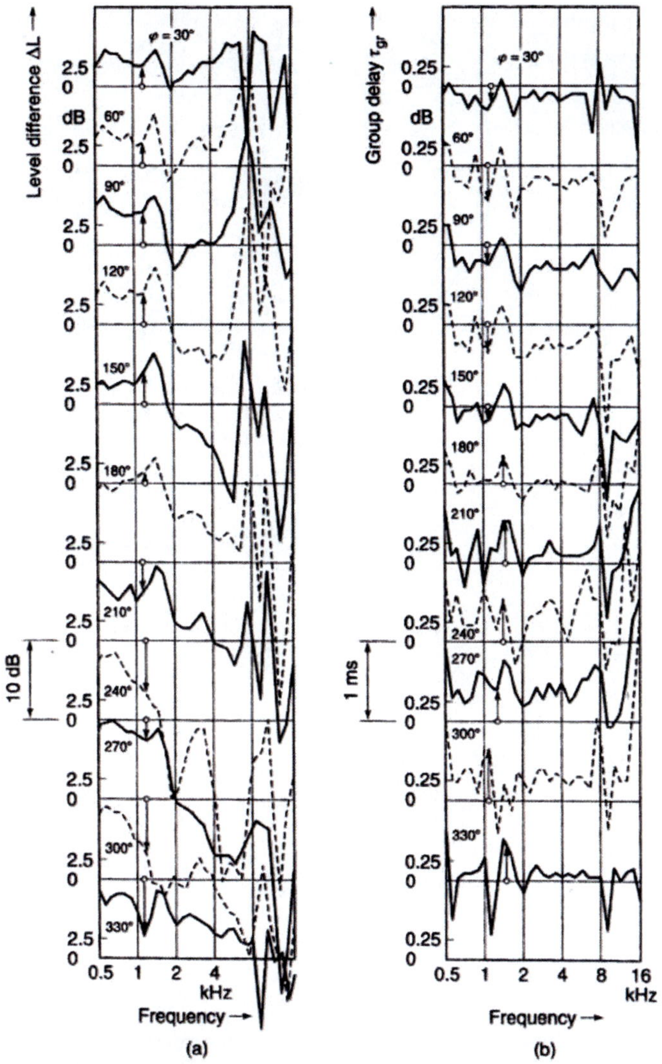

Fig. 2.4 Monaural transfer-function at the left ear for several directions in the horizontal plane, (front: $\varphi = 0°$, acoustically damped room. Loudspeaker-distance 2 m (6–7 ft.), 25 listeners, complex mean **a** level-difference, **b** time-difference [from Rumsey (2001), p. 24, after Blauert (1997)]

ILDs. In analogy to this front-back ambiguity there exists also a so-called 'elevation-ambiguity' for sound source positions x (up) and y (down).

With natural hearing (as opposed to listening via headphones) two additional, very helpful mechanisms come into play:

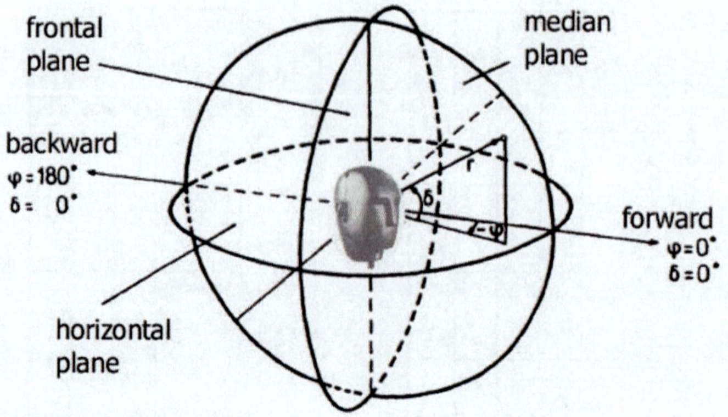

Fig. 2.5 A head-related coordinate system, as being used for listening-experiments; definition of median-, frontal- and horizontal-plane; r is the distance to the sound-source, φ is the azimuth, δ is elevation (after Blauert 1997)]

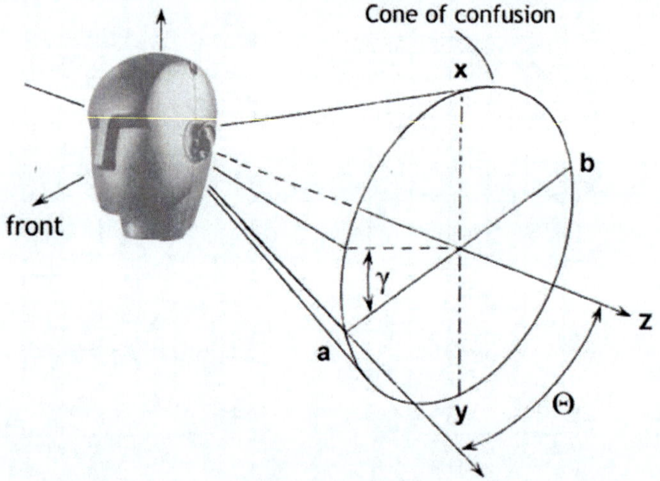

Fig. 2.6 'Cone of confusion': a mere lateral localization of a sound source needs only determination of angle θ, while for a precise localization also γ needs to be detected by the human hearing mechanism [graphic after Hall (1980), p. 342, German edition; after von Hornbostel and Wertheimer (1920)]

3. the frequency response of monaural signals ('pinnae frequency response') usually helps to achieve higher localization accuracy within the 'cone of confusion' (see Figs. 2.4, 2.5 and 2.6)

Already Mach was convinced that the pinnae had a certain 'directivity' and therefore should be important also for localization. In the 1960s D. W. Batteau published his research on the effects of the pinnae-reflections in the time-domain (see Batteau

1967). In his view, the pinnae were serving as directional reflectors, which generated characteristic echo-patterns depending on the angle of incidence (azimuth, elevation) of the sound. In his measurements he was able to detect ITDs in the range from 10 to 300 μs. According to Batteau, it would then be the task of the inner ear to analyze the echo patterns and derive the corresponding angle-of-incidence. However, in order for this mechanism to function, the human ear would need to have a much higher accuracy in the time-domain, than it actually has.

Hebrank and Wright (1974) found that Batteau's pinnae-reflections result in corresponding spectral 'colorations', which indeed can be decoded by the brain. Later examinations by various researchers came to the conclusion that the resulting spectral filtering due to the pinnae is especially important for localization in the median plane as well as for front-back distinction.

4. In addition small head-turns and the changes in the ITDs, which go along with them, help humans to detect the position of a sound source, if in doubt (Blauert 1997) (Fig. 2.7).

5. Furthermore humans are very sensitive to inter-aural signal coherence [see Boehnke et al. (2002) and the references cited therein, as well as Tohyama and Suzuki (1989)], which was acknowledged as an important factor for human listening and sound source localization in reverberant rooms and acoustic environments with several simultaneous sound sources (see Faller and Merimaa 2004).

All of the above-mentioned features are of rather individual nature and depend strongly on the size and shape of the head, the pinnae, as well as chest and shoulders. These inter-individual differences can be analyzed best by measuring a listener's HRTF ('head-related-transfer-function').

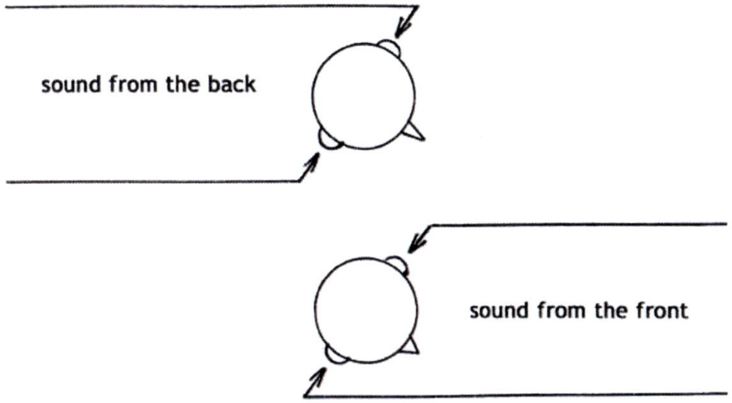

Fig. 2.7 The change of ITDs—caused by a head-turn—are of opposite sense for sound from the front or from the back (after Blauert 1997, p. 180)

The resolution of human hearing in the time- and frequency-domain was researched extensively for monaural hearing [see for example Moore (1997)]. For binaural listening the frequency resolution seems to be quite similar to that of monaural (Moore 1997; van der Hejden and Trahiotis 1998), even though for some of the test signals higher bandwidths were found (Kollmeier and Holube 1992; Holube et al. 1998). Therefore it seems valid to use the ' Equivalent Rectangular critical Band' (ERB) resolution, which was found accurate for monaural listening, also for hearing in general (Glasberg and Moore 1990) (see Fig. 2.8). On this occasion it should also be mentioned that not all frequencies seem to have the same importance in terms of contributing to localization: the 'Weighted-Image Lateralization Model' by Stern et al. (1988) shows a dominant region around 600 Hz, and Wightman and Kistler (1992) have managed to prove the high importance of ITD's of the LF-band it this respect.

To determine the temporal resolution of human hearing is more difficult: it was possible to prove that human listeners are capable of detecting spatial movement of sound sources, which cause sinusoidal fluctuations of ITDs an ILDs, but only up to a fluctuation frequency of about 2.4–3.1 Hz (see Blauert 1972). Despite this fact (Grantham and Wightman 1978) were able to show that their test listeners were able to discern ITD fluctuations up to 500 Hz, but primarily based on the resulting broadening

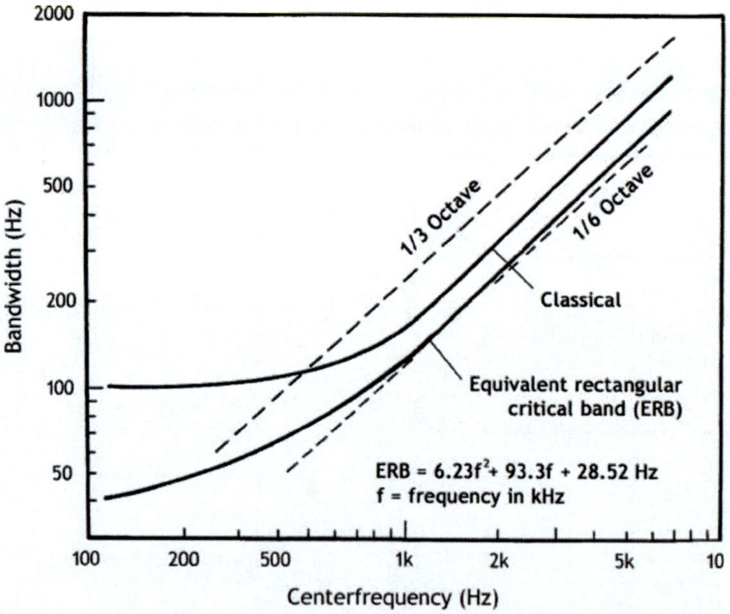

Fig. 2.8 ERB—'equivalent rectangular critical band': shown is a comparison of the bandwidths of 1/3rd and 1/6th octave-band, the 'critical bandwidth' of human hearing, as well as the 'Equivalent Rectangular critical Band' (ERB), which was calculated according to the formula shown in the graphic (after Everest 1994)

of the sound source's ASW (Apparent Source Width). Several Studies on this subject came to the conclusion that binaural perception apparently functions similar to signal processing with a double-sided exponential, rounded exponential or Gaussian time-window [see Holube et al. (1998); Grantham and Wightman (1979); Kollmeier and Gilkey (1990); Culling and Summerfield (1999); Akeroyd and Summerfield (1999), as well as Breebart et al. (2002)]. The window-lengths which were found in these studies are quite varied, but the mean 'Equivalent Rectangular Duration' (ERD), i.e. the mean integration time (or time-constant τ) of human hearing seems to lie in the order of 100 ms. However, also much shorter time-constants were found by some researchers (see Akeroyd and Bernstein 2001) with an ERD of 10 ms, as well as Bernstein et al. (2001) with an ERD of 13.8 ms). A possible explanation was given in that these varied time-constants may correspond to different parts of the binaural 'hearing mechanism' (Kollmeier and Gilkey 1990; Akeroyd and Bernstein 2001; Bernstein et al. 2001).

The so-called 'Precedence-Effect' [or 'Haas-Effect', a.k.a. 'Law of the first wave-front'; see Haas (1951)] also plays a very important role in connection with localization. Of two closely following sound events arriving at the listener from different directions, the first one will determine the apparent localization direction, as long as both sound events are separated only by a few milliseconds (Blauert 1997; Litovsky et al. 1999). Upon replay of a mono test signal via headphones, it seems to shift or 'wander' from the center ('in head localization') to the ear at which the sound arrives first, for ITDs from 0 to 0.6 ms. For the time-range of Δt equal 0.6–35 ms the sound is being localized at the 'advanced' ear; for Δt larger than 35 ms, two distinct sound events will be perceived in the form of an echo.

2.3.1 HRTF Phase-Characteristics

Time-of-arrival differences Δt can also be looked at as phase differences if pure sinusoidal signals are concerned. It is a—quite common—misbelieve (among sound-engineers, but not only) that we (humans) cannot detect phase differences, unless we have the case of 'reversed polarity' of one loudspeaker in a (2-channel) stereo replay system, in which the cable was attached 'out of phase' at one of the loudspeaker terminals. Quite to the contrary, there is evidence that the human ear is capable of detecting phase differences at least at low frequencies up to about 700 Hz. Around this frequency the head diameter becomes equal to half the wavelength of sound and hence the 'phase detection' mechanism of human hearing starts to falter, as a clear detection at which ear the sound arrives first is no longer possible. In addition, phase differences between signals at both ears can also be misleading, if they are caused by room modes (which is more likely when listening in small rooms than in large rooms), or reflections.

Figure 2.9 shows the interaural phase-difference IPD for a listener of a sound-event impinging from 0° to 150° in steps of 30° in the horizontal plane. The thick lines show the corresponding interaural time-delays Δt at 0.25, 0.5 and 1.0 ms.

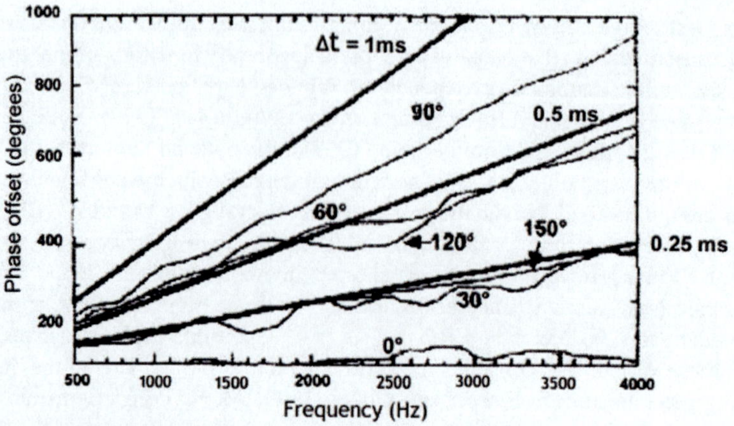

Fig. 2.9 Continuous interaural phase-difference IPD (after Begault 1994)

As mentioned above, localization based on ITDs works very well up to approximately 1.5 kHz. Above this frequency the phase relationship starts to be ambiguous to our perception, but the human hearing mechanism is nevertheless capable of analyzing the ITDs of the amplitude-envelope of a sound event, if it is appropriately structured (i.e. a non-continuous sound).

Apart from the outer ear, which has the largest spectral influence, also head, shoulders and torso interact with sound and cause slight sound colorations. In Fig. 2.10 the direction-dependent and direction-independent contributors to sound coloration within the human hearing system are pictured according to their rank-order (top to bottom).

The 'Cavum Conchae' is the central cavity and the biggest area of resonance of the pinna at the entrance to the ear canal. The outer ear canal leads from the cavum conchae to the tympani and is about 2.5 cm long with a diameter of 7–8 mm. Due to these physical dimensions the ear canal is characterized by a strong acoustic resonance usually around 4 kHz (see Fig. 2.11).

According to the current state of science it is assumed that it is actually the ear canal resonance which is responsible for 'externalization' of sound perception [as opposed to 'in-head-localization'; see Rumsey (2001)]. Due to this reason, efforts were made to excite the ear canal from various directions (when sound is replayed via headphones) in order to overlay this direction-specific EQ, which is different and characteristic for each individual listener, before sound arrives to the tympanum, in an attempt to achieve better 'externalization' (Tan and Gan 2000).

The spectral influence of the shoulders is in the order of ± 5 dB, while for the torso it is only about ± 3 dB.

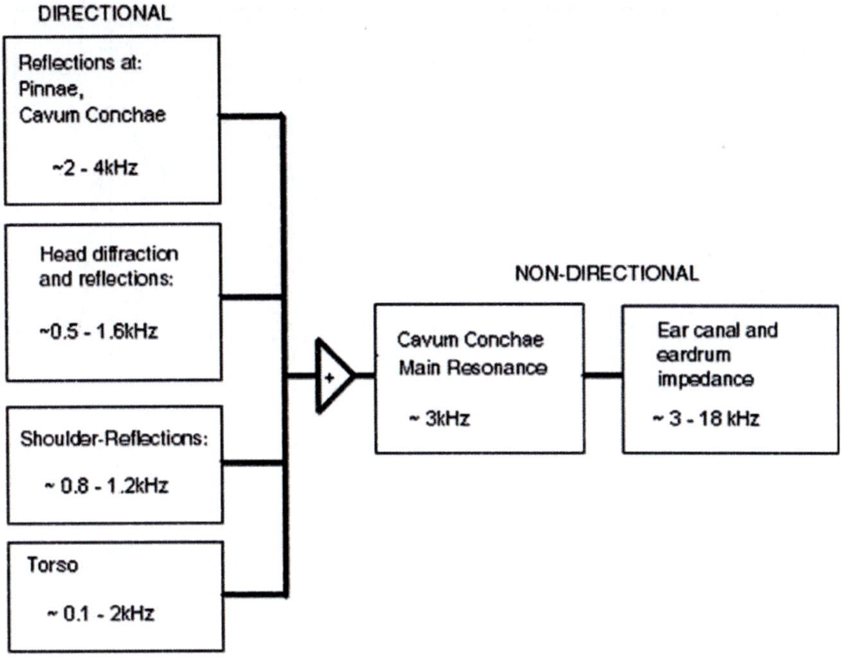

Fig. 2.10 HRTF-block diagram after Blauert (adapted from Fellner and Höldrich 1998b)

2.3.2 *Localization and HRTFs*

One of the most important results of Blauert's research in the 1970s was the discovery of the 'directional bands', responsible for localization in human hearing (Blauert 1997).

As we have already seen in Fig. 2.4 with the individual HRTF-curves there are characteristic peaks and dips in the frequency response, dependent on the sound source position relative to the human head. This is effectively a kind of 'frequency coding' of sound source positions, which the human brain seems to learn at a young age. Typically sound perception for sources behind the listener is characterized by less amplitude at high frequencies, which is due to the shadowing effect of the pinna.

As can be seen in Fig. 2.12 the frequency range around 8 kHz is mainly responsible for 'above' localization, while 'front' localization is being determined by signal content in the frequency bands from 300 to 600 Hz, as well as 3000–6000 Hz. 'Back' localization is related to the frequency bands around 1200 Hz as well as 12 kHz. Similar results were found also by Hebrank and Wright (1974), as well as Begault (1994). It is interesting to note that—looking at Blauert's findings—the relevant frequency bands seem to be at a distance of one decade from each other for both front-, as well as rear-localization mechanisms.

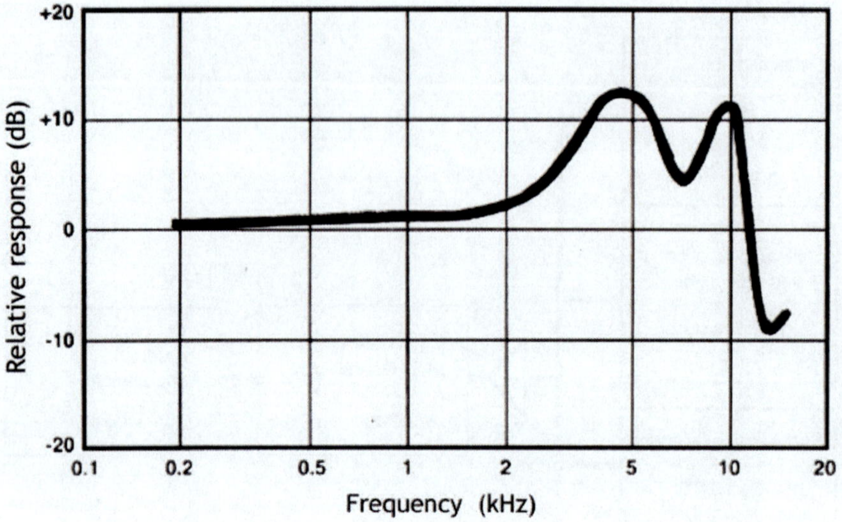

Fig. 2.11 A typical transfer function of the auditory canal—which can be considered to act as a resonant pipe. The transfer function of the ear canal is a fixed entity that is superimposed on each of the highly variable transfer functions of the outer ear. This affects the character of what we hear and the directional cues as well [adapted from Streicher and Everest (2006), after Mehrgardt and Mellert (1977)]

Fig. 2.12 Directional bands: pictured is the relative statistical frequency of test listeners, who are giving one of the answers 'back', 'above', 'front' more often than the other two taken together. Bands drawn at a 90% confidence level; colored background areas indicate most probable answers (after Blauert 1974)

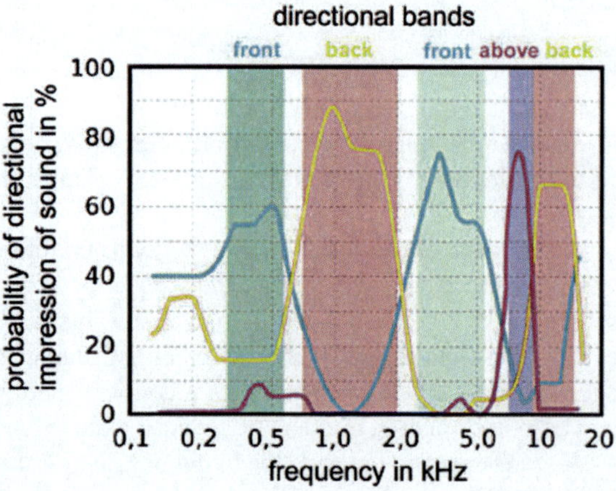

2.4 Mechanisms of Distance Perception

The following explanations are excerpts taken from Nielsen (1993), with additions from Fellner and Höldrich (1998a), as well as Rumsey (2001).

2.4.1 *Sound Intensity*

For humans, the most important 'anchor' for distance perception is the sound intensity, which is transformed into the individual perception of 'loudness' by our hearing mechanism. Loudness is the individually perceived amplitude or level of a sound event, which is also frequency dependent, according to the sensitivity of the human ear (see the Flechter-Munson curves) (Fig. 2.13).

Under free-field conditions sound pressure suffers a reduction in level by 6 dB_{SPL}, when doubling the distance to the sound source. A 'halving' of the loudness perceived by the listener—which is marked in 'Sone'—corresponds to a loss in sound pressure level by 10 dB. (As human hearing is 'non-linear' in its behavior in many respects, the exact amount of attenuation (in dB) necessary to achieve the impression of 'half the loudness' is largely signal dependent, but the SPL level-drop by 10 dB is usually true at least for the frequency range from 400 Hz to 5 kHz and sound pressure levels in-between 40 dB and approx. 100 dB). In this context it needs to be added that usually the estimation of distance of a sound source in a natural environment almost always happens in connection with a corresponding visual stimulus, which is also why it is easier to judge the distance of a sound-source in an environment that is well-known. Whether the perceived loudness is a meaningful indicator of distance for the listener, or not, depends largely on the question whether he or she knows the 'normal' loudness of the sound source. (see Nielsen 1993).

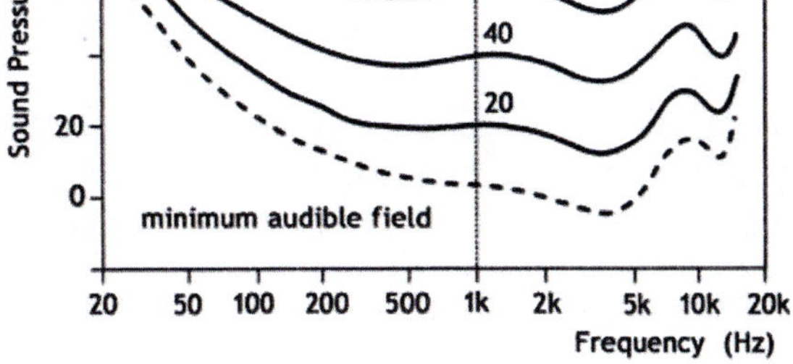

Fig. 2.13 Equal loudness contours of hearing, sound pressure level in dB re 20 $\mu N/m^2$ (0 dB_{SPL} is about the threshold of hearing) versus frequency in Hz, from ISO 226 (after Fletcher-Munson). These curves show that at no level is the sensation of loudness flat with frequency; more energy is required in the bass range to sound equally loud as the mid-range

2.4.2 Diffuse Sound

Due to the energy contained in diffuse sound, the level drop when doubling the distance to the sound source in an average room is not 6 dB, as would be the case in a completely damped room or under free-field conditions, but less. The direct-/diffuse-sound ration seems to be an even stronger indicator for distance perception than the mere sound pressure level at the listener position. According to a hypothesis by Peter Craven, the perceived distance of a sound source is determined by the relative time-delay and the relative amplitude of the early reflections in relation to the direct sound signal (see Gerzon 1992).

If the acoustic properties of the listening room are not known to the listener (as can be the case under test-conditions), apparently the human hearing mechanism is capable of extracting the acoustic information about the room from what is being heard (Nielsen 1993). In their experiments (Mershon and Bowers 1979) found that it made no difference whether their test listeners were already familiar with the acoustic conditions of the listening room before the test, or not. However, the diffuse sound triggered off by a sound stimulus seems to be an important source of information for the human listener: in a sound-deadened room, test listeners did not manage to establish a relationship between the perceived and the real physical distance to the sound source (Nielsen 1993).

2.4.3 Frequency Response

As the HF-part of a sound signal is more heavily damped when propagating through air than LF signals (air absorption effect), for the human listener also the frequency-spectrum of an acoustic sound event contains distance information. A prerequisite for this is that the listener is familiar with the regular spectral content of the sound event. In a room with diffuse sound components (reverb) the spectral balance of the reverb changes due to the frequency dependent absorptive behavior of the surrounding surfaces, as well as due to the air absorption effect. The absorption coefficient of air depends on temperature, as well as relative humidity, which can lead to relevant changes in sound especially with live outdoor sound-reinforcement systems. Also, local climate factors (wind, etc.) can lead to a change of the frequency response or spectral balance.

2.4.4 Binaural Differences

In connection with sound sources that are close to the listener, there can be relevant changes of the HRTFs of the listener. Especially the ILDs can become much larger for low and high frequencies (see Huopaniemi 1999).

2.5 Spatial Impression

The resulting 'spatial impression' or 'spaciousness' for a human listener is primarily based on two components: the so-called 'Apparent Source Width' (ASW), i.e. the perceived base width of a sound event, as well as 'Listener EnVelopment' (LEV), i.e. the acoustic (or musical) envelopment of a listener—the feeling of being 'surrounded by sound'. In the time domain, the acoustic properties of a room can be represented by the 'Impulse Response' (IR) and the 'Reflectogram', which can be derived from the IR (Fig. 2.14).

When analyzing the temporal aspects of a diffuse sound field in a room, acousticians distinguish between the 'early reflections' (ER) which occur during the first 80 ms right after the sound event, and the 'late reflections' (LR), which happen afterwards. While the first (single) reflections from the walls, ceiling and floor of a venue can help a listener to get an impression of his (or her) position in the room, the later reflections are of much more diffuse nature and also much weaker in level, as they result from multiple reflections at the borders of the room and the objects contained within.

Based on psychoacoustic research, which takes into account and distinguishes between the early reflections (which happen within the first 80 ms) and the late reflections, the acoustic measure of 'Clarity' (C) was defined by Reichardt and Kussev in the early 1970s (see Reichardt and Kussev 1972)

$$C_{80} = 10 \log \frac{\int_0^{80\text{ ms}} p^2(t)\mathrm{d}t}{\int_{80\text{ ms}}^{\infty} p^2(t)\mathrm{d}t} (\mathrm{dB}) \tag{2.1}$$

with p ... sound pressure.

The above mentioned ASW is primarily influenced by the sound reflections which arrive during the first 80 ms at the listener, a context in which the frequency range from 1 to 10 kHz is most important (see Fig. 2.15). On the other hand, for LEV it

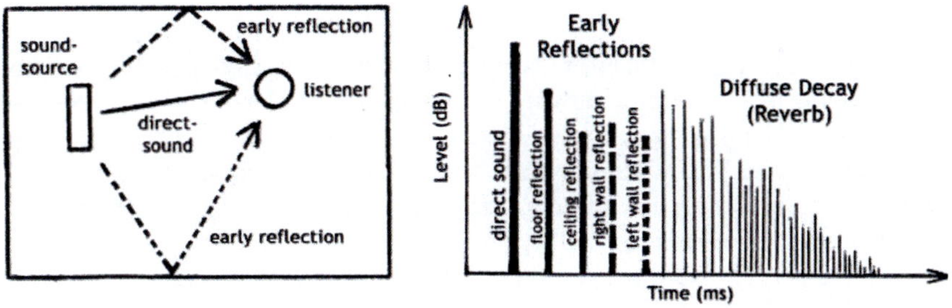

Fig. 2.14 The response of an enclosed space to a single sound impulse: (left) The direct path from source to listener is the shortest, followed by early reflections from the nearest surfaces. (Right) The impulse response in the time domain shows the direct sound, followed by some discretely identifiable early reflections, followed by a gradually more dense reverberant tail that decays exponentially

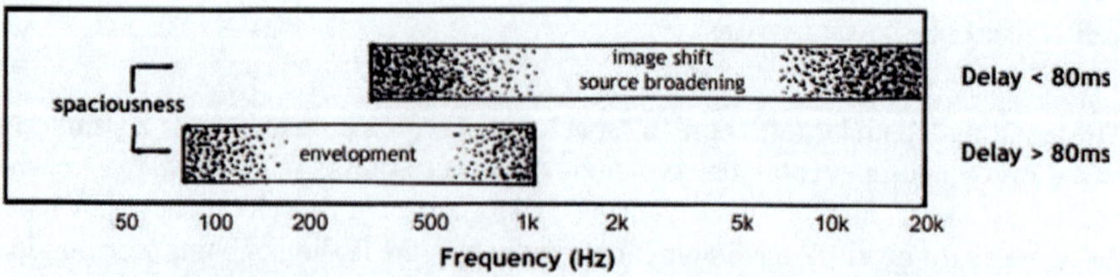

Fig. 2.15 The approximate frequency- and delay-ranges over which reflected sound contributes to the perception of different spatial aspects (ASW and LEV) (after Toole 2008, Fig. 7.1, p. 97)

is the low-frequency signal components (mainly below 200 Hz and up to 500 Hz) arriving after 80 ms and from various directions, which are most important for the listener.

According to research by Hidaka et al. (1997) ASW is mainly influenced by reflections in the octave-bands around 500, 1000 and 2000 Hz, which cause low Inter-Aural Cross Correlation (IACC). As defined in Hidaka et al. the $IACC_{E3}$ is the mean value of the IACCs in these three octave-bands measured during the first 80 ms (E = early), which is directly related to the impression of ASW. Ando (1977) had already showed that in respect to the angle of incidence of a single reflection apparently there is a listener-preference for 60°, which finds its correspondence in a reduced IACC around that angle-range (see Fig. 2.16). The findings of this study are reinforced through the outcome of later research by Barron and Marshall (1981).

Research by Griesinger has shown that the ideal angle of incidence to evoke the sensation of envelopment in listeners is frequency dependent: while LF signal-components under 700 Hz should ideally arrive laterally (at ± 90° from the listener), at higher frequencies LEV can also be achieved with sounds much closer to the median plane (see Griesinger 1999).

As visible in Fig. 2.17 for a single reflection of a 1000 Hz signal arriving at 45° there is a high fluctuation of ITD, resulting in a strong sensation of acoustic envelopment (LEV). For frequencies around 2000 Hz arriving in the same angular segment (40–45°) there is much less fluctuation and hence a much lower sensation of envelopment.

Unfortunately it seems that there is not really a consensus among scientists on the exact meaning of acoustical terms used especially in the realm of spatial perception. While Toole (2008) essentially declares 'spaciousness' and 'envelopment' to be synonymous (ibid. p. 99), Rumsey also uses the words 'room impression' and—rightly so—sees a connection between listener envelopment and the 'externalization' of a sound-event (see Rumsey 2001, p. 38).

In Lehnert (1993) 'spaciousness' and 'spatial impression' are defined as follows (after Kuhl 1978):

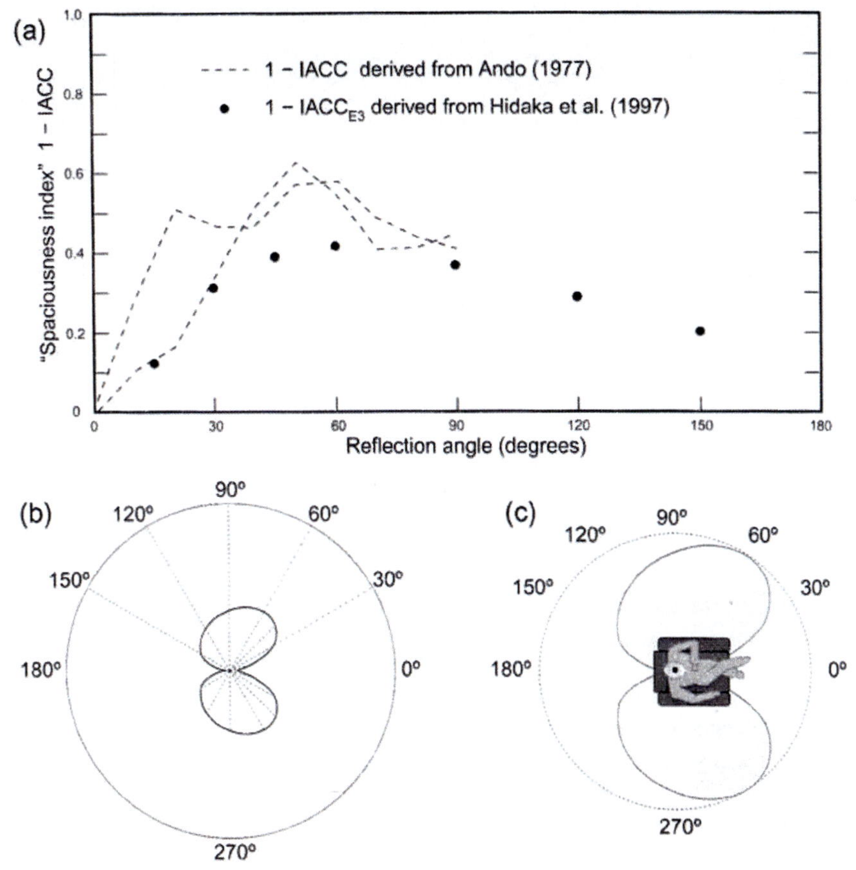

Fig. 2.16 'Spaciousness-Index' (1-IACC) depending on the angle of incidence of a reflection; (various forms of graphical representation (**a**, **b**, **c**) from Toole (2008), Fig. 7.5, p. 106). **a** Both curves from Ando (1977) and Hidaka et al. (1997) reach a maximum around 60°. **b** The 1-$IACC_{E3}$ data from Hidaka et al. (1997) reformatted to a polar plot. **c** The intention is made even more clear with the addition of a listener: a 'spatial-effect balloon' surrounding a listener showing estimates of ASW and image broadening generated by early reflections incident from different directions

- Auditory Spatial Impression (German: 'Raumeindruck'): The concept of the type and size of an actual or simulated space to which a listener arrives spontaneously when he/she is exposed to an appropriate sound field. Primary attributes of auditory spatial impression are auditory spaciousness and reverberance.
- Auditory Spaciousness (German: 'Räumlichkeit'): Characteristic spatial spreading of the auditory events, in particular the apparent enlarged extension of the auditory image compared to that of the visual image. Auditory spaciousness in mainly influenced by early lateral reflections.

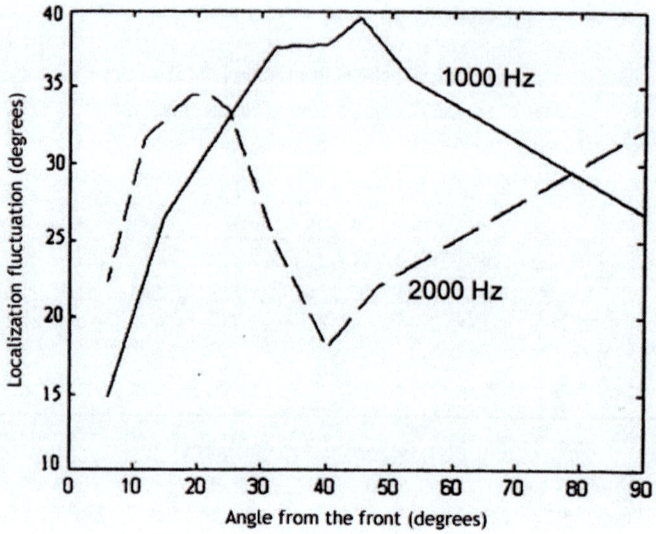

Fig. 2.17 Fluctuations of the ITD for a single reflection depending on the angle of incidence plotted for two frequencies (computer-simulation). This shows that the resulting spatial impression is both angle- and frequency-dependent (after Griesinger 1997)

Sarroff and Bello have proposed and tested two quantitative models for the estimation of the spatial impression contained in a stereo recording: (a) (for ASW) based on the analysis of stereophonic width (or ASW) of several panned single sound sources, and (b) (for LEV) the analysis of the quantification of reverb as part of the overall signal (see Sarroff and Bello 2008).

In Avni and Rafaely (2009) it is analyzed if and how spatial impression can be explained by the connection between IACC and spatial correlation in diffuse sound fields via 'spherical harmonics' (see Avni and Rafaely 2009).

2.6 Physical Measures in Relation to Spatial Impression

In the previous sub-chapter it has already been reported that the individual impression of ASW is best represented through measuring the $IACC_{E3}$ (see Hidaka et al. 1997).

Already back in 1975 *R*, the 'Measure of Spatial Impression' was defined by Lehmann (1975) (see also Reichardt et al. 1975). For the practical measurement of *R* a 40° funnel around the 0° axis of main orientation was defined. In this context, sound signals arriving with more than 80 ms delay within the funnel, as well as those arriving laterally (i.e. outside of the funnel) with a delay of 25–80 ms are regarded to increase spatial impression. Not in favor of increased spatial impression are lateral reflections arriving within the first 25 ms, as well as frontal reflections within the time window of 25–80 ms.

If we denote the sound-energy components, which are positive for spatial impression with E_R, and the sound-energy components which are negative for spatial impression with E_{NR}, then R can be described as:

$$R = \lg \frac{E_R}{E_{NR}} (\mathrm{dB}) \tag{2.2}$$

The practical implementation of measuring R requires the use of a directional microphone in addition to the pure pressure transducer (omni-pattern), and R can be obtained by performing the following calculation:

$$R = 10 \log \frac{\int_{25\,\mathrm{ms}}^{\infty} p_{omni}^2(t)\mathrm{d}t - \int_{25\,\mathrm{ms}}^{80\,\mathrm{ms}} p_{shotgun,front}^2(t)\mathrm{d}t}{\int_{0\,\mathrm{ms}}^{25\,\mathrm{ms}} p_{omni}^2(t)\mathrm{d}t + \int_{25\,\mathrm{ms}}^{80\,\mathrm{ms}} p_{shotgun,front}^2(t)\mathrm{d}t} (\mathrm{dB}) \tag{2.3}$$

… with 'shotgun, front' denoting the output-signal of a highly directional (shotgun) microphone, directed towards the sound source.

For the sake of completeness also R, the 'Reverb-Measure' (German: 'Hallmaß' or ratio of reverberant sound energy to early sound energy), as defined by Beranek and Schultz (1965) should be mentioned:

$$R = 10 \log \frac{\int_{50\,\mathrm{ms}}^{\infty} p^2(t)\mathrm{d}t}{\int_{0\,\mathrm{ms}}^{50\,\mathrm{ms}} p^2(t)\mathrm{d}t} (\mathrm{dB}) \tag{2.4}$$

Concerning listener envelopment LEV, different measures such as 'Lateral Fraction' (LF), as well as 'Lateral Gain' (LG_{80}) were proposed.

Research by Marshall (1968) and Barron (1971) shows the importance of lateral reflections for spatial impression for the listener in the concert hall. The 'Lateral energy Fraction' (LF) was used as a measure for ASW in Barron and Marshall (1981) and is defined as follows:

$$\mathrm{LF} = 10 \log \frac{\int_{0\,\mathrm{ms}}^{80\,\mathrm{ms}} p_F^2(t)\mathrm{d}t}{\int_{0\,\mathrm{ms}}^{80\,\mathrm{ms}} p_O^2(t)\mathrm{d}t} (\mathrm{dB}) \tag{2.5}$$

with $p_F(t)$ being the sound pressure of the impulse response, which was measured in the concert hall with a side-oriented figure-8 microphone (i.e. 'on-axis' is oriented at 90° from the sound source), while $p_O(t)$ is the sound pressure at the same point in the concert hall, but measured with an omni-directional microphone. Usually signals of both microphones are integrated over the first 80 ms for the calculation of LF.

The definition of 'Lateral Gain' can be found in Bradley and Soulodre (1995), which sets the energy of the late-arriving (i.e. after 80 ms) lateral diffuse-sound in relation to the energy picked up by an omni-directional microphone, both at the listener position in the hall. In Soulodre et al. (2003) LG_{80} is defined as:

$$\mathrm{LG}_{80} = 10\log\frac{\int_{80\,\mathrm{ms}}^{\infty} p_F^2(t)\mathrm{d}t}{\int_{0\,\mathrm{ms}}^{\infty} p_A^2(t)\mathrm{d}t}(\mathrm{dB}) \tag{2.6}$$

with $p_F(t)$ being measured by a side-oriented figure-8 microphone and $p_A(t)$ with an omni-directional microphone, both at a distance of 10 m from the sound-source under 'free-field' conditions. In this paper it is also analyzed that the restriction to a fixed time-interval of 80 ms (independent of the frequency band concerned) is not sufficient if one wants to take into account the frequency-dependent 'inertia' (or time-constant) of human hearing. This is why Soulodre et al. propose an integration time of 160 ms up to 500 Hz, above that a shorter integration time (75 ms at 1000 Hz) and only 45 ms above 2000 Hz.

Further they plead for a new objective measure for LEV which they denote with $\mathrm{GS}_{\mathrm{perc}}$

$$\mathrm{GS}_{\mathrm{perc}} = 0.5\mathrm{G}_{\mathrm{perc}} + \mathrm{S}_{\mathrm{perc}}(\mathrm{dB}) \tag{2.7}$$

in which $\mathrm{G}_{\mathrm{perc}}$ is a gain-component and $\mathrm{S}_{\mathrm{perc}}$ is a component of spatial distribution. $\mathrm{G}_{\mathrm{perc}}$ is essentially the acoustic measure of 'Strength' from concert hall acoustics, also known as 'Strength factor' (G) (see Beranek 2004, p. 617), which was first defined by Lehmann in 1976 (see Lehmann 1976):

$$\mathrm{G}_x = 10\log\frac{\int_{x\,\mathrm{ms}}^{\infty} p_O^2(t)\mathrm{d}t}{\int_{0\,\mathrm{ms}}^{\infty} p_A^2(t)\mathrm{d}t}(\mathrm{dB}) \tag{2.8}$$

According to the proposal of Soulodre et al. $\mathrm{G_x}$ is a measure for the relative sound pressure level of the late energy, while—depending on the octave band which is being measured—the integration times $\times$ of Table 2.1 are used. With this new measure for listener envelopment—which can be considered a refinement of LG_{80}—it was possible to further the correlation of LEV by another 4%, reaching a total of 0.98. In connection with the search for a physical measure for the spatial distribution $\mathrm{S}_{\mathrm{perc}}$ Soulodre argues that a measure based on IACC may be considered, as this is generally used to quantify the spatial distribution of a sound field (see Soulodre et al. 2003, p. 838, 2nd paragraph).

In this context it is interesting to note that IACC can be used not only in connection with the first component of spatial impression, namely ASW (as pointed out above), but also a part of LEV (i.e. the component which concerns the spatial distribution $\mathrm{S}_{\mathrm{perc}}$) could be represented through a measure related to IACC. These are both strong arguments for taking a further look at IACC as a measure for spatial impression with human listeners.

In relation to the acoustics of small rooms (Griesinger 1996) proposes the 'Lateral Early Decay Time' (LEDT) as a measure for spaciousness:

$$\mathrm{LEDT}_{350} = \frac{60\ 350\,\mathrm{ms}}{(\mathrm{S}(0) - \mathrm{SD}(350))1000\,\mathrm{ms/s}} \tag{2.9}$$

Table 2.1 Frequency dependent integration times, adapted to human psychoacoustics (after Soulodre et al. 2003)

Octave bands (Hz)	Integration limit × (ms)
–	–
63	–
125	160
250	160
500	160
1000	75
2000	55
4000	45
8000	45

with S(t) being the Schröder-integral of the impulse response and SD(t) being the Schröder-integral of the interaural level-difference (ILD).

A multitude of factors contributes to spatial impression: research by Griesinger (1997) has shown that it is the fluctuations of ITD and ILD, which contribute significantly to the spaciousness of a sound event. Research by Mason and Rumsey (2002) seems to confirm this: measurements based on the $IACC_E$ showed the highest correlation to the perceived source width as well as environmental width (i.e. the acoustically perceived width of the performance space).

ASW and IACC are very useful in connection with acoustic measurements of concert halls, but not equally suited for the measurement of smaller rooms. This is why (Griesinger 1998) has proposed two additional measures: The 'Diffusefield Transfer Function' (DTF) as a measure of envelopment, which is equally suited for small and large rooms, and the 'Average Interaural Time Delay' (AITD) as a measure for externalization.

As can be seen from all the information provided above, the individual listening impression in respect to spatial impression is determined by ASW and LEV, which are both caused by reflected sound, which arrives at the listener in various time-slots and from different directions. The result for both is a decreased IACC at the ears of the listener (due to lateral reflections), which is important as research has shown that in respect to the perceived acoustic quality of concert halls listener preference increases with lowering IACC (see Hidaka et al. 1995; Beranek 2004). In this context also the 'Binaural Quality Index' BQI was defined in Keet (1968). According to Beranek (2004) it is "one of the most effective indicators of acoustic quality of concert halls" (ibidem, p. 506) and defined as follows:

$$\mathrm{BQI} = (1 - \mathrm{IACC}_{E3}) \tag{2.10}$$

The subindex $_{E3}$ denotes the early sound energy in the time window from 0 to 80 ms in the octave-bands with center-frequencies at 500 Hz, 1 kHz and 2 kHz. Figure 2.18 clearly shows the correlation between hall ranking and BQI.

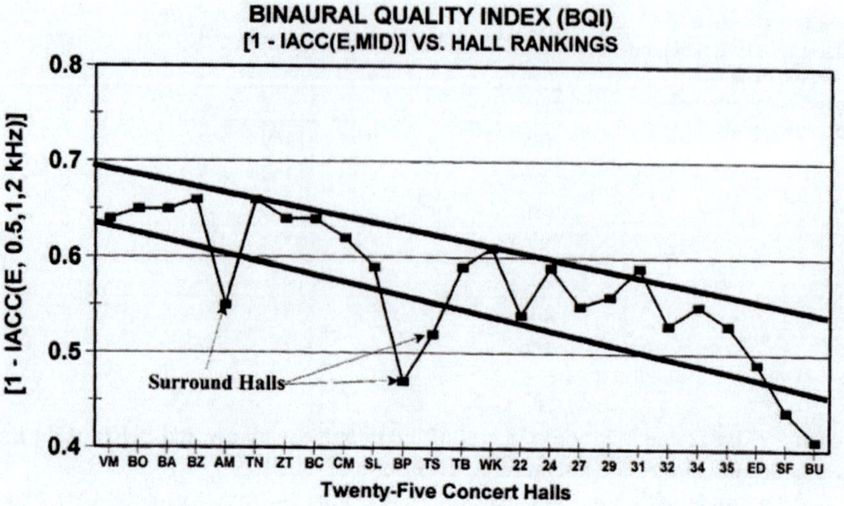

Fig. 2.18 Binaural Quality Index (BQI) for 25 concert halls, measured when unoccupied, plotted versus the subjective rank orderings of acoustical quality; average standard deviation 0.11 s (from Beranek 2004, p. 509)

A newly defined "Binaural Quality Index of Reproduced Music" (BQI_{rep}) has been proposed in Pfanzagl-Cardone (2011), which has been derived from the "Binaural Quality Index", as defined in Keet (1968) [see also Chap. 1, Sect. 1.5 in this book for more details, or Pfanzagl-Cardone (2012)].

In analogy to the BQI, the BQI_{rep} is defined as:

$$BQI_{rep} = (1 - IACC_3) \tag{2.11}$$

with $IACC_3$ being the mean value of the cross-correlation coefficients of the octave bands 500 Hz, 1 kHz and 2 kHz measured between the L and R binaural signals (no time-window applied). (Rem.: MATLAB-code for calculating the BQI_{rep} from binaural samples can be found in Appendix C, p. 404 of Pfanzagl-Cardone (2020), free for download here: https://link.springer.com/content/pdf/bbm%3A978-3-7091-4891-4%2F1.pdf). The BQI_{rep} is intended to serve as a qualitative measure of the spatial impression contained in recordings and is mainly intended for use with music recordings (see also Chap. 10, Sects. 10.9.6–10.9.8 in this respect).

2.7 The Influence of Loudspeaker Quality and Listening Room Acoustics on Listener Preference

When undertaking the task of evaluating the quality and various aspects of microphone techniques, it must also be assumed that the acoustical quality of the reproduction loudspeakers as well as the acoustical properties of the listening room will have an influence on the result. Changes in the signal from the source (e.g. musical instrument) to the recipient (ear of human listener) are unavoidable due to technical limitations in the form of qualitative limits of the electronics and electroacoustic transducers involved (microphones, loudspeakers, AD- and DA-converters, etc.) as well as limitations of the transmission channel or storage medium.

In connection with the selection of appropriate loudspeakers the question arises whether their directivity index (DI) would have an influence on listener evaluations of test material. Various researchers seem to have divergent answers to this: (Kates 1960) favors loudspeakers with higher directionality, as he thinks that it is of advantage to avoid unnecessary room-reflections. Also Zacharov (1998) shares this opinion, but encounters strong criticism by important leaders within the audio-engineering community (see—among others—Holman 2000). To underline his arguments, Holman points out the results of his own research (Holman 1991), in which he was able to show that a higher directivity index of loudspeakers is helpful for better localization, but a smaller directivity index is in favor of better 'envelopment' for the listener (especially with a surround-replay setup). Likewise (see Toole 2008, p. 137)—based on the analysis of his previous listening tests (Toole 1985, 1986)—arrives at the conclusion that the majority of listeners apparently prefers loudspeakers with a broader dispersion angle (i.e. smaller directivity index). In his case this also had to do with the fact that he was using listening rooms with acoustically untreated side-walls which caused relevant reflections due to the wide dispersion angle of the loudspeakers. In turn these reflections were responsible for a lowering of the IACC at the listeners which lead to an increase in perceived ASW. This resulted in a higher listener preference for loudspeakers which had a wider dispersion characteristic.

That a lager base width or ASW is preferred by listeners coincides with the findings of Berg and Rumsey (2001), as well as Bech (1998). In the case of Bech's research, the variation in base-width of the sound source had been achieved on the reproduction side by means of a varied spacing of the (stereo-) front-speakers (see Fig. 2.19). Additional research which arrived to similar results can be found in Moulton et al. (1986), Moulton (1995) and Augspurger (1990).

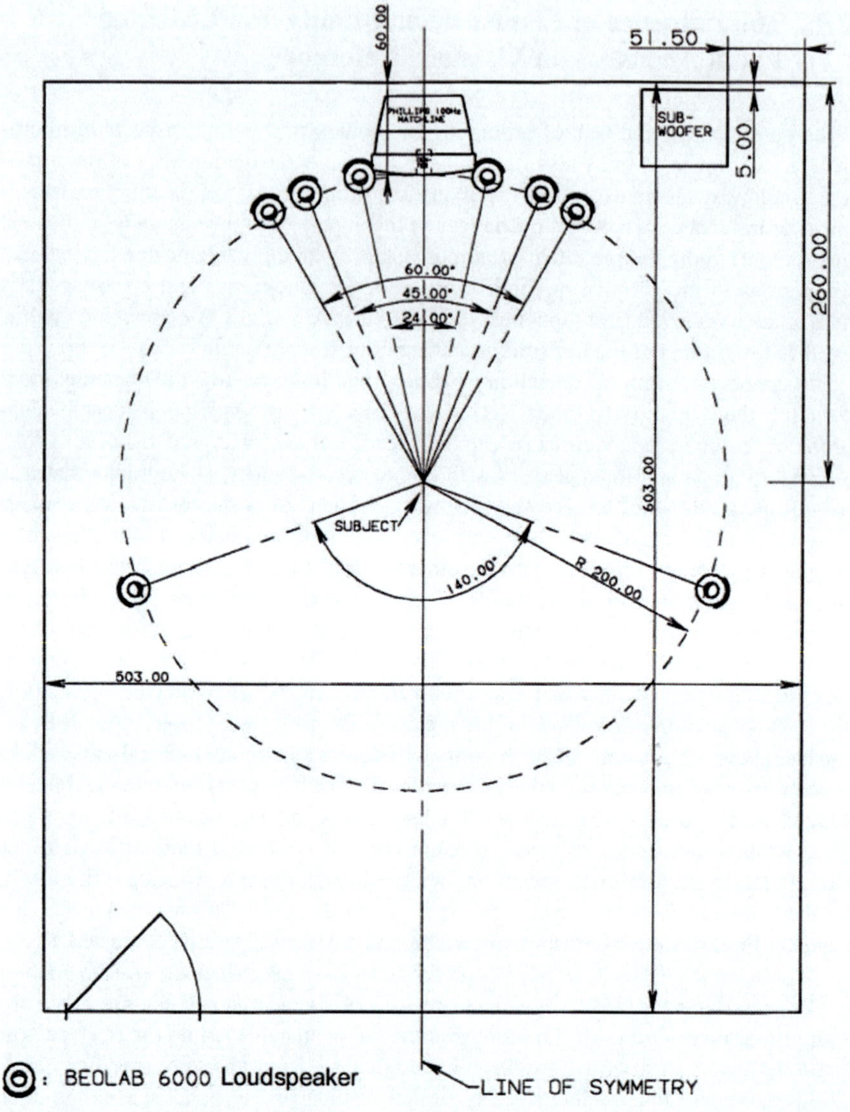

Fig. 2.19 Loudspeaker and TV setup in the listening room for research on the influence of base width on perceived audio quality (from Bech 1998)

2.8 Psychoacoustic Effects Concerning Localization and Spatial Impression with Loudspeaker Reproduction

When playing back sound via a standard stereo loudspeaker setup (i.e. speakers at − 30° and + 30°) sufficiently large level or time differences (ILDs and ITDs) between the two signals will essentially lead to localizing the stimulus only in one of the two speakers. In order for this to happen, the level difference needs to be about 15 dB or the time difference about 1.1 ms, or a suitable combination of level- and time-difference values in between (see Fig. 2.20). First research in that direction can already be found in de Boer (1940), and also in later publications such as Theile et al. (1988), Gernemann (1994). In his research (Sengpiel 1992) arrives at slightly different results with 18 dB and 1.5 ms respectively.

When comparing the results from various research, it becomes clear that the deviation between values is much larger for time-of-arrival based localization than for level-based localization. The big discrepancies for time-of-arrival based stereophony is also backed by the findings of a study by Wittek and Theile (2002).

As is visible in Table 2.2, the differences in perceived localization width (regarding the resulting recording angle) can be more than 100% (!) (compare the results of Wittek with those of Sengpiel for a small AB-pair with a capsule spacing of 50 cm, or roughly 2.7 ft.). In Knothe and Plenge (1978) various studies are named which make it clear that the effect of frequency-dependence of sound source localization

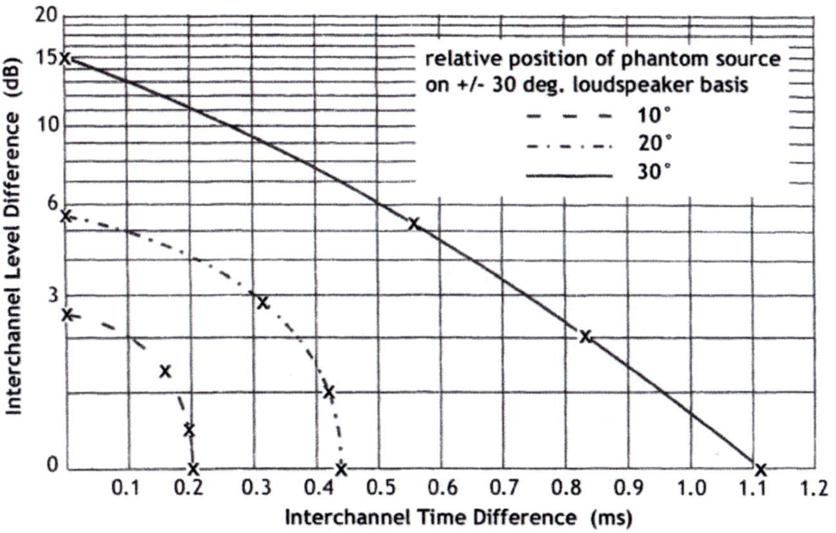

Fig. 2.20 Localization curves after Williams (1987) and Simonsen (1984)

Table 2.2 Recording angles as found in literature Hugonnet and Walder (1998), Sengpiel (1992), Williams (1987), Wittek and Theile (2000), Wittek (2002); after Wittek and Theile (2002)

Setup	Hugonnet	Sengpiel	Williams	Wittek
AB omnis 50 cm	130°	180°	100°	74°
AB omnis 100 cm	–	62°	–	36°
ORTF cardioids 110°, 17 cm	90°	96°	100°	102°

was known at least since 1934. More recent research on this topic can be found in Griesinger (2002), among others.

Reasons for the differences between results of the studies examined in Wittek and Theile (2002) may be the following:

- difference in the test signals which were used for the various studies,
- varying acoustical characteristics of the loudspeakers used for the different studies [concerning this topic see (Gernemann 1998)],
- large inter-individual differences in acoustic perception among the test listeners, in combination with localization-distortion which happens at mid-frequencies [see Benjamin and Brown (2007), as well as Fig. 2.17)] (Fig. 2.21).

If one looks at the differences of more than 100%, as pointed out in the study of Wittek and Theile (2002), it seems clear that the results achieved through small-AB

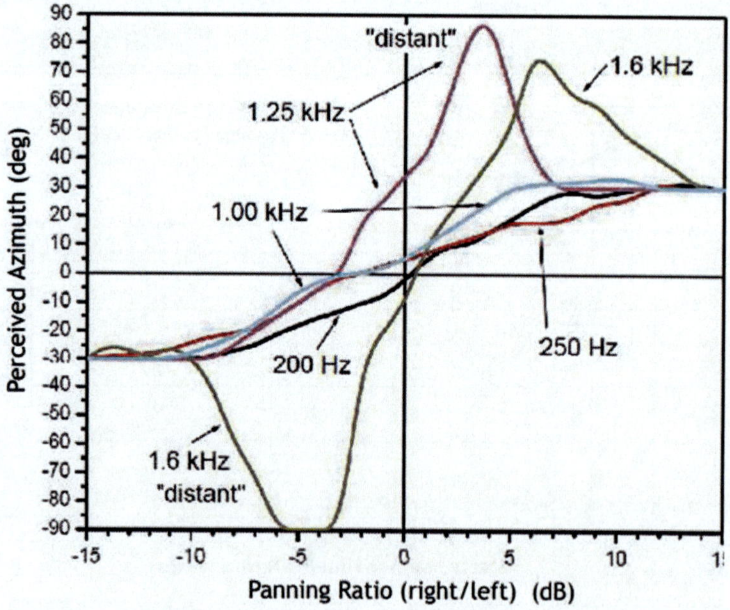

Fig. 2.21 Perceived azimuth at various frequencies for signals (narrow-band Gaussian sine-bursts), which were panned between the stereo-channels by means of level-difference (from Benjamin 2006)

based recording techniques are not consistent enough in terms of localization (or stereophonic 'base width') so that inter-individual differences between listeners (as well as qualitative differences between loudspeakers) could be neglected. Therefore small-AB cannot be recommended as a reliable microphone technique what concerns the reproduction of localization and spatial distribution of sound sources.

Through experiments regarding the localization of sound sources replayed through a 5.1 loudspeaker-setup according to (ITU-R BS.775.1 2012) (Martin et al. 1999) it was found that already a time-difference of 0.6 ms was sufficient to cause full localization of the sound source in just one of the two rear speakers. Most likely this has to do with the much larger distance (or greater opening angle of 120°) between the rear speakers (in contrast to the 60° between the L and R front speaker), relative to the listener (see Rumsey 2001, p. 32). Another conclusion of this research was that level-based localization (ILD) results in a more stable listener impression than time-of -arrival based (ITD) localization.

2.8.1 *Frequency Dependent Localization-Distortion in the Horizontal Plane*

As was pointed out in Knothe and Plenge (1978), localization of a panned mono sound-source is strongly frequency dependent (see Fig. 2.22). The level-difference ΔS, achieved on the summing bus due to panning, needs to be larger at lower frequencies in order to achieve the same impression of position in terms of localization than with higher frequencies. The more lateral the perception of the sound source is supposed to be on the $\pm$ 30° stereo base-width, the larger ΔS needs to be: for a localization impression at 10° the difference in necessary ΔS between high and low frequencies is about 1–2 dB, for 25° azimuth the necessary difference is already up to 5 dB (compare ΔS at 400 Hz and 7 kHz in Fig. 2.22).

Also Griesinger (2002) arrives at similar conclusions in respect to frequency dependent localization both for stereo, as well as surround signals. For spectral components above 600 Hzlocalization is heavily 'biased' towards the loudspeaker to which the signal gets panned. According to Griesinger's research this is due to interference of the direct signal from one loudspeaker with the signal from the other loudspeaker, which diffracts around the listener's head.

2.8.2 *Frequency-Dependent Localization in the Vertical Plane*

As research by Ferguson and Cabrera (2005) has shown, also in the vertical plane localization distortion between signals of high and low frequencies takes place: while high frequency sound sources are normally localized at their correct physical position,

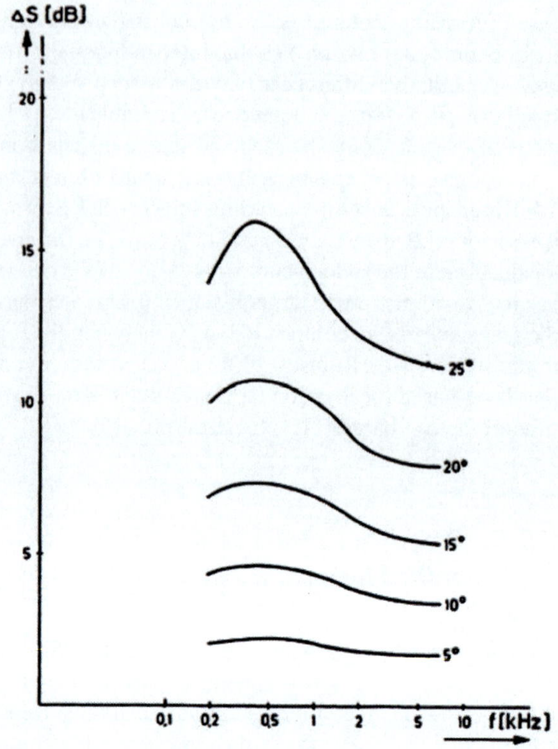

Fig. 2.22 Frequency-dependent localization of a sound-source and ΔS of relative level difference between L and R channel on a stereo loudspeaker system (from Knothe und Plenge 1978)

this is not the case for low frequency sound sources. These are usually localized below their real physical position. In the case of a broadband sound signal localization is dominated by the high frequency signal components (see also Morimoto et al. 2003).

2.8.3 *Effects Concerning the Reproduction of Spaciousness*

In practical recording situations the frequency dependent localization of sound signals can be especially difficult for instruments that are not only broadband in terms of frequency spectrum, but also broad or wide in a physical sense—like a piano for example: for the human listener the source width of the piano seems much smaller for low frequencies, than for high frequencies; at least for microphone techniques which are purely level-based, which is the case for the coincident "Blumlein-Pair", consisting of two crossed figure-8 microphones. In this context methods for 'spatial equalization' were proposed by various researchers [see e.g. Gerzon (1986), as well as Griesinger (1986)].

The change in terms of spaciousness towards low frequencies can be checked by listening selectively to isolated frequency bands to see (or better: hear) which spatial impression a microphone technique is able to provide at different frequency ranges. In this respect the entire frequency band below approx. 800 Hz is especially important, as the human head is not yet effective as a baffle and sound signals bend around it. Above 800 Hz the shadowing effect of the human head becomes more and more evident and thus human hearing is based mainly on ILD, while at low frequencies it is mainly based on an analysis of phase- and time-differences (see Fig. 2.23).

In this context research by Yost et al. (1971) should be mentioned, which has shown that the low-frequency components of transient binaural signals are of highest importance for localization: the highpass-filtering of clicks (similar to Dirac impulses) with a cutoff-frequency at 1500 Hz lead to a clear deterioration of localization, while lowpass-filtering with the same frequency resulted only in a minimal change (deterioration) of localization.

Research by Hirata (1983) deals with the phenomenon of localization distortion of the low-frequency components of a stereo signal upon loudspeaker playback. He proposes PICC, a "Perceptual Interaural Cross-correlation Coefficient":

$$\mathrm{PICC} = \mathrm{DR_0} + (1 - \mathrm{D})\mathrm{R_E} \tag{2.12}$$

with D (Deutlichkeit), as it was defined by R. Thiele (see Thiele 1953)

$$\mathrm{D} = \frac{\int_0^{50\,\mathrm{ms}} p^2(t)\mathrm{d}t}{\int_{0\,\mathrm{ms}}^{\infty} p^2(t)\mathrm{d}t} \tag{2.13}$$

with:

$\mathrm{R_0}$ the interaural cross-correlation coefficient of direct sound (which is 1 for sound from 0°),

$\mathrm{R_E}$ the interaural cross-correlation coefficient of diffuse sound, which is defined as:

$$\mathrm{R_E} = \frac{\sin \mathrm{kr(f)}}{\mathrm{kr(f)}} \tag{2.14}$$

and the wave-number k as: $\mathrm{k} = 2\pi\mathrm{f/c}$

with c, the speed of sound and

r(f) the effective acoustic distance between the human ears, which is 30 cm [see Yanagawa et al. (1976), as well as Suzuki and Tohyama (1981)].

In addition he defines ASI, an "Index of Acoustic Spatial Impression", as follows:

$$\mathrm{ASI} = 100(1 - \mathrm{D})(\%) \tag{2.15}$$

Total spatial impression signifies ASI = 100%, while the complete absence of spatial impression is ASI = 0%.

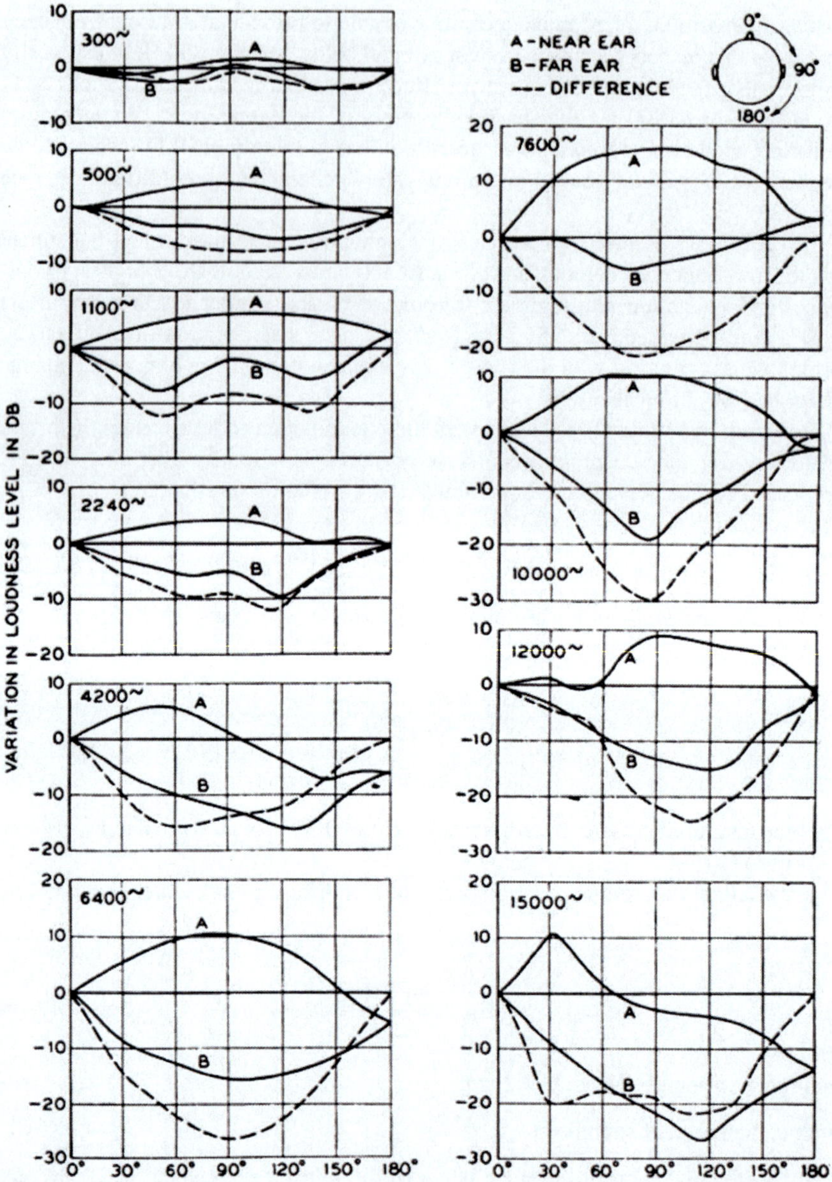

Fig. 2.23 Interaural level difference over source angle for different frequencies [graphic from Steinberg and Snow (1934)]

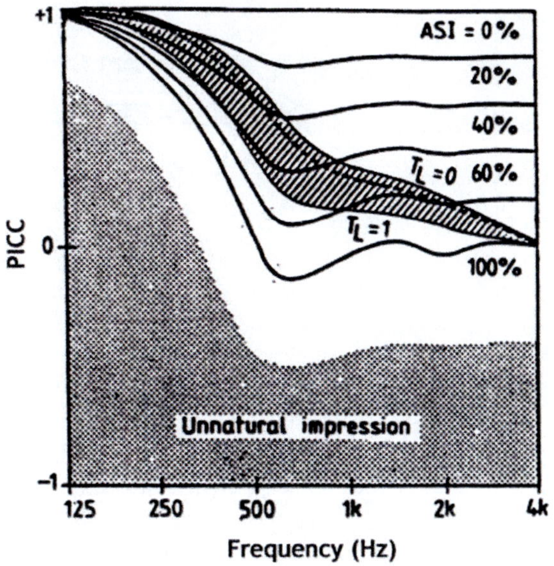

Fig. 2.24 The PICC-curves for stereo-reproduction in a listening room with a reverb time T_L (0–1 s) show small ASI values at low frequencies compared to an ASI = 60% for the seats in the middle section of a concert hall. The dashed curve stands for $T_L = 0.3$ s (from Hirata 1983)

Figure 2.24 shows that in a standard listening room ($RT_{60} = 0.3$ s) ASI is small for frequencies below 800 Hz, but high for frequencies above 800 Hz in comparison to ASI at a seat in the concert hall, for which ASI equals 60%.

Research by Griesinger concerning listener envelopment (Griesinger 1999) has shown that for rising frequency the ideal loudspeaker position moves towards the median plane. For frequencies below 700 Hz instead, a speaker layout is ideal which enables maximal lateral separation between the transducers, i.e. at a position left and right of the listener at ± 90° (see Fig. 2.25).

The above mentioned research by Hirata and Griesinger shows that the standard loudspeaker layout for stereo [as well as 5.1 surround, see ITU specification according to (BS.775-1)] with loudspeakers at ± 30° are far from ideal for the reproduction of low frequency signal components.

For this reason it is very important to choose—already at the stage of the recording process—a microphone technique, which captures the sound signal in a de-correlated manner over the whole frequency range, as a deterioration (in respect to a 'forced' increase in correlation) upon replay has to be expected anyway (see Hirata 1983).

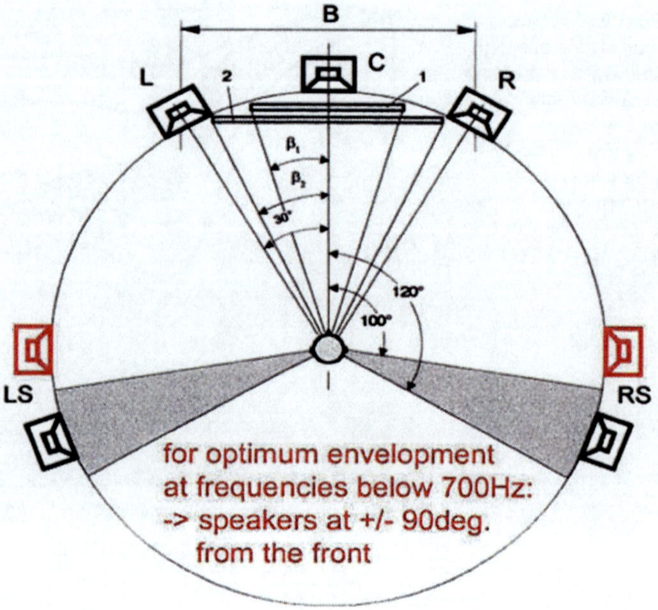

Fig. 2.25 A '5.2' arrangement with subwoofers at the sides at ± 90°, optimized for the spatial reproduction of low frequencies, according to the recommendation by Griesinger (1999)

References

Akeroyd MA, Summerfield AQ (1999) A binaural analog of gap detection. J Acoust Soc Am 105:2807–2820

Akeroyd MA, Bernstein LR (2001) The variation across time of sensitivity to interaural disparities: behavioural measurements and quantitative analyses. J Acoust Soc Am 110:2516–2526

Ando Y (1977) Subjective preference in relation to objective parameters of music sound fields with a single echo. J Acoust Soc Am 62:1436–1441

Augspurger GL (1990) Loudspeakers in control rooms and listening rooms. Paper presented at the audio engineering society 8th international conference

Avni A, Rafaely B (2009) Inter-aural cross correlation in a sound field represented by spherical harmonics. J Acoust Soc Am 125(4):2545

Barbour JL (2003) Elevation perception: phantom images in the vertical hemisphere. In: Proceedings to the 24th audio engineering society international conference: multichannel audio. The New Reality

Barron M (1971) The subjective effects of first reflections in concert halls—the need for lateral reflections. J Sound Vibr 15:475–494

Barron M, Marshall AH (1981) Spatial impression due to early lateral reflections in concert halls: the derivation of a physical measure. J Sound Vibr 77:211–232

Batteau DW (1967) The role of the pinna in human localization. Proc Roy Soc B168(1011):158–180

Bech S (1998) The influence of stereophonic width on the perceived quality of an audiovisual presentation using a multichannel sound system. J Audio Eng Soc 46(4):314–322

Begault D (1994) 3-D sound for virtual reality and multimedia. Academic Press, USA

Benjamin E (2006) An experimental verification of localization in two-channel stereo. Paper 6968 presented at the 121st audio engineering society convention
Benjamin E, Brown R (2007) The effect of head diffraction on stereo localization in the mid-frequency range. Paper 7018 presented at the 122nd audio engineering society convention, Vienna
Beranek L (2004) concert halls and opera houses: music, acoustics and architecture, 2nd edn. Springer, New York
Beranek LL, Schultz TJ (1965) Some recent experiences in the design and testing of concert halls with suspended panel arrays. Acustica 15:307
Berg J, Rumsey F (2001) Verification and correlation of attributes used for describing the spatial quality of reproduced sound. Paper presented at the audio engineering society 19th international conference
Bernstein LR, Trahoitis C, Akeroyd MA, Hartung K (2001) Sensitivity to brief changes of interaural time and interaural intensity. J Acoust Soc Am 109:1604–1615
Blauert J (1972) On the lag of lateralization caused by interaural time and intensity differences. Audiology 11:265–270
Blauert J (1974) Räumliches Hören. S. Hirzel Verlag, Stuttgart
Blauert J (1997) Spatial hearing. The MIT Press
Boehnke SE, Hall SE, Marquadt T (2002) Detection of static and dynamic changes in interaural correlation. J Acoust Soc Am 112:1617–1626
Bradley J, Soulodre G (1995) Objective measures of listener envelopment. J Acoust Soc Am 98:2590–2597
Breebart J, van der Par S, Kohlrausch A (2002) A time-domain binaural signal detection model and its predictions for temporal resolution data. Acta Acustica-Acustica 88:110–112
Corteel E, Pesce D, Foulon R, Pallone G, Changenet F, Dejardin H (2016) An open 3D audio production chain proposed by the Edison 3D project. Paper 9589 presented at audio engineering society 140th international convention in Paris, France
Culling JF, Summerfield AQ (1999) Measurement of the binaural temporal window using a detection task. J Acoust Soc Am 103:3540–3553
de Boer K (1940) Plastische Klangwiedergabe. Philips Tech Rdsch 5(4)
de Keet VW (1968) The influence of early lateral reflections on spatial impression. In: 6th international congress on acoustics, Tokyo
Eaton C, Lee H (2022) Subjective evaluations of three-dimensional, surround and stereo loudspeaker reproductions using classical music recordings. Acoust Sci Tech 43(2):149–161
Eska G (1997) Schall und Klang: wie und was wir hören. Birkhäuser Verlag
Everest FA (1994) The master handbook of acoustics, 3rd edn. TAB Books McGraw-Hill
Faller C, Merimaa J (2004) Source localization in complex listening situations: selection of binaural cues based on interaural coherence. J Acoust Soc Am 116:3075–3089
Fellner M, Höldrich R (1998a) Physiologische und psychoakustische Grundlagen des räumlichen Hörens. IEM-Report 03 KUG: Univ f Musik u darst Kunst, Graz
Fellner M, Höldrich R (1998b) Außenohr-Übertragungsfunktion—Messung und Datensätze. IEM-Report 04, KUG: Univ f Musik u darst Kunst, Graz
Ferguson S, Cabrera D (2005) Vertical localization of sound from multiway loudspeakers. J Audio Eng Soc 53(3):163–173
Geluso P (2012) Capturing height: the addition of Z microphones to stereo and surround microphone arrays. Paper 8595 presented at the 132nd audio engineering society convention
Gernemann A (1994) Summenlokalisation im Stereodreieck—Überlegungen zu psychoakustischen Untersuchungen mit dynamischem Testsignal und hochpräzisen Schallwandlern. Manus, Düsseldorf
Gernemann A, Rösner T (1998) Die Abhängigkeit der stereophonen Lokalisation von der Qualität der Wiedergabelautsprecher. In: Proceedings to the 20. Tonmeistertagung des VDT, Karlsruhe, p 828
Gerzon M (1986) Stereo shuffling: new approach, old technique, pp 122–130. Studio Sound

Gerzon M (1992) Psychoacoustic decoders for multispeaker stereo and surround sound. Paper 3406 presented at 103rd audio engineering society convention, San Francisco

Glasberg BR, Moore BCJ (1990) Derivation of auditory filter shapes from notched-noise data. Hear Res 47:103–138

Grantham DW, Wightman FL (1978) Detectability of varying interaural temporal differences. J Acoust Soc Am 63:511–523

Grantham DW, Wightman FL (1979) Detectability of a pulsed tone in the presence of a masker with time-varying interaural correlation. J Acoust Soc Am 65:1509–1517

Gribben C, Lee H (2014) The perceptual effects of horizontal and vertical interchannel decorrelation using the Lauridsen decorrelator. Paper 9027 presented at the 136th audio engineering society convention

Griesinger D (1986) Spaciousness and localization in listening rooms and their effects on the recording technique. J Audio Eng Soc 34(4):255–268

Griesinger D (1996) Spaciousness and envelopment in musical acoustics. In: Proceedings to the 19. Tonmeistertagung des VDT, pp 375–391

Griesinger D (1997) Spatial impression and envelopment in small rooms. Paper 4638 presented at the 103rd audio engineering society convention

Griesinger D (1998) General overview of spatial impression, envelopment, localization and externalization. Proceedings to the audio engineering society 15th international conference on small rooms

Griesinger D (1999) Objective measures of spaciousness and envelopment. Paper 16–003 presented at the audio engineering society 16th international conference on spatial sound reproduction

Griesinger D (2002) Stereo and surround panning in practice. Paper 5564 presented at the 112th audio engineering society convention, Munich

Haas H (1951) The influence of a single echo on the audibility of speech (German). Acoustica 1(2)

Hahn E (2018) Musical emotions evoked by 3D audio. Paper presented at the conference on spatial reproduction of the audio engineering society, Tokyo, Japan

Hall DE (1980) Musical acoustics. Brooks/Cole Publication Company, California. German edition: Musikalische Akustik—ein Handbuch, Schott-Verlag

Hamasaki K (2003) Multichannel recording techniques for reproducing adequate spatial impression. In: Proceedings to the audio engineering society 24th international conference on multichannel audio. The New Reality, Banff, Canada

Hebrank J, Wright D (1974) Spectral cues in the localization of sound sources on the median plane. J Acoust Soc Am 56(3):1829–1834

Hidaka T, Beranek L, Okano T (1995) Interaural cross-correlation, lateral fraction, and low- and high-frequency sound levels as measures of acoustical quality in concert halls. J Acoust Soc Am 98(2)

Hidaka T, Beranek L, Okano T (1997) Some considerations of interaural cross correlation and lateral fraction as measures of spaciousness in concert halls. In: Ando Y, Noson D (eds) Music and concert hall acoustics. Academic Press, London

Hirata Y (1983) Improving stereo at L.F. Wireless World, pp 60

Holman T (1991) New factors in sound for cinema and television. J Audio Eng Soc 39:529–539

Holman T (2000) Comments on the 'subjective appraisal of loudspeaker directivity for multichannel reproduction.' J Audio Eng Soc 48(4):314–317

Holube I, Kinkel M, Kollmeier B (1998) Binaural and monaural auditory filter bandwidths and time constants in probe tone detection experiments. J Acoust Soc Am 104:2412–2425

Hugonnet C, Walder P (1998) Stereophonic Sound Recording. John Wiley & Sons

Huopaniemi J (1999) Virtual acoustics and 3D sound in multimedia signal processing. Dissertation, Helsinki University of Technology

ITU-R Recommendations BS.2159-4 (2012) Multichannel sound technology in home and broadcasting applications. Int Telecommun Union

ITU Recommendation ITU-R BS.775-3 (2012) Multichannel stereophonic sound system with and without accompanying picture. Int Telecommunications Union. 08-2012

Kates JM (1960) Optimum loudspeaker directional patterns. J Audio Eng Soc 28:787–794
Kim S, Howie W (2021) Influence of the listening environment on recognition of immersive reproduction of orchestral music sound scenes. J Audio Eng Soc 69(11):834–848
Knothe J, Plenge G (1978) Panoramaregler mit Berücksichtigung der frequenzabhängigen Pegeldifferenzbewertung durch das Gehör. In: Proceedings to the 11. Tonmeistertagung des VDT, Berlin
Kohlrausch A (1988) Auditory filter shape derived from binaural masking experiments. J Acoust Soc Am 84:573–583
Kollmeier B, Gilkey RH (1990) Binaural forward and backward masking: evidence for sluggishness in binaural detection. J Acoust Soc Am 87:1709–1719
Kollmeier B, Holube I (1992) Auditory filter bandwidths in binaural and monaural listening conditions. J Acoust Soc Am 92:1889–1901
Kuhl W (1978) Räumlichkeit als eine Komponente des Höreindrucks. Acustica 40:167–168
Lee H (2011) The relationship between interchannel time and level differences in vertical sound localization and masking. Paper 8556 presented at the 131st audio engineering society convention
Lee H, Gribben C (2014) Effect of vertical microphone layer spacing for a 3D microphone array. J Audio Eng Soc 62(12):870–884
Lehmann U (1975) Untersuchung zur Bestimmung des Raumeindrucks bei Musikdarbietungen und Grundlagen der Optimierung. Dissertation, TU Dresden
Lehmann P (1976) Über die Ermittlung raumakustischer Kriterien und deren Zusammenhang mit subjektiven Beurteilungen der Hörsamkeit. Dissertation, TU Berlin
Lehnert H (1993) Auditory spatial impression. In: Proceedings of the audio engineering society 12th international conference on the perception on reproduced sound, pp 40–46
Lipshitz SP (1986) Stereo microphone techniques: are the purists wrong? J Audio Eng Soc 34:717–743
Litovsky RY, Colburn HS, Yost WA, Guzman SJ (1999) The precedence effect. J Acoust Soc Am 106:1633–1654
Marshall AH (1968) Acoustical determinants for the architectural design of concert halls. Arch Sci Rev 11:81–87
Martin G, Woszczyk W, Corey J, Quesnel R (1999) Sound source localization in a five channel surround sound reproduction system. Paper 4994 presented at the 107th audio engineering society convention, New York
Mason R, Rumsey F (2002) A comparison of objective measurements for predicting selected subjective spatial attributes. Paper 5591 presented at the 112th audio engineering society convention, Munich
Mehrgardt S, Mellert V (1977) Transformation characteristics of the external human ear. J Acoust Soc Amer 61(6):1567–1576
Merimaa J, Pulkki V (2005) spatial impulse response rendering I: analysis and synthesis. J Audio Eng Soc 53(12)
Mershon DH, Bowers JN (1979) Absolute and relative cues for the auditory perception of egocentric distance. Perception 8:311–322
Moore BCJ (1997) An introduction to the psychology of hearing, 4th edn. Academic Press, London UK
Morimoto M, Yairi M, Iida K, Itoh M (2003) The role of low frequency components in median plane localization. Acoust Sci Technol 24:76–82
Moulton D (1995) The significance of early high-frequency reflections from loudspeakers in listening rooms. Paper 4094 presented at the 99th audio engineering society convention
Moulton D, Ferralli M, Hebrock S, Pezzo M (1986) The localization of phantom images in an omnidirectional stereophonic loudspeaker system. Paper 2371presented at the 81st audio engineering society convention
Nielsen SH (1993) Auditory perception in different rooms. J Audio Eng Soc 41(10)
Otani M, Shigetani H (2019) Reproduction accuracy of higher-order Ambisonics with Max-rE and/or least norm solution in decoding. Acoust Sci Tech 40(1):23–28

Otani M, Shigetani H, Mitsuishi M, Matsuda R (2020) Binaural Ambisonics: its optimization and applications for auralization. Acoust Sci Tech 41(1):142–150

Otani M, Hirahara T, Morikawa D (2021) Origin of frequency dependence of interaural time difference. Acoust Sci Tech 42(4):181–192

Pfanzagl E (2002) Über die Wichtigkeit ausreichender Dekorrelation bei 5.1 Surround-Mikrofonsignalen zur Erzielung besserer Räumlichkeit. In: Proceedings to the 21. Tonmeistertagung des VDT, Hannover

Pfanzagl-Cardone E (2011) Signal-correlation and spatial impression with stereo- and 5.1 surround-recordings. Dissertation, University of Music and Performing Arts, Graz, Austria. https://iem.kug.ac.at/fileadmin/media/iem/altdaten/projekte/dsp/pfanzagl/pfanzagl_diss.pdf. Accessed Oct 2018

Pfanzagl-Cardone E (2012) 'Naturalness' and related aspects in the perception of reproduced music. In: Proceedings to the 27. Tonmeistertagung des VTD, Köln

Pfanzagl-Cardone E (2020) The art and science of surround and stereo recording. Springer-Verlag GmbH Austria.https://doi.org/10.1007/978-3-7091-4891-4

Rayleigh (1907) On our perception of sound direction. Phil Mag 13

Reichardt W, Kussev A (1972) Zeitschrift elektr Inform u Energietechnik 3(2):66, Leipzig (rem.: without title citation (see (Cremer und Müller, 1978), footnote 2, p. 345)

Reichardt W, Abdel Alim O, Schmidt W (1975) Definitionen und Messgrundlage eines objektiven Maßes zur Ermittlung der Grenze zwischen brauchbarer und unbrauchbarer Durchsichtigkeit bei Musikdarbietung. Acustica 32:126

Roffler SK, Buttler RA (1968) Factors that influence the localization of sound in the vertical plane. J Acoust Soc Am 43(6):1255–1259

Rumsey F (2001) Spatial audio. Focal Press (Elsevier)

Rumsey F, Lewis W (2002) Effect of rear microphone spacing on spatial impression for omnidirectional surround sound microphone arrays. Paper 5563 presented at the112th audio engineering society convention

Rumsey F, Segar P (2001) Optimisation and subjective assessment of surround sound microphone arrays. Paper 5368 presented at the 110th audio engineering society convention, Amsterdam

Sarroff A, Bello JP (2008) Measurements of spaciousness for stereophonic music. Paper 7539 presented at the 125th audio engineering society convention

Sengpiel E (1992) Grundlagen der Hauptmikrophon-Aufnahmetechnik—Skripten zur Vorlesung (Musikübertragung). Hochschule der Künste, Berlin. www.sengpielaudio.de. Accessed 2004

Simonsen G (1984) Master's thesis, Technical University of Lyngby, Denmark (no title information available)

Soulodre GA, Lavoie MC, Norcross SG (2003) Objective measures of listener envelopment in multichannel surround systems. J Audio Eng Soc 51(9)

Steinberg JC, Snow WB (1934) Auditory perspective—physical factors. Electr Eng 53(1):12–15

Stern RM, Zeiberg AS, Trahoitis C (1988) Lateralization of complex binaural stimuli: a weighted image model. J Acoust Soc Am 84:156–165

Streicher R, Everest A (2006) The new stereo soundbook, 3rd edn. Audio Engineering Associates

Suzuki A, Tohyama M (1981) Interaural cross-correlation coefficient of Kemar head and torso simulator. IECE Japan, Tech Rep EA80-78

Tan CJ, Gan WS (2000) Direct Concha excitation for the introduction of individualized hearing cues. J Aud Eng Soc 48(7/8):642–653

Theile G (1978) Weshalb ist der Kammfilter-Effekt bei Summenlokalisation nicht hörbar? In: Proceedings to the 11. Tonmeistertagung des VDT

Theile G, Wittek H (2011) Principles in surround recordings with height. Paper 8403 presented at the 130th audio engineering society convention

Theile G et al (1988) Raumbezogene Stütztechnik—eine Möglichkeit zur Optimierung der Aufnahmequalität. In: Proceedings to the 15. Tonmeistertagung des VDT

Thiele R (1953) Richtungsverteilung und Zeitfolge der Schallrückwürfe in Sälen. Acustica 3:291–302

Tohyama M, Suzuki A (1989) Interaural cross-correlation coefficients in stereo-reproduced sound fields. J Acoust Soc Am 85(2). Reprinted in: Rumsey F (ed) (2006) An anthology of articles on ,spatial sound techniques—part 2: multichannel audio techniques. Audio Engineering Society, New York

Toole FE (1985) Subjective measurements of loudspeaker quality and listener performance. J Audio Eng Soc 33(1/2):2–32

Toole FE (1986) Loudspeaker measurements and their relationship to listener preferences. J Audio Eng Soc 34:227–235

Toole FE (2008) Sound reproduction—loudspeakers and rooms. Focal Press (Elsevier)

Van Daele B, Van Baelen W (2012) Productions in Auro-3D: professional workflow and costs. White paper by Auro-Technologies

Van der Hejden M, Trahiotis C (1998) Binaural detection as a function of interaural correlation and bandwidth of masking noise: implications for estimates of spectral resolution. J Acoust Soc Am 103:1609–1614

von Hornbostel EM, Wertheimer M (1920) Über die Wahrnehmung der Schallrichtung. Report to the Academy of Sciences, Berlin, pp 388–396

Wallis R, Lee H (2014) Investigation into vertical stereophonic localization in the presence of interchannel crosstalk. Paper 9026 presented at the 136th audio engineering society convention

Wendt F, Florian M, Zotter F (2014) Amplitude panning with height on 2, 3, and 4 loudspeakers. Proceedings to the 2nd international conference on spatial audio

Wightman FL, Kistler DJ (1992) The dominant role of low-frequency interaural time differences in sound localization. J Acoust Soc Am 91:1648–1661

Williams M (1987) Unified theory of microphone systems for stereophonic sound recording. Paper 2466 presented at the 82nd audio engineering society convention

Williams M (2013) The psychoacoustic testing of the 3d multiformat microphone array design, and the basic isosceles triangle structure of the array and the loudspeaker reproduction configuration. Paper 8839 presented at the 134th audio engineering society convention

Williams M (2016) Microphone array design applied to complete hemispherical sound reproduction—from integral 3D to comfort 3D. Paper presented at the 140th audio engineering society convention, Paris

Williams M (2022a) MMAD 3D audio—designing for height—practical configurations. Paper presented at the 152nd convention of the audio engineering society

Williams M (2022b) MMAD. Sounds of Scotland, France

Wittek H (2002) Image Assistant V2.0. http://www.hauptmikrofon.de. Accessed 24 June 2008

Wittek H, Theile G (2000) Investigations into directional imaging using L-C-R stereo microphones. In: Proceedings to the 21. Tonmeistertagung des VDT, p 432–454

Wittek H, Theile G (2002) The recording angle—based on localisation curves. Paper 5568 presented at the 112th audio engineering society convention, Munich

Yanagawa H, Higashi H, Mori S (1976) Interaural correlation coefficients of the dummy head and the feeling of wideness. Acoust Soc Jap Tech Rep H-35-1

Yost WA, Wightman FL, Green DM (1971) Lateralisation of filtered clicks. J Acoust Soc Am 50:1526–1531

Zacharov N (1998) Subjective appraisal of loudspeaker directivity for multichannel reproduction. J Audio Eng Soc 46(4):288–303

Zollner M, Zwicker E (2003) Elektroakustik, 3rd edn. Springer

Zwicker E, Fastl H (1990) Psychoacoustics. Springer, Berlin

Chapter 3
The 'AURO-3D®' System and Format

Abstract With its introduction at the AES Conventions in 2006 by inventor Wilfried Van Baelen, Auro-3D was the first true two-layered 3D audio system that became fast established in all audio markets. Additionally, Van Baelen introduced the new generic terms 'Immersive Sound' (related to the audible experience) and 'Immersive Audio' (related to the technical components) to replace 'Surround Sound with Height' (see SMPTE Standard ST 2098-2:2019, in Immersive audio bitstream specification (2019) Immersive audio bitstream specification standard for Digital Cinema). Auro-3D is not dependent on the technology behind, being either channel-, object- or scene-based. Next to 'quality' and 'efficiency', 'compatibility' is a fundamental element of the concept behind the format as Auro-3D is accordant with any existing workflows, delivery formats and available bandwidth, as well as with the existing standards for stereo and 5.1 Surround. Auro-3D is a 'hybrid' format, which means it uses both channel- and object-based technology, which currently can cater for loudspeaker layouts up to '26.1' discrete channels and some interactive solutions for the end-consumers. Crucial questions such as speaker positioning and time alignment are answered and details of the post-production and mixing process for 3D audio revealed. Auros 'Creative Tool Suite' functionality in terms of 3D panning, mixing, encoding, etc. is now available as software plug-in for the most common DAWs like Cubase, LogicPro, Nuendo and Black Magic. Finally, the encoding of the Auro-3D mix for a DCP (Digital Cinema Package) format or for a consumer 'home' format (e.g. Blu-ray Disc) or downloads and streaming are explained.

Keywords Auro · 3D · Immersive · Audio · Sound · Codec · DCP

Preamble: large parts of the text in this chapter are cited from a whitepaper by Auro Technologies NV 2015 with kind permission of the authors.

E. Pfanzagl-Cardone, *The Art and Science of 3D Audio Recording*,
https://doi.org/10.1007/978-3-031-23046-2_3

3.1 Historic Development of Auro-3D and Competitive Systems

Wilfried Van Baelen started the developments of his Auro-3D format at Galaxy Studios in April 2005 with '3D music' in mind. As a pioneer for music productions in 5.1 surround and as an experienced mixing engineer for movies, he envisioned a system that adds the missing dimension 'height' as efficiently as possible with optimum usage in all audio markets. This means that 'maximum mono compatibility' is required in order to experience the content everywhere as closely as possible to the intention of the content creators of movies, music, sports, games etc.

After having first developed the most efficient speaker layouts for home for playback of 'immersive sound' using two or three layers (which he introduced at the Paris and San Francisco AES Conventions in 2006), Van Baelen expanded that concept in 2010 to public commercial theaters (marketed as Auro 11.1 by Barco) in order to get movie content. Due to Auro-Codec [by inventors Wilfried Van Baelen & Guido Vandenberghe (see Van Den Berghe and Van Baelen 2008)] the existing DCP (Digital Cinema Package) could be used for distribution without any change in the specifications, including the obliged watermarking on the audio as part of the DCP standard (see Fig. 3.1).

In summer of 2011, the worldwide leading post-production facility Skywalker Sound installed the first Auro-3D Studio System in their Mix Studio A. Only a few months later, on November 1, 2011, George Lucas announced that his next film 'Red Tails' would be released in Auro 11.1 as the first film in history with immersive sound (see Fig. 3.2).

The movie was mixed at Galaxy Studios in Mol using the Auro Creative Tool Suite and the first ever AMS-Neve Digital Film Console (DFC 3D) with 3D panning and 3D routing which was custom made for Galaxy Studios (see Fig. 3.3). The reactions of exhibitors were overwhelming, and immediately there was a sizable demand to install the Auro 11.1 by Barco system in hundreds of theaters all over the world.

This is a substantial difference to the Dolby 'Atmos' system, which was originally developed for Digital Cinema only, as Dolby stated publicly in 2012, when

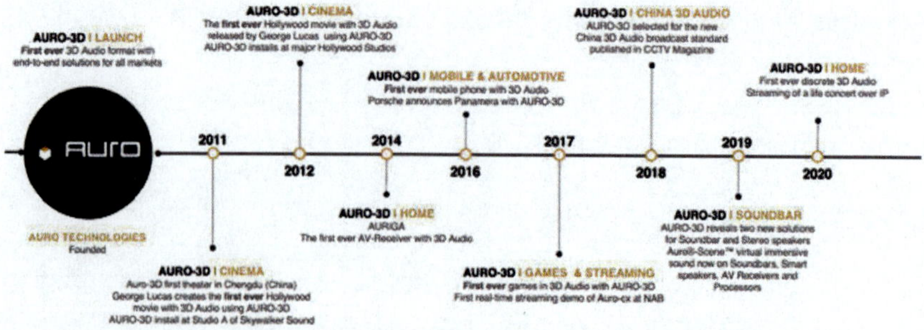

Fig. 3.1 Timeline of Auro-3D milestones since 2010 (courtesy Auro Technologies)

Fig. 3.2 George Lucas 'Red Tails'—first film mixed in Auro-3D

launching their format at the Cinemacon convention in Las Vegas where Auro-3D had their introduction a year earlier together with their partner Barco (leader in Digital Cinema Projectors). But after Auro's successful introduction at the CES convention at Las Vegas in 2014 from their AURIGA (= the first ever AV Receiver for home with 3D audio) developed by Auro Technologies, Dolby announced a few months later to adopt their Atmos-System for home-theatres as well, which they did with a different, non-compatible speaker layout and with a different non-compatible delivery technology in order to avoid any infringement of the worldwide patented 'Auro' technology.

3.2 Auro-3D—Basic Concept and System Description

The concept of the Auro-3D format is based on an end-to-end system for all audio markets which includes specific speaker layouts and patented technologies in order to be able to deliver anywhere "the same sound experience—as intended by the content creators" (see Figs. 3.4 and 3.5). Part of the concept is the easy integration and easy

Fig. 3.3 The first ever AMS-Neve Digital Film Console (DFC 3D) at Galaxy Studios in Mol, Belgium with Auro-3D inventor Wilfried Van Baelen (photo courtesy of Van Baelen)

distribution to all audio markets which was made possible by complete compatibility of the Auro-Codec with PCM, which allowed the use of existing distribution formats without any change in the specs.

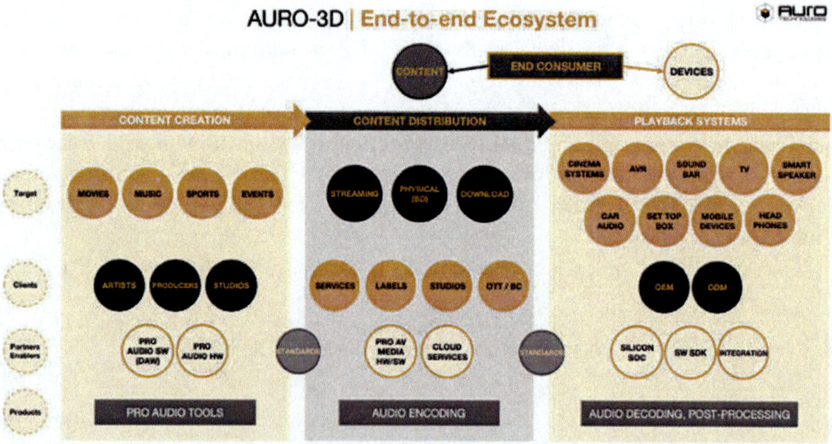

Fig. 3.4 Auro-3D—end-to-end 'Ecosystem' in terms of content creation, distribution and playback systems (courtesy Auro Technologies)

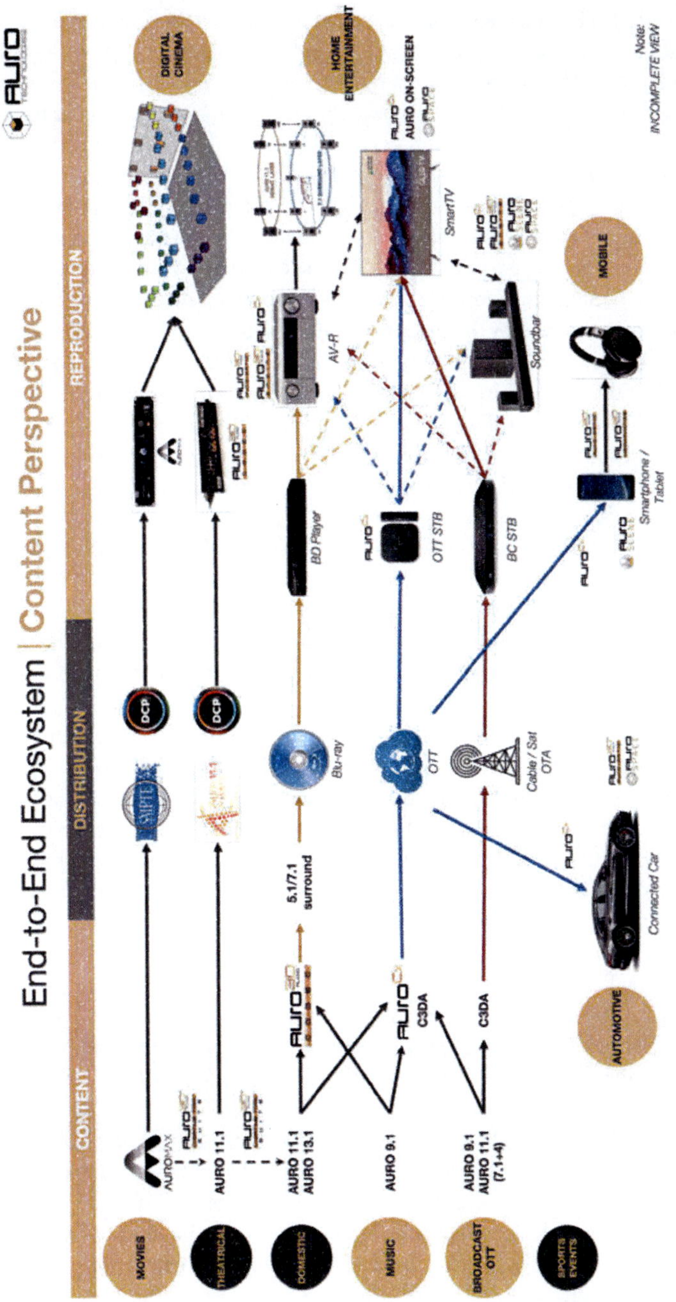

Fig. 3.5 Auro-3D—end-to-end 'Ecosystem' in terms of content perspective (courtesy Auro Technologies)

The Auro-3D format also contains virtual technology that creates a similar immersive sound experience wherever the Auro-3D speaker layouts are not present like in soundbars, headphones, mobiles, computers etc. Auro's virtual technology 'Auro-Matic' is able to generate in real time a natural 3-dimensional sound experience in real time over speakers, soundbars or over headphones from existing legacy content (mono, stereo, surround).

The aim behind the Auro-3D format is to be able to create 3D audio content with ease and bring this new sound experience to all audio markets, independent of content (movies, music, sports, games…) and playback system (cinema, home, mobile, cars, headphones, soundbars, earbuds etc.) by using existing distribution formats without the need for any change of the technical specifications or extra bandwidth to add the missing dimension 'height' all around the audience (see Figs. 3.4 and 3.5).

The addition of the third dimension 'height' was not new, but before the introduction of Auro-3D, no other system was able to establish itself successfully, due to various reasons. As early as 1999, the THX 10.2 system (see THX10.2) by Tomlinson Holman had already had two height front speakers behind the screen but did not manage to make a successful entry to the Digital Cinema market. Also the 22.2 system by Kimio Hamasaki, launched in 2005 (see Hamasaki et al. 2005; Hamasaki 2011; Hamasaki and Van Baelen 2015; Matsui 2015; Oode et al. 2011; Sawaya et al. 2011, 22.2 System) with eight height channels and the speakers positioned at an elevation of + 45° (above ear-level) did not become the commercial success as envisioned. Both systems did not have an end-to-end solution in terms of compatibility and delivery to all audio markets.

In some way German Tonmeister Werner Dabringhaus has anticipated part of the basic principle of Auro-3D's height loudspeakers by proposing his '2 + 2 + 2' system of surround sound playback in 2000 (see Dabringhaus 2000). Instead of adhering to the regular 5.1 surround layout he remained with only four speakers for the horizontal ear-level layer (by leaving out the center speaker) and added two height speakers—above the front speakers L, R—lifted up half the distance of the front loudspeaker's base width. But this was a first step only on the way to real 3D audio, as it is not possible to create a fully convincing 3D 'sound impression' with two height or 'overhead' speakers only.

Wilfried von Baelen based his Auro 9.1 system on the existing 5.1 surround main standard to which he added a 'quadraphonic height layer' with an elevation between 25° and 35° in order to get a "*vertical coherent stereo field all around the audience*" (see Fig. 3.6).

Using the three orthogonal axes (x, y, z) the Auro-3D speaker setups reproduce true three-dimensional sound as a hemisphere all around the audience. It enhances and augments their emotional impact substantially compared to stereo or surround (as commented by Morten Lindberg; "*the emotional component becomes a tenfold compared to surround or stereo formats*" (see Genelec 2021).

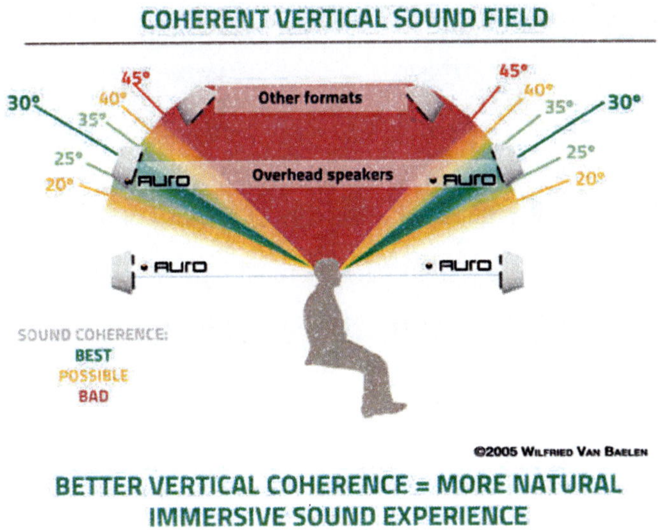

Fig. 3.6 Auro-3D concept: achieving a coherent vertical sound-field by use of additional overhead speakers (graphic courtesy and © Van Baelen)

3.2.1 Auro-3D Engine

However, the challenge was to bring an audio format to the market with more than eight channels as—from production to distribution and playback—everything was limited to eight channels (7.1 surround sound, see Fig. 3.7). First of all, tools needed to be developed in order to be able to move sounds all around the audience in '3D space' (i.e. 3D panning) and 3D routing to the different speakers and channels.

Also, a solution was needed for the distribution to consumer markets since Blu-ray and HDMI interface were both limited to eight channels of PCM (uncompressed audio). This major hurdle was solved by the Auro-Codec, the technical enabler that made this historic revolution in sound possible because it could make use of existing delivery formats (e.g. Blu-ray) without the need of any change in the technical specifications.

The **Auro-Codec** allows for mixing and 'un-mixing' (decoding) *while staying in the PCM domain* (i.e. the uncompressed audio format), delivering up to high-resolution audio (e.g. 96 kHz-24bit) in all encoded and decoded channels (see Fig. 3.8). As such, Auro-3D can be played back from a standard DCP (Digital Cinema Package) or from any Blu-ray player in the market without the need of change of the specifications allowing for a fast integration without any compromise in quality, compatibility, and efficiency in order to create, deliver and play back as intended by the creators.

Early in 2006, Van Baelen had the idea to make use of the least significant bits in the PCM stream and developed this idea further together with Guido Van Den Berghe

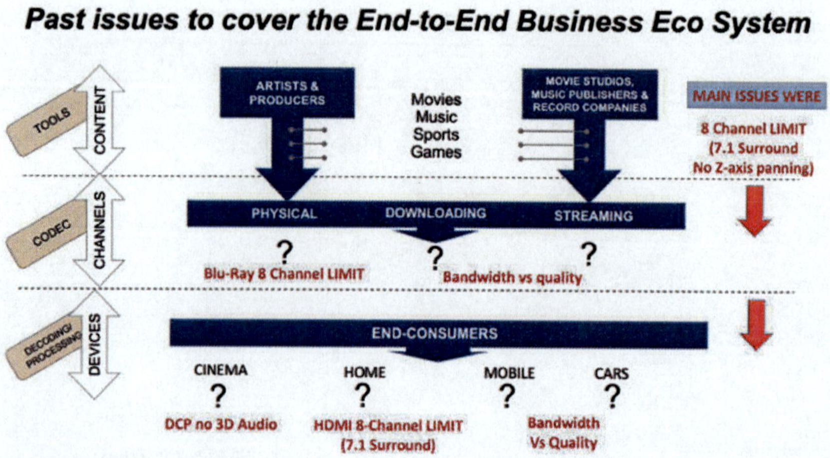

Fig. 3.7 Past limitation of eight audio channels only for production, distribution and home playback (graphic courtesy and © Van Baelen)

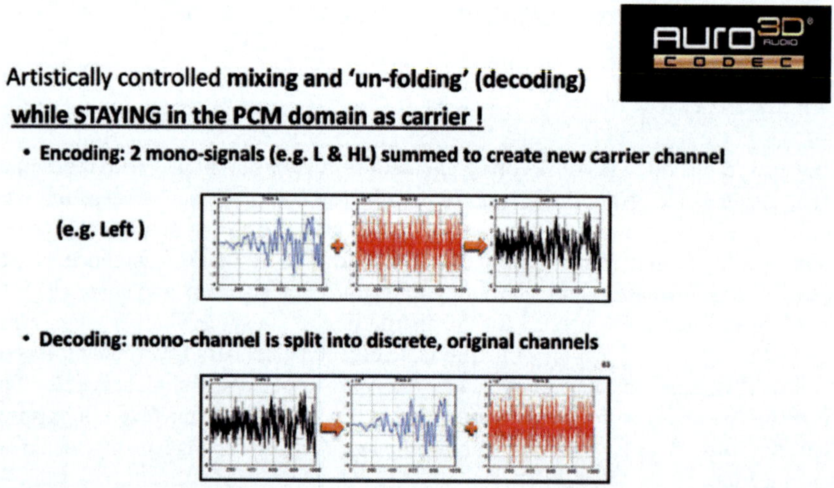

Fig. 3.8 Basic coding and decoding scheme of the Auro-3D Coder (graphic courtesy of Auro Technologies)

(see Van Den Berghe and Van Baelen 2008). It took about three years of development to have it tested and perfectly working in 48 and 96 kHz with 24-bit resolution. The principle behind this technology is based on a pure mathematical solution as used for lossless codecs, so there is no psycho-acoustic optimization as with Dolby Atmos, suffering from noticeable artifacts using their 'Spatial Coding Technology' which is not lossless. This is also the reason why encoding the same content with Dolby

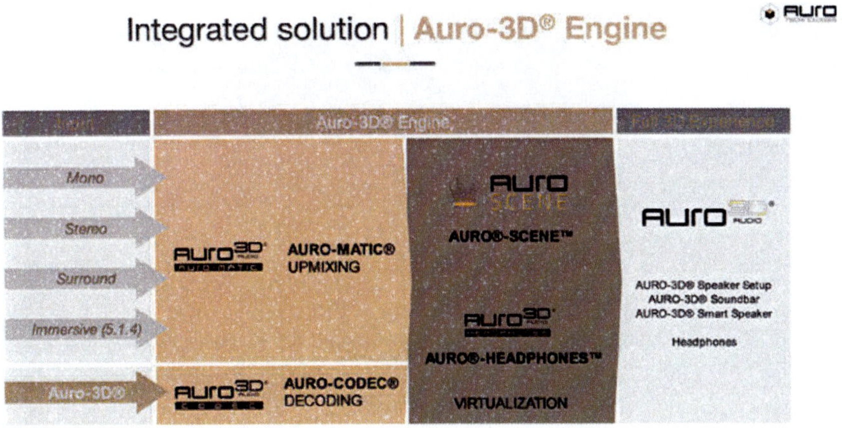

Fig. 3.9 Auro-3D engine for integration of Auro-Codec, Auro-Matic, Auro-Scene and Auro-Headphones (graphic courtesy of Auro Technologies)

Atmos and Auro-3D sounds different. Auro-3D delivers the same sonic quality as the original master, as it does not suffer from artifacts like some competitor formats.

The **Auro-Matic** algorithm turns all legacy content—whether it is mono, stereo or surround sound—into a natural, immersive sound field over either one of the Auro-3D speaker layouts, or even a standard pair of headphones by using **Auro-Headphones**, which is an algorithm that is able to process a binaural output for headphone or earbud reproduction from a 3D master in real time.

Those technologies are combined in the **Auro-Engine** which recognizes if the original source content is in Auro-3D format, or in a legacy audio format and will use the corresponding technology to deliver a natural 3D sound experience over loudspeakers or headphones. Also, the newest **Auro-Cx** streaming codec will be integrated in the Auro-3D Engine and available as an update in most AV-Receivers that have Auro-3D on board (Fig. 3.9).

3.2.2 *A Vertically Coherent Soundfield*

The introduction of the vertical axis in sound recording and reproduction brings about a completely different experience for human beings compared to what they are used to hear in the horizontal plane only. There is a clearly perceivable improvement in transparency, depth and natural sound coloration when adding—by means of the so called 'height layer'—what Van Baelen refers to as a '*vertical stereo field*' in front and all around the listener. Research at Galaxy Studios in 2005 by Van Baelen has led to the discovery of an ideal vertical angle between 25° and 35° for the position of the height layer—creating a (quadraphonic) field all around the listener—to be installed

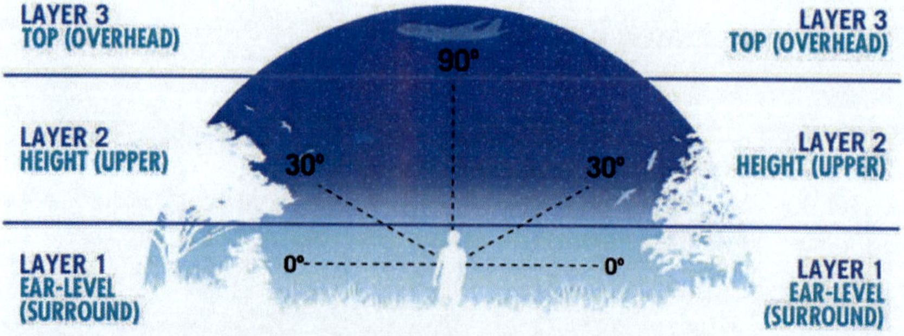

Fig. 3.10 Three layer proposal by Van Baelen for natural hearing (graphic courtesy and © Van Baelen)

above the 2D surround Layer at ear-level as defined by the ITU-recommendation (ITU-R BS.775) (2012) (see Fig. 3.6).

Van Baelen's substantial research in 2005 showed that if the vertical angle between the ear-level speaker and its corresponding vertical height speaker is larger than 40°, the natural coherence gets lost and our brain is not able to perceive it as a vertical phantom image but will experience it as two individual sound sources. From a psycho-acoustic point of view this is completely different to perception in the horizontal plane, where the included angle for phantom image creation can reach up to approx. 70° between left and right speaker. Van Baelen's conclusion is that the 'sound-perception' part of our brain is somehow 'horizontally oriented' as we do not have an ear on the top of our head to analyze any vertical time-of-arrival differences in sound (see Fig. 3.10).

In larger home theatres or larger rooms like professional theaters, an additional third layer, recreating a natural space is required to fill the gap directly above the audience. Contrary to what many people would believe, this 'Voice of God' channel—situated directly over the head of the listeners—is not a crucial element in creating the most natural immersive sound experience. Its use is actually limited to fly-overs and other special effects, as found in sound-mixing for cinema (Fig. 3.11).

3.2.3 3D-Reflections Around Sound Objects

A large part of what we hear in a natural sound field are the 3D-reflections of sound, which happen around or close to their sources. Object based technology like Dolby Atmos typically uses 'mono sources' which do not contain the original 3D-reflections and are therefore not reproducing a natural sound experience. Also, the typical 'Doppler effect' (which occurs with moving sound-sources) can only be experienced by capturing and reproducing the original 3D-reflections which are missing when only 'mono' sound objects are moving around the audience. The

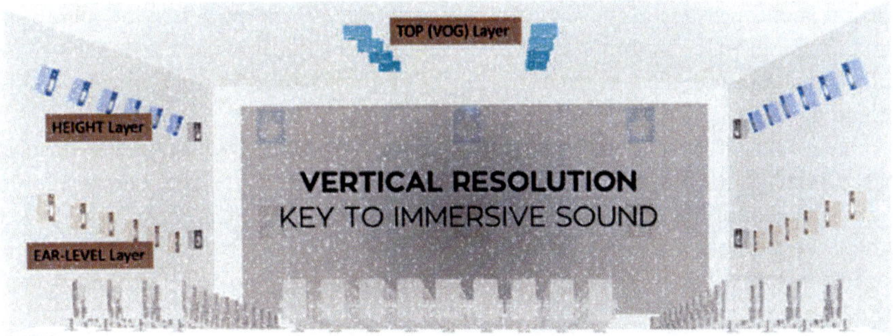

Fig. 3.11 Auro-3D three layer loudspeaker arrangement for the cinema theatre including ear-level layer, height layer and top layer ('Voice of God') (graphic courtesy and © Wilfried Van Baelen)

renderer at playback is not able to reproduce such native Doppler effects with 3D-reflections related to the number of speakers installed. The advantage of the Auro-3D system is that it can capture and reproduce original 3D reflections around fixed or moving sources in a coherent way resulting in a much more natural sound impression compared to formats that need object-based technology to create a 3D-spatial image around the audience (e.g. Dolby Atmos).

3.2.4 Efficiency as a Key Element

Many people think that more channels and more speakers allow for a more natural reproduction of sound but "… even with a million speakers around our ears, we will not be able to get a natural reproduction of sound", Tomlinson Holman stated at the ICTA Convention in 2014. The art is to get that illusion of a natural sound reproduction in the most efficient way, which is the concept behind the Auro-3D speaker layout and its related technologies.

The goal of Auro-3D is to use a minimum amount of channels to achieve a good 'sound spread' in the upper hemisphere. Every additional channel can become a possible cause for problems with phase, workflow, distribution, bandwidth etc.… Therefore the goal of the Auro-3D format is to deliver the most immersive listening experience with the minimum amount of channels/speakers. It provides a scalable solution from Auro 9.1 up to Auro 13.1 for home cinemas (more playback channels are possible using the **AuroMax** system, with up to 26.1 discrete channels, but this is usually not implemented in home cinemas).

There is a discussion about the inclusion of ceiling speakers in the layout, with the alternative suggestion of 'beaming systems' using the ceiling as a reflective surface. However, at the 2014 CEDIA convention Dr. Floyd Toole explained that such systems are based on psycho-acoustic tricks, and thus not capable of reproducing a natural sounding space. These psycho-acoustic effects are quite easy to achieve in the horizontal plane within surround sound systems, but this becomes much more difficult when using the vertical axis, since the human hearing system is horizontally oriented.

The resulting 'sweet spot' by using such psycho-acoustic tricks apparently is also rather small. The Auro-3D format, on the other hand, is supposed to create an even larger sweet spot than the 5.1/7.1 surround standard formats can provide. This was proven on a scientific base with measurements that show a better coverage of the spectrum from all signals by an Auro-3D setup versus a 5.1 or 7.1 surround system (see Röhbein et al. 2019).

3.3 Auro-3D Listening Formats for Home, Music and Broadcast Applications

The Auro 9.1 and Auro 10.1 listening formats were already presented by their inventor Wilfried Van Baelen at the AES convention in 2006 in Paris and in San Francisco. He impressed many people and it inspired the audio industry to move forward integrating the missing dimension 'height' in order to create a 3D space around the listener. In 2007 he expanded the system up to Auro 13.1 and envisioned that above that number of channels, object-based technology could be used to create more 'zones' within the same speaker install, the basic concept behind AuroMax.

The Auro-3D format has speaker layouts with many more playback channels (e.g. AuroMax 26.1) which are typically used in large theaters with more than 20 discrete playback channels. Such systems with a large numbers of speakers are typically not installed at home and do require a different approach since they are based on the use of 'speaker arrays' like the ones we are used to have in cinema theaters since many decades in order to create a large 'sweet zone' and provide—for every seat in the theater, if possible—the same surround experience as intended by the creators of the movie.

Note that the sound systems in cinema theaters are 'non-linear' in terms of frequency response, while systems at home are based on the reproduction of a 'linear' frequency spectrum (i.e. preferably no frequency-related amplitude deviations between input and output signal) which is not the case in cinema theaters which are based on the 'Academy curve' also called the X-Curve as described in the SMPTE standard recommendations (see ISO 2969 and SMPTE 202 M).

Content in Auro-3D can be presented in various channel configurations. The figure below gives an overview of the main configurations used for home, music studios and broadcast in which Auro-3D content will be delivered through various media such as Blu-ray Disc, downloads and even streaming.

The most efficient speaker layout that can reproduce a 3D space is the Auro 8.0 system: two quadraphonic layers above each other, creating a vertical stereo field around the listener. Adding a center channel and LFE delivers Auro 9.1. Auro 10.1 is

Configuration	Icon	L	R	C	LFE	Ls	Rs	Lb	Rb	HL	HR	HC	T	HLs	HRs
Auro 8.0	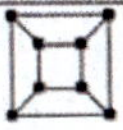	x	x			x	x			x	x			x	x
Auro 9.1		x	x	x	x	x	x			x	x			x	x
Auro 10.1	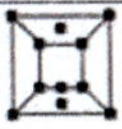	x	x	x	x	x	x			x	x		x	x	x
Auro 11.1	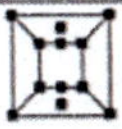	x	x	x	x	x	x			x	x	x	x	x	x
Auro 11.1 (7+4)		x	x	x	x	x	x	x	x	x	x			x	x
Auro 13.1	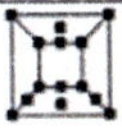	x	x	x	x	x	x	x	x	x	x	x	x	x	x

Fig. 3.12 Auro-3D listening formats and according speaker layout schematic (from Auro Technologies NV 2015)

adding the third layer (over the head 'top' speaker 'T' or 'Voice of God') and is the most efficient 3-layered speaker layout with 5.1 backwards compatibility (Fig. 3.12).

'Auro 9.1' is the most efficient two-layered speaker layout able to reproduce a 3D space and 5.1 compatible. Its 'vertical stereo field' around the listener, which is part of each native Auro-3D speaker layout, is key to getting a natural immersive sound experience (Fig. 3.13).

'Auro 10.1' is the most efficient 3-layered speaker layout to reproduce a 3D sound field with 5.1 compatibility. To achieve a natural spread of sound in the vertical axis in larger spaces, at least three layers are needed in order to have a hemispherical coherent sound field (Fig. 3.14).

'Auro 11.1' has two different layouts, one with two layers and one with three layers. Auro-3D's very popular speaker layout used in Digital Cinema theaters contains 3-layers. It is based on Auro 10.1 with the addition of a height center channel, providing a total of six front channels spread over two vertical planes delivering the best screen sound experience of all sound formats on the market (as of 2016). Auro 11.1 became video-beamer manufacturer Barco's favorite choice because it delivers the most immersive sound experience in small as well as large cinemas in the most efficient way which makes it very scalable and easy to install on top of the already existing 5.1 installed cinema system (Fig. 3.15).

To maximize the compatibility of movie content made for 'Digital Cinema' to 'home cinema', this Auro 11.1 speaker layout for cinema theaters is using 'speaker arrays', which means that one channel is represented by more speakers. This dates back to the seventies when surround sound became popular in cinema theaters. The

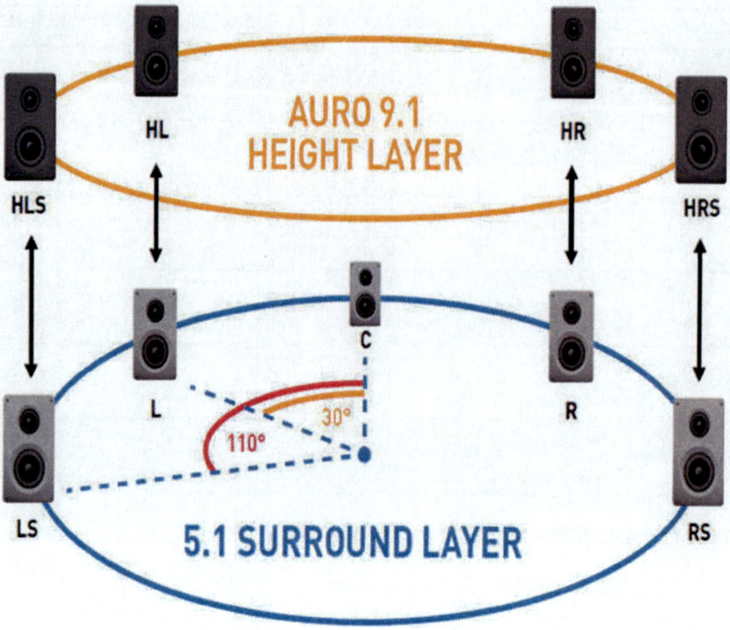

Fig. 3.13 The Auro-3D 9.1 speaker layout (from Auro Technologies NV 2015)

use of more speakers for the same surround channel is to—ideally—provide the same surround sound experience to each seat in the theater. That same concept of speaker arrays is used for Auro-3D's height layer and top layer (= Voice of God Layer, or VOG) to maintain a large 'sweet-spot'.

The VOG layer is typically even split in two different rows in order to enable for every seat the same 'overhead experience' (see Fig. 3.16); something that is much less the case with a 'stereophonic overhead layer' like in the Dolby Atmos system. In case cinema theaters have a low ceiling or a very wide distance between left and right, more rows are installed in the overhead layer (e.g. in Nagoya in Japan there are four overhead rows in order to create a natural spread of sound in the vertical plane).

'Auro 13.1' is the most efficient 3-layered speaker layout based on the 7.1 surround standard with six screen-loudspeaker channels distributed over two vertical planes (similar to the Auro 11.1 system in Fig. 3.17). The Auro 13.1 is the channel-based part (the fixed 'audio beds') of the AuroMax system (see Fig. 3.18)

One important note to make is that in theory, there should be no audible difference using object-based technology versus channel-based technology when having the same number of speakers installed. Scalability of the immersive sound experience with object-based technology depends also very much on the available speaker system and is often less scalable compared to Auro's channel-based system. While object-based 3D audio seems more flexible in terms of scalability, this—at times—comes at the expense of localization accuracy or localization precision for (moving)

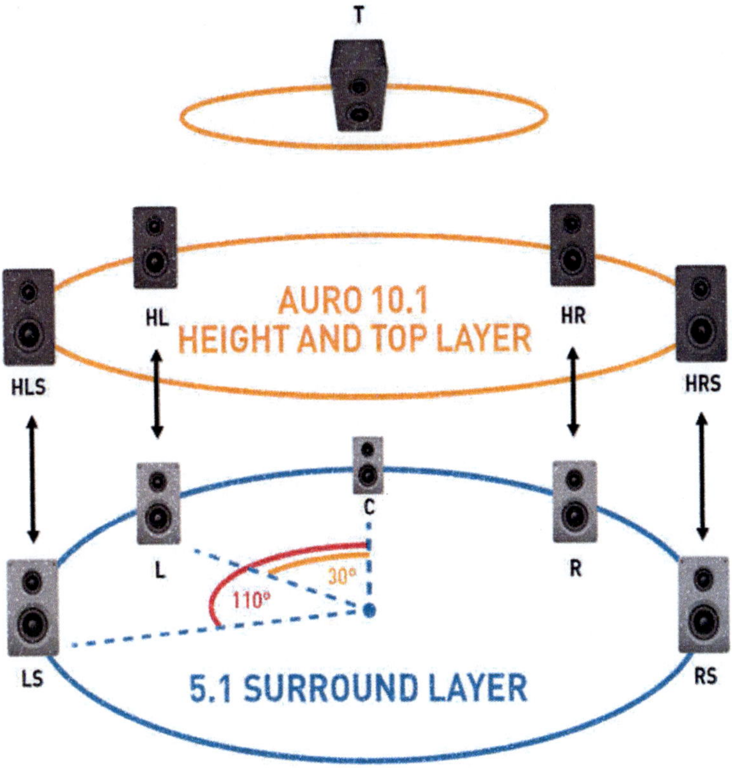

Fig. 3.14 The Auro-3D 10.1 speaker layout (from Auro-Technologies NV 2015)

3D audio objects, because the final sonic result in the listening room depends very much on the quality of the playback rendering algorithm, while in channel-based 3D audio the signal relationship (i.e. signal correlation, including amplitude and phase relationships between any pair of channels) is 'fixed' already during the process of mixdown and therefore less 'tangible' or 'fragile' in its final outcome.

3.4 Room Acoustics and Practical Speaker Setup for Auro-3D

3.4.1 Auro-3D Screensound

Another innovative element in the Auro-3D audio concept is the 'Auro 3D Screensound': the speakers behind the screen are spread in two vertical layers allowing for a

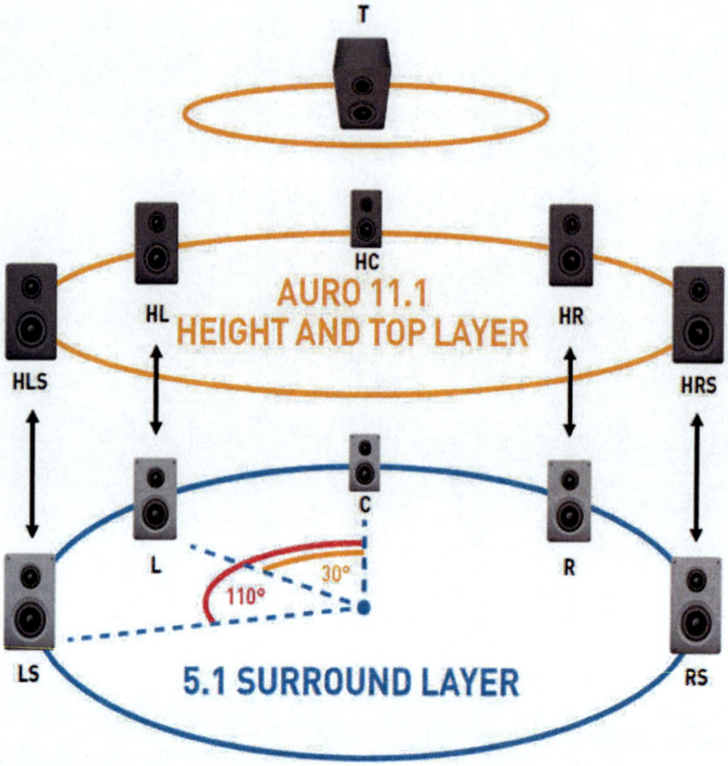

Fig. 3.15 The Auro-3D 11.1 speaker layout showing the discrete channels (from Auro-Technologies NV 2015)

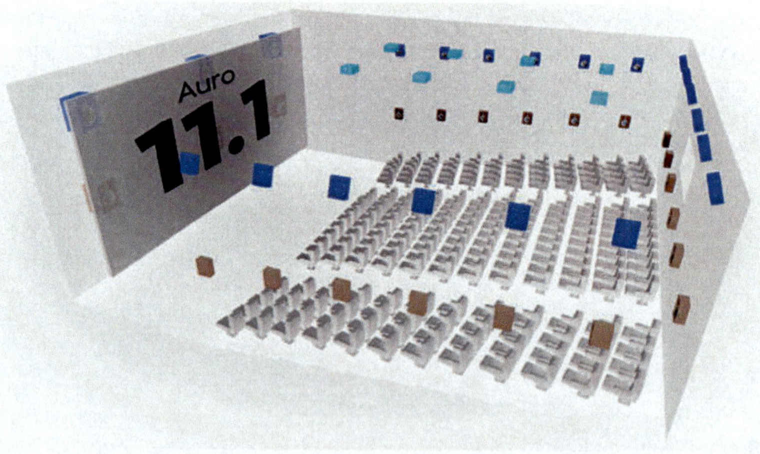

Fig. 3.16 The Auro-3D 11.1 concept as implemented in a cinema theatre (graphic courtesy and © Van Baelen)

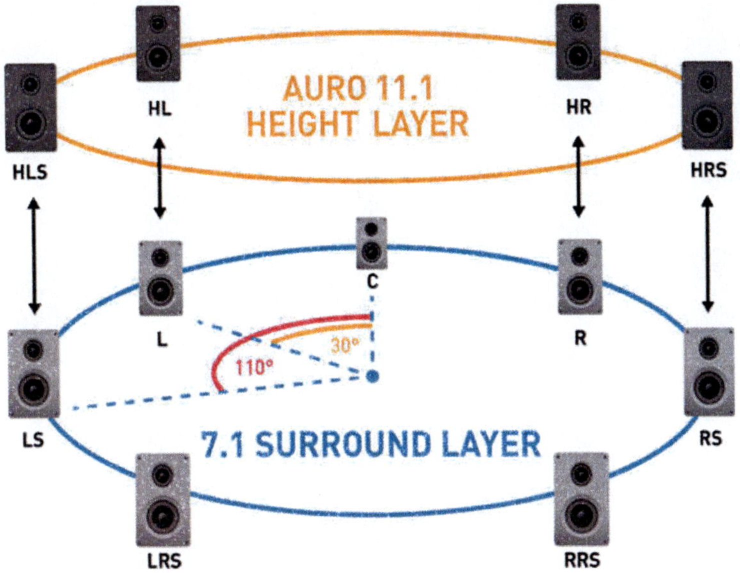

Fig. 3.17 The Auro-3D 11.1 speaker layout for home based on 7.1 surround (without a height layer) (from Auro Technologies NV 2015)

much better sound from the front, more depth in the sound, as well as a better connection to the visual objects on the screen. Because the substantial majority of sound is coming from the screen channels (typically around 80%), having the ability to spread them over more speakers in a different elevation is resulting in a much higher transparency and also leaves more (acoustic) 'space' for the dialogue, which enhances the intelligibility, which is the backbone of every movie soundtrack (Figs. 3.19 and 3.20).

This is the reason why the same immersive sound can be experienced at home where Auro-3D has the same vertical quadraphonic field around the screen as basic part of their native Auro-3D speaker layouts. Since most of the sound energy is coming from the screen channels, it makes sense to have those parts of the system optimized, which is also a key differentiator for Auro-3D versus competitors which are using only one horizontal layer behind the screen.

3.4.2 *Further Aspects in Relation to Room Acoustics and Practical Speaker Setup*

As with any other control room the basic design considerations are room size and geometry, acoustics, equipment, wiring, etc. Especially the room height and acoustics will be affected by the introduction of the additional speakers. As usual, it is always

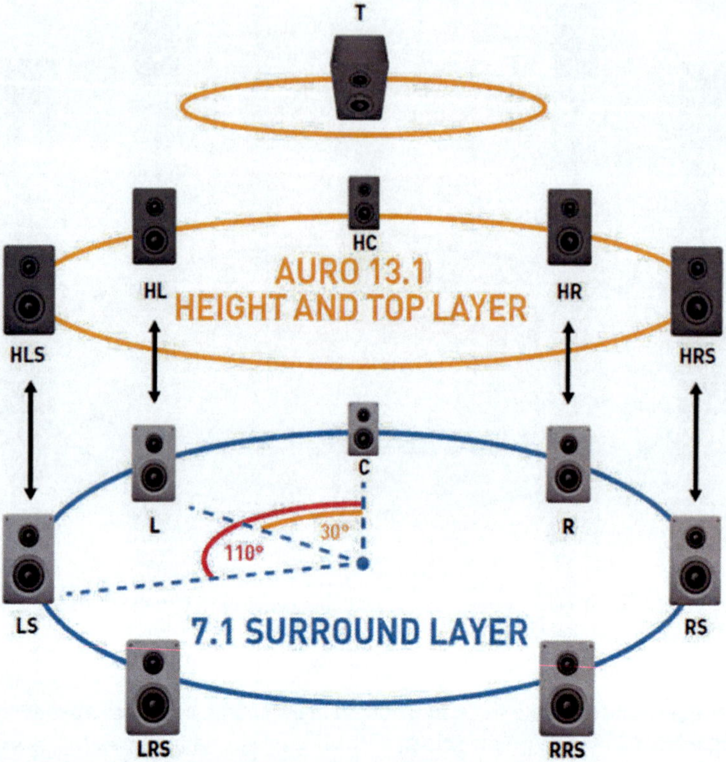

Fig. 3.18 The Auro-3D 13.1speaker layout (from Auro Technologies NV 2015)

Fig. 3.19 'Auro-3D Screensound' two vertical layers for the screen speakers in the cinematic setup (graphic courtesy and © Van Baelen 2022)

Fig. 3.20 'Auro-3D Screensound' two vertical layers of speakers also 'around' the TV-screen in the home setup (graphic courtesy and © Van Baelen 2022)

a good idea to consult a specialized acoustician for advice. Care should be taken, both when building a new room or retrofitting an existing one, that the acoustics do not influence the overall performance of the system negatively. An acoustician can measure and point out problems, especially when e.g. positioning the height speakers close to a (too) reflective ceiling or placed in corners. Such anomalies should be eliminated to ensure a proper listening environment for Auro-3D productions.

As we have already seen, there are many possible room layouts for Auro-3D installations. They range from smaller setups, typically configured as Auro 9.1 targeting music productions (see Fig. 3.21), to large home theater setups (see Figs. 3.22 and 3.23), geared for configurations up to Auro 13.1. In all cases, a space is required that can allow for the placement of at least nine speakers around the main listening position, plus a subwoofer for the reproduction of the LFE (Low Frequency Effects) channel (20–120 Hz). The spectral range for the loudspeakers depends very much on the use of bass management or not (see Sects. 3.4.6 and 3.4.7 concerning Subwoofer Placement and Bass Management).

Figures 3.21 and 3.22 give a schematic representation of the standard Auro 9.1 and Auro 13.1 setups for small to medium rooms. These setups use discrete channels only. Larger rooms typically use multiple speakers for the surround speakers, both in the Lower and height layer. This not only allows for a better coverage of the listening area in larger rooms, but also gives the sound the diffuse character as heard in a cinema theater. It is advisable to use the same number of surround speakers in both the lower and height layer to maintain equal power and spread in both layers (Fig. 3.23).

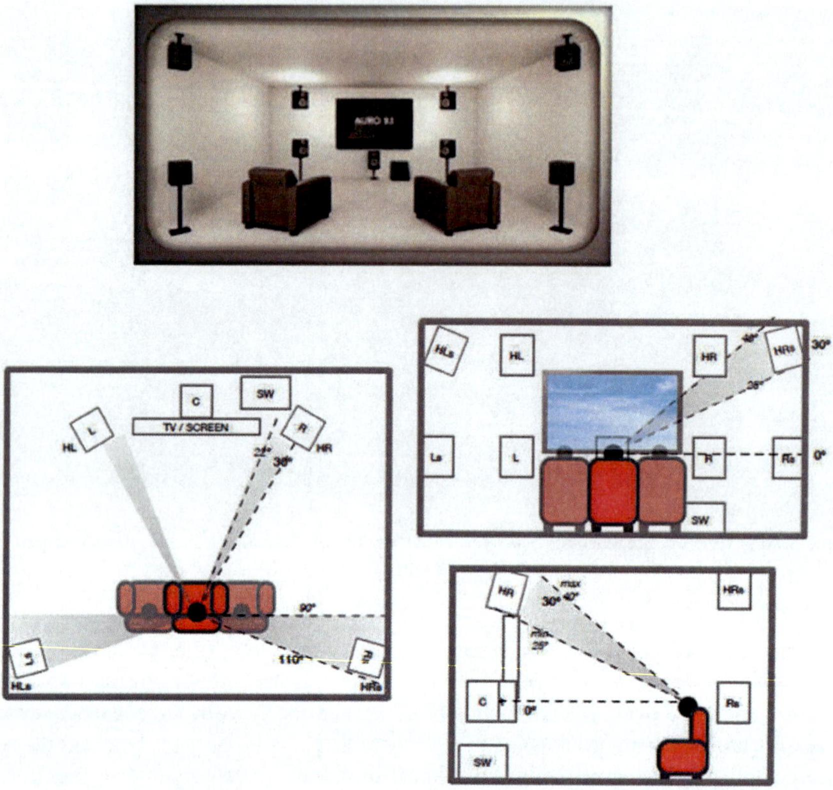

Fig. 3.21 Typical Auro-3D 9.1 speaker setup at home (from Auro Technologies NV 2015)

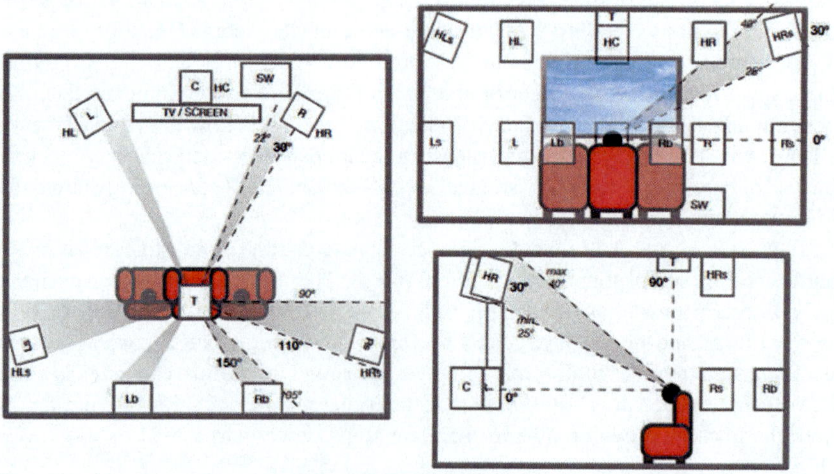

Fig. 3.22 Auro-3D 13.1 speaker setup for medium sized rooms (from Auro Technologies NV 2015)

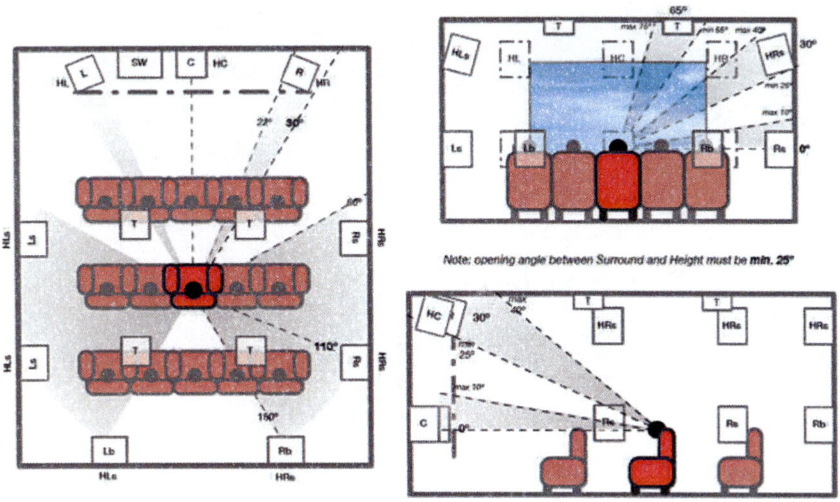

Fig. 3.23 Auro-3D 13.1 speaker setup for a large home theater (from Auro Technologies NV 2015)

3.4.3 Speaker Positioning and Time Alignment in Home Setups

It is important to know that the specifications for a Digital Cinema (following the recommendations by the SMPTE standards) are different from the specifications used for home theaters. The specifications for Digital Cinema will be described in the sections of AuroMax (see Sect. 3.5).

The specifications for an Auro-3D listening room layout 'for home' are based on the recommendations provided by the ITU-R1 (Rec. ITU-R BS.775-2 (2012)). This means that in theory, all speakers should be equidistant from the main listening position. This is often not the case especially in home theaters where people are usually closer to the surround speakers than the front speakers. Adjustments can be made in different ways and most AV-Receivers that have Auro-3D on board have the ability to make time alignment of the surround channels to avoid that the signals of those channels are getting to the ears of the listener faster than those of the front channels. The same goes for the volume, which can be adjusted to have the same volume from all speakers.

The acoustic centers of the lower speakers should be in the horizontal plane at ear-level. The speakers in the height layer should be elevated to an angle of about 30° (ideally between 25° and 35°) and tilted in such a way that the acoustical center is aimed at the listener's head when standing. The top speaker, for configurations that have one, should be positioned right above the listener, at 90° (see Fig. 3.24).

It is important to note that the placement of the speakers, with the corresponding angles as described above, is measured from the center of the circle (the place from which every speaker is equidistant) even if the audience will afterwards be seated on

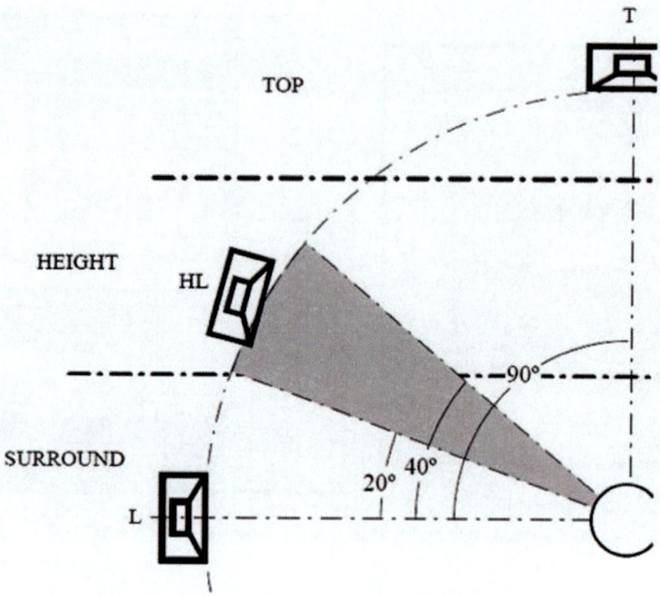

Fig. 3.24 Auro-3D elevation angles of height and top speakers (from Auro-Technologies NV 2015)

a seat that is closer to the surround speakers. Because if the horizontal and vertical angles are measured from the seat closer to the surround speakers, the angles would be different and will not match the standard setup as used by the content creators and therefore deliver a different sound experience.

If the center speaker needs to be positioned above or below a video monitor and as a result the acoustic centers of the front speakers are not aligned, it is important to try to position the tweeters in as close to a horizontal line, as possible. If the center, surround, height and/or top speakers are not equidistant with the L/R pair, a signal delay may be applied to obtain coincident arrival of the sound at the listening position.

Setup for Auro 13.1, Auro 11.1, Auro 10.1 and Auro 9.1:

The center speaker is positioned directly in front of the listener, with its acoustical center at ear-level, and is used as reference for the angles with the remaining speakers.
The left and right speakers are positioned at ± 30° from center, forming a 60° angle, the surround speakers at ± 110° off-center.

The surround layer is normally placed at ear level (0° elevation). However, in some larger rooms these speakers may need to be slightly elevated to provide sufficient distribution of the sound power to all listening positions. In such cases, the surround layer should be placed as low as possible, not exceeding an elevation of 10°, while always maintaining a minimum angle of 25° between the surround and height layers.

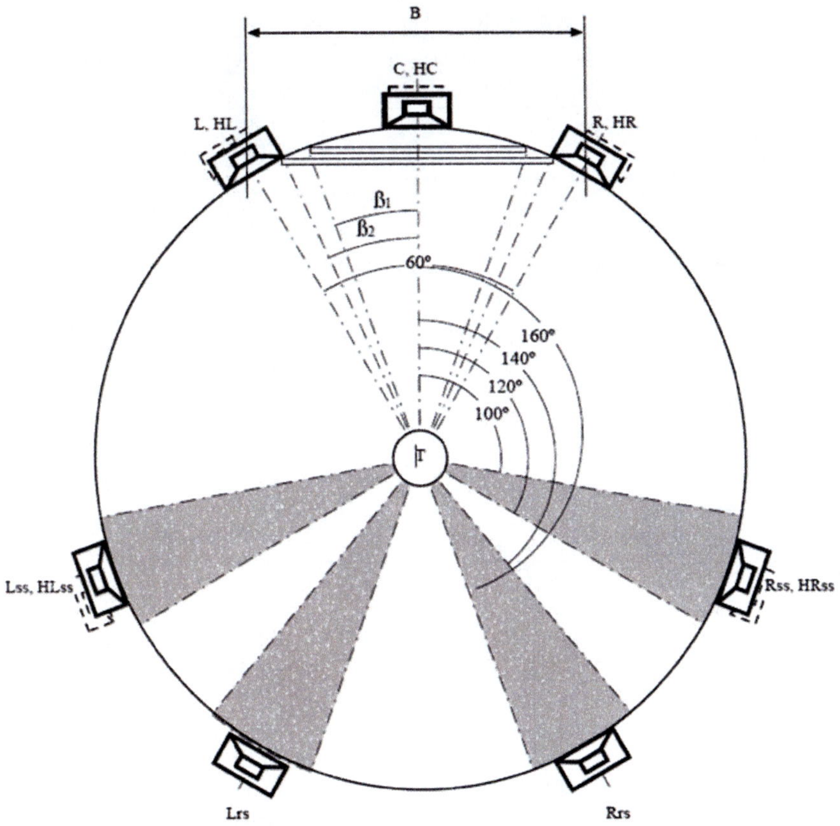

Fig. 3.25 Auro-3D and horizontal speaker positions for configurations based on 7.1 surround (from Auro Technologies NV 2015)

This means that the height layer should be elevated to at least 35° if the surround layer is positioned at the maximum elevation of 10° (see Sect. 3.4.4 for further details).

Some Auro-3D configurations are based on a 7.1 surround lower layer. For these configurations, the horizontal placement of the speakers follows the ITU-R Recommendation for 7.1 surround speaker setups in the Lower layer, and that of the 5.1 surround setup in the height layer (Auro 13.1 and Auro 11.1(7 + 4) are examples of this) (see Fig. 3.25 and Table 3.1).

3.4.4 *Tilt of the Height Speakers*

During the hundreds of tests that Wilfried Van Baelen and his team were doing in 2005 in order to find the best results for the placement of the height speakers, it

Table 3.1 Normative speaker positions for Auro-3D (from Auro Technologies NV 2015)

Speaker	Abbrev	Azimuth (deg.)			Elevation (deg.)		
		Min	Nom	Max	Min	Nom	Max
Left	L	– 22	– 30	– 40	–	0	10
Center	C	–	0	–	–	0	10
Right	R	22	30	40	–	0	10
Left surround	Ls	– 90	– 110	– 135	–	0	10
Right surround	Rs	90	110	135	–	0	10
Left back	Lb	– 135	– 150	– 155	–	0	10
Right back	Rb	135	150	155	–	0	10
Height left	HL	– 22	– 30	– 40	25	30	40
Height center	HC	–	0	–	25	30	40
Height right	HR	22	30	– 40	25	30	40
Height left surround	HLs	– 90	– 110	– 135	25	30	40
Height right surround	HRs	90	110	135	25	30	40
Top	T	–	–	–	65	90	100

became clear that there is a more natural spatial impression if the height speakers are not tilted so that the main axis of the speaker is directed straight to the listener. Tilting the speaker in such a way that the crossing point is about one or two feet above the listener creates a more natural 'spatial effect'. Audio engineers typically think that being 'off axis' from the height speaker gives a different timbre, but in reality this is not the case because most speakers maintain a similar sound spectrum (assuming typical dispersion characteristics of professional loudspeakers) even when making small adjustments by tilting the axis of the speakers a bit above the listener (see Figs. 3.26 and 3.27).

Van Baelen's recommendation how to align the tilt for height speakers: "A very practical way to achieve this tilt of the height speakers can be done as follow; stand up and direct the axis of all height speakers to your head and leave it like that. When sitting down, the crossing point of all height speakers will be one to two feet above the listener's head which is giving the best spatial impression. Additionally, this method delivers also a larger sweet-spot".

The listing in Table 3.1 gives an overview of the speaker positions for standard, discrete setups for Auro-3D. Note that the opening angle between the surround and height layers should be at least 25° for the surround speakers and at least 22° for the height screen channels.

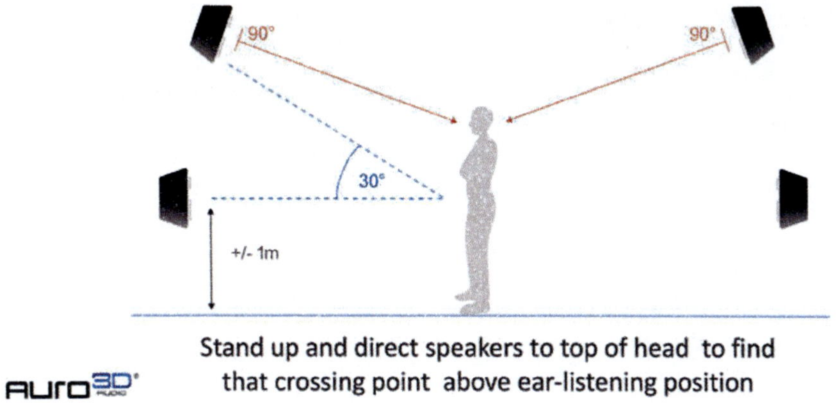

Fig. 3.26 Suggested tilt alignment for height layer speakers in an Auro-3D home setup (graphic courtesy and © Van Baelen)

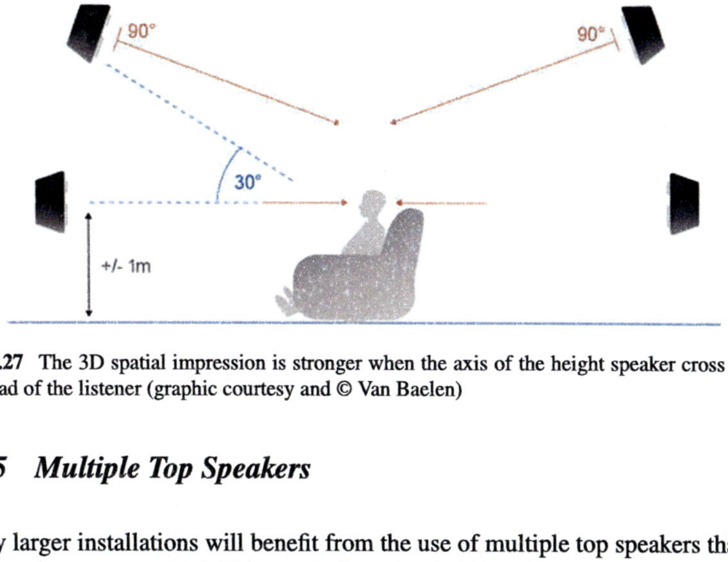

Fig. 3.27 The 3D spatial impression is stronger when the axis of the height speaker cross above the head of the listener (graphic courtesy and © Van Baelen)

3.4.5 Multiple Top Speakers

Many larger installations will benefit from the use of multiple top speakers that are then used as an array (a single group of speakers). This will ensure an equal spread of the sound throughout the room and provide an equal experience to everyone in the room. This is also especially true for rooms with lower ceilings, as this will also help to minimize the proximity effect of a too closely positioned ceiling speaker. The number of speakers used in such configuration depends on the room and is done in pairs. When using two top speakers, they should be positioned as a L/R overhead pair at 90° elevation. Larger rooms can benefit of using four or six top/overhead speakers, positioned in two rows above the audience.

3.4.6 *Subwoofers*

It is important to include at least one subwoofer in the speaker system, which is used for the LFE channel. (Fig. 3.28)

When some or all of the speakers are not capable of reproducing the lowest frequencies of a soundtrack or music recording, bass management (cross-over filters, mixing of the bass with the LFE channel at the correct ratio) should be used. Various products are available that redirect the bass from any channel that is not reproduced by the speakers to the subwoofer(s). Positioning the subwoofer(s) can be a tedious task and the positions will not be the same in all rooms. Finding the right spot(s) requires some experimentation, especially when retrofitting any existing room. While playing familiar program material with significant low frequency content, several subwoofer locations can be tried. The locations providing the smoothest bass response are the best choice for final subwoofer placement.

3.4.7 *Bass Management*

Although all channels in all Auro-3D configurations are specified to be full-range (20–20 kHz), many home theater systems and even smaller monitoring setups will not allow for full-sized speakers in all positions. For this reason, the larger front speakers or even a dedicated subwoofer uses a bass management system to reproduce the frequencies that are otherwise lost. With up to 14 channels in Auro 13.1, there are many possible combinations, ranging from large speakers for all channels, not requiring any bass management at all, to only small speakers and one or more subwoofers that reproduce all low frequencies. Crossovers, subwoofer and main speakers should work together to reproduce a flat frequency response for all installed channels. Most home installs do use bass management, ideally with a crossover frequency as low as possible (like 80 Hz). This means that the main speakers on the surround layer (i.e. 'lower layer') should be able to reproduce a spectrum from 50 Hz to 20 kHz. Ideally the same speakers are used for the height layer but in practice those speakers are often of a smaller size (preferably from the same brand) and should be able to reproduce a frequency range of 100 Hz–20 kHz.

3.4.8 *Polarity*

All speakers should have the same polarity. It is also highly recommended to maintain the same electrical polarity throughout the complete reproduction system.

3.4.9 Signal Delay

It is important that the sound from each speaker not only has the correct level, but also arrives at the listening position at the right time. In an ideal room setup, all speakers are positioned equidistant from the mixing position so those common sounds coming from two or more speakers are heard as one. If one or more speakers are not placed at the same distance, signal delay is required to achieve the intended result. The speaker positioned furthest from the listener is used as reference point to determine the required delays. The difference in distance between this speaker and each individual speaker determines the delay time, which is calculated as approximately 3 ms per meter or 1 ms per foot.

3.4.10 Room Equalization

Various AV-processors provide some means of equalization to compensate for possible problems with the room's acoustics. Some of these systems also have automatic algorithms, often using a measurement microphone in one or more positions in the room. Whatever system is used, it is important to realize that many of these systems introduce phase problems when applying too narrow, or too strong equalization. The following guidelines will help to achieve a natural sounding result.

- Avoid too many EQ bands (< 5 bands if possible)
- Limit the equalization to a frequency range below 500 Hz
- Avoid gains larger than 6 dB or attenuations larger than 10 dB
- Use wide EQ bands as much as possible.

3.5 Installs for Digital Cinema and Dubbing Stages

3.5.1 'AuroMax'—Concept and Speaker Layouts

'AuroMax' had already been designed by Wilfried Van Baelen before Dolby's 'Atmos' system appeared on the market. AuroMax is the newest member of the Auro-3D format and combines its object-based technology (up to 128 objects) with Auro-3D's unique 3-layered channel-based technology (up to Auro 13.1 beds). It is the ultimate cinema sound system within the SMPTE 2098 standard (i.e. the worldwide standard about immersive systems for Digital Cinema) (see Fig. 3.29).

AuroMax delivers the highest resolution and sound precision of all immersive sound systems on the market: with at least 20 individually amplified channels, it makes sense to use Object-Based technology to further enhance the spatial resolution. But such large numbers of speakers and channels are typically only found in professional cinema theaters or in 'high-end' home cinemas.

Fig. 3.28 Auro 11.1 Certified Home Cinema install at demo room of Stassen Hifi in the Netherlands (from Auro Technologies NV 2015)

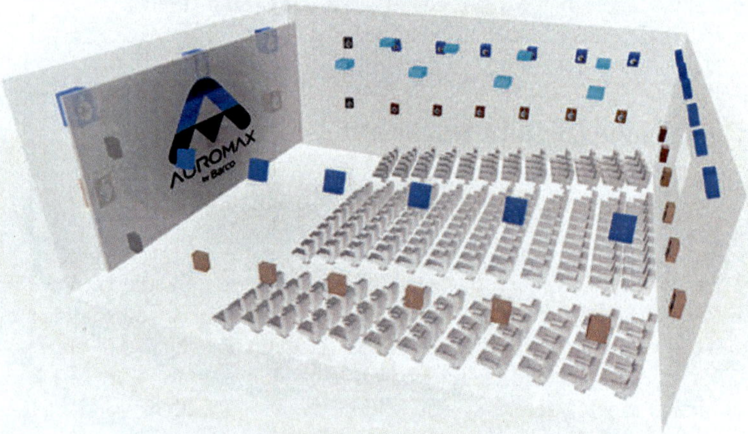

Fig. 3.29 The AuroMax speaker layout (graphic courtesy and © Van Baelen)

The vision behind the AuroMax system is to enhance the resolution and sound precision in a 3D space beyond the three layered Auro 13.1 system while keeping the advantages of the larger sweet spot. A system with more individual channels does not guarantee that the sound gets more natural. Additionally, it creates another level of complexity in respect to downward compatibility to systems with less channels (more comb filtering, leveling issues etc.) which is a key element of the Auro-3D concept. AuroMax is combining—in the best possible way—the 'pros' and 'cons' for cinema systems with more than 15 discrete audio channels in order to achieve the ultimate sound reproduction system with the largest possible sweet spot.

The idea behind AuroMax is to expand the Auro 11.1 system with object-based technology in order to create more 'zones' of sound around and above the audience. This means that 'objects' will be played back via a predefined speaker array. The advantages are again that the sweet spot can be maintained. If sound objects in a large theater are only played back by one speaker, the dispersion of that sound in the theater is limited to a small number of seats and therefore causes a much smaller sweet spot. That is the reason why AuroMax is using 'zones' which can reach all seats in the theater more easily and maintain the large sweet spot of the Auro-3D system.

The AuroMax speaker layout is based on the 3-layered Auro 11.1 system and adds four proscenium speakers (also called 'Wide Screen Channels'). Those proscenium speakers are filling the (acoustic) gap between the left or right screen channel and the first surround speaker (typically installed at the sides of first row of seats). This will also happen in the height layer which means that the AuroMax system has ten front speakers in total divided over two vertical layers, aiming to deliver the best screen sound on the market.

There is a total of 16 zones around and above the audience (see Fig. 3.30). The speaker layout of AuroMax is described in the SMPTE 2098 standard. AuroMax is the ultimate immersive sound system in the cinema market, and at the time of this writing many hundreds of installs can be found all around the globe.

3.6 Content Creation

3.6.1 Workflow Considerations

One strength of the Auro-3D concept seems to be a unified workflow concept, in which there is one mix session for all kinds of delivery formats, i.e. simultaneous mixing of 3D, 2D and stereo using standard mixing workflows which everybody understands—independent of the delivery format. Whether it is channel-based Auro-3D (e.g. Auro 9.1, 7.1, 5.1) or object-based as AuroMax with support for open standards. This approach is very economic because no extra time is needed to create the main audio formats (stereo, surround, immersive) simultaneously (see Figs. 3.31, 3.32 and 3.33).

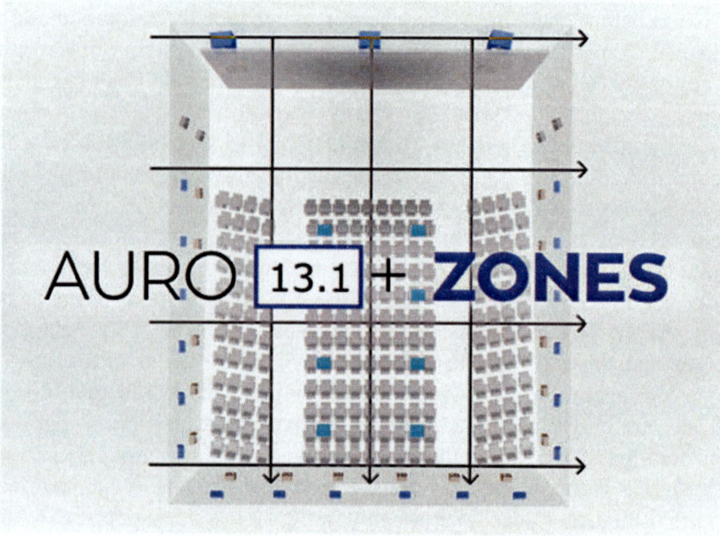

Fig. 3.30 AuroMax 13.1 speaker layout, zone distribution (graphic courtesy and © Van Baelen)

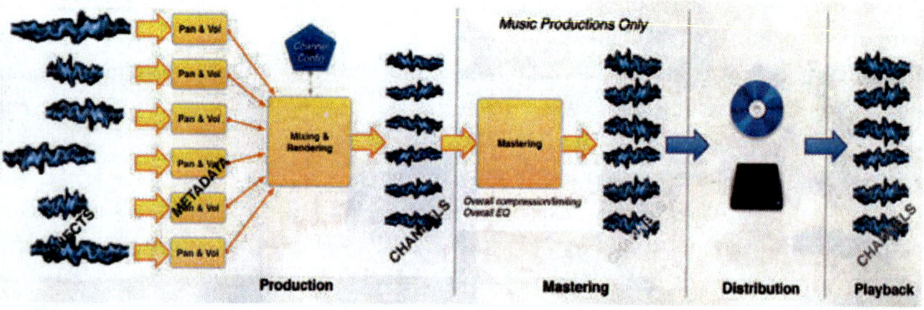

Fig. 3.31 Schematic of Auro-3D channel-based production chain, e.g. stereo, surround, -3D audio (graphic from Van Daele 2016)

3.6.2 Auro-3D Music and Broadcast Production as a Linear Based Workflow

In contrary with the audio systems at cinema theaters, music production studios as well as broadcast and games use a linear sound reproduction: the same sound spectrum that goes in, should be heard in the output. This is totally different from digital cinema, which uses the X-Curve (also called Academy Curve) for which the reproduction of the frequency spectrum of the output is very different from the input (see Sect. 3.6.3).

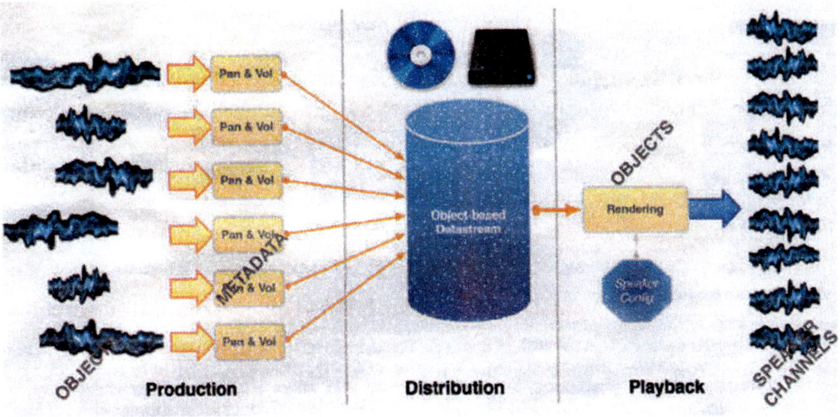

Fig. 3.32 Schematic of an object-based production chain (e.g. IOSONO) (graphic from Van Daele 2016)

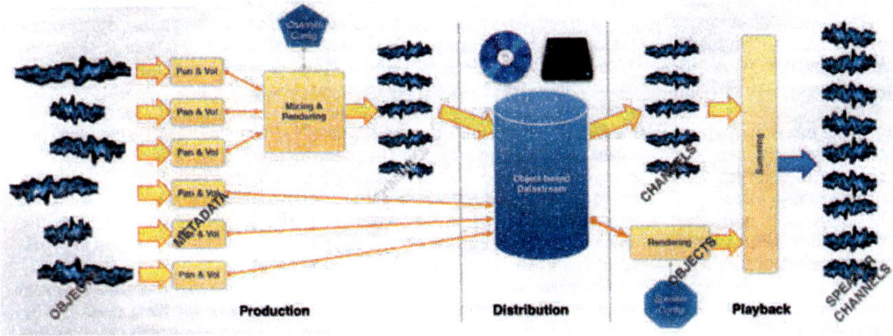

Fig. 3.33 Schematic of Auro-3D hybrid production chain (e.g. AuroMax, Auro-Cx, Dolby Atmos, DTS:X) (graphic from Van Daele 2016)

In music production, the established workflow consists of the following major parts:

A Recording/Editing
B Mixing
C Mastering.

At the 'Recording' stage, signals are either captured acoustically (via microphones) or derived from electronic sound sources (synthesizers, programming, electric guitars, online sources, etc.) and this is always a channel-based process. During the 'Editing' process the best parts of the recorded takes are brought together so that the final edited version sounds as a fluent 'one-take' performance.

The 'Mixing' process is typically used in pop music in which the many recorded channels are mixed, which means that each recorded channel is going to be optimized in timbre (by equalizer), volume, localization in the auditory field and often automation is needed to dynamically control all the changes. This process can be channel-based, object-based or both (hybrid). Mixing engineers are using specific speaker layouts in order to control the corresponding channel layout for each delivery format (stereo, surround and immersive).

The 'Mastering' Process is the last and also very important step and has two main tasks;

1. Create the master which is needed for duplication or direct distribution (streaming)
2. Optimize the overall sound of the mix in order to give it a 'gloss' which is often achieved by use of very specific compression and some 'spatial tricks'. After the mastering process, no further technical or aesthetic changes are allowed, therefore this is the final step from a sound creation point of view.

One big advantage of Auro-3D is that it is the only immersive sound format that can maintain the traditional mastering process. Systems like Dolby Atmos do need workarounds because the renderer at playback cannot reproduce the results achieved in a traditional channel-based mastering process. This is because in hybrid formats, the objects and channels are split in the delivery master, and will be rendered again before playback for which the renderer takes the specific speaker layout into account. But not having access to the compression and spatial techniques used in the traditional mastering process, the renderer at playback cannot reproduce the artistic intent of the mastering engineer during the traditional mastering process.

3.6.3 Post-production and Mixing for Auro-3D (X-Curve Based Workflow)

Through many decades, the film-industry for theatrical releases has evolved into a system that creates an improved compatibility between small and large theaters. Additionally, the so-called 'X-Curve' is also a part of the noise reduction system (see ISO 2969 (1987) and SMPTE 202 M (1998)) which was needed due to the use of analogue tapes. In small rooms (like mixing rooms, which are called 'dubbing rooms' in the film industry) there is a good control over mid and high frequencies acoustically, but less so for the low frequencies, because the rooms are not large enough to allow for recreation of the full wavelength at low frequencies. Mixing engineers are typically sitting in one part of a specific frequency's wavelength and therefore cannot correctly hear (or 'monitor') the energy of the entire frequency spectrum as it will sound later on, when played back in a large room (with many room modes, which are creating an uneven frequency response at such low frequencies).

In large theaters we almost have the opposite effect, as the modes appearing at low frequencies can be reasonably well controlled, but the high frequencies will

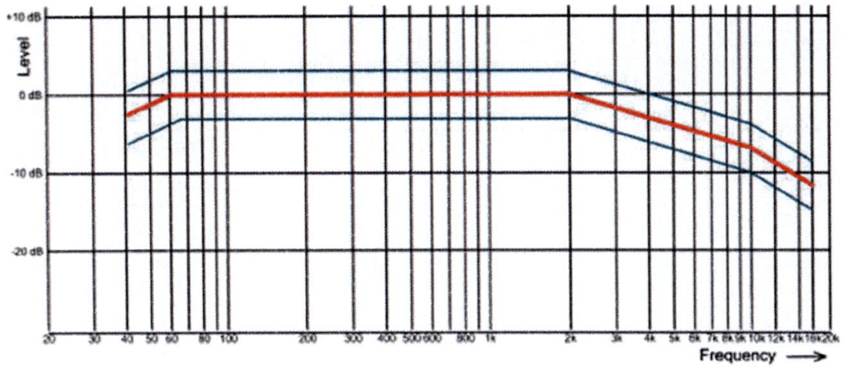

Fig. 3.34 The 'X-Curve' of spectral distribution as suggested for movie theatre sound (also called 'Academy Curve') (graphic courtesy of Van Baelen)

suffer attenuation due to air absorption. This means when mixing in a large room and playing it back in a small room, the sound experience is pretty different and the same happens vice versa when mixing in a small room and hearing a pretty different result in a large room.

In order to compensate this effect, the 'X-Curve' was defined, which means that pink noise, reproduced through each speaker, needs to result in a spectral distribution, following the X-Curve (Fig. 3.34).

Any difference that a measurement is showing needs to be corrected via an equalizer (typically 30-band 1/3 octave graphic equalizer). This means that in a small mixing room, the biggest part of corrective equalizing—in order to arrive at the desired frequency response in accordance with the X-Curve—will have to be applied from 1 kHz up to 20 kHz, while much less correction is needed in a larger mixing room or cinema theatre.

Therefore, the sound of the master mixed for cinema is not linear at all and cannot be used for playback at home. Additionally, the dynamic range used for the cinema mix is established with 85 dB (A) reference level at the reference listening place (at 2/3 of the distance between the screen speakers and rear wall loudspeakers) of a −20 dB pink noise signal.

Large parts of the following section are cited—in a slightly altered form—from Van Daele and Van Baelen (2012a, b), Van Daele (2016):

"... The mixing stage is where all the multichannel audio content comes together to create the complete 3-dimensional cinematic experience (see Table 3.2). As a preparation, it is important that all this material is organized to match the track layout on the mixing desk. Source materials may be:

- Auro-3D multichannel recordings from the set (dialogue, ambience) and orchestral scoring stage.

Table 3.2 Table of multichannel sound formats for the Auro-3D mixing stage and Auro-3D stem layout (from Van Daele and Van Baelen 2012a, b)

	Lower layer			Height layer		Comment
	Stem	Width		Stem	Width	
Dialogue	A	5.0	+	G	4.0	In case no dialogue in height top (HT/VoG) is needed
		5.0			5.0	Auro-3D 10.1: quad height + HT
		5.0			6.0	Auro-3D 11.1: quad height (incl. height center) + HT
Rem.: Typically, the most important channels for Dialogue are L, C and R. The other channels are normally only used to create spatial reflections						
Ambiances	B	5.1	+	H	4.0	Minimum configuration: Auro-3D 9.1
		5.1			5.0	Auro-3D 10.1: quad height + HT
		5.1			6.0	Auro-3D 11.1: 5.0 height (incl. height center) + HT
		6.1			6.0	Auro-3D 12.1: incl. center surround channel
		6.1			7.0	Auro-3D 13.1 + HT
		7.1			6.0	When a 7.1 surround mix is needed
		7.1			7.0	7.1 surround + back height center (for mixing Auro-3D 13.1 content in compatibility with 7.1)
FX	C	5.1	+	I	6.0	Minimum configuration: Auro-3D 11.1
		6.1			6.0	Auro-3D 12.1: incl. center surround channel
		7.1			6.0	When a 7.1 surround mix is needed

(continued)

Table 3.2 (continued)

	Lower layer			Height layer		Comment
	Stem	Width		Stem	Width	
		7.1			7.0	7.1 surround + back height center (for mixing Auro-3D 13.1 content in compatibility with 7.1)
Music	D	5.1	+	J	4.0	Minimum configuration: Auro-3D 9.1
		5.1			5.0	Auro-3D 10.1: quad height + HT
		5.1			6.0	Auro-3D 11.1: 5.0 height (incl. height center) + HT
Rem.: As the basic standard for music is recording 5.1 surround, most music will be recorded and mixed in Auro-3D 9.1. However, in some occasions the music channels are also used to create certain effects, making it useful to also provide the HC and HT channels for such situations						
Spare	E		+	K		
Rem.: The AMS-Neve Gemini II console with Encore 2 automation enables 12 stems, providing two additional stems to create an extra Lower and height layer when needed						
SUM	F	5.1	+	L	6.0	Final Mix in Auro-3D 11.1
		6.1			6.0	Final Mix in Auro-3D 12.1
		6.1			7.0	Final Mix in Auro-3D 13.1
		7.1			7.0	Final Mix in Auro-3D 13.1 with 7.1 surround compatibility

- While complete multitracks can be used immediately, already Auro-encoded material will have to be decoded first using the *Auro-3D Decoder* (hardware unit or decoder plug-in), before entering the final mix.
- Existing 5.0 (surround sound) multichannel or stereo recordings (e.g. from sound libraries), can be upmixed to Auro-3D 9.1 or 11.1 with the *Auro-Matic Pro* plug-in.

As part of the complete Auro-3D concept a mixing template was developed that allows for the simultaneous mixing to Auro-3D 11.1 and 5.1 surround sound, using the *Auro-3D Encoder* plug-in that is inserted in the final stem.

When creating separate mixes for two formats (e.g. stereo and surround), many mixing engineers have confirmed that both formats gain in quality by checking and switching between those formats on a regular basis. The Auro-3D Encoder plug-in allows dynamically mixing (with automation) the height channels into the lower channels during the mixing process. This means that the engineer—if he wishes to do so—can always hears the final result of the surround mix while listening through the Auro-3D plug-in. This way the optimal 5.1 surround mix will be created while mixing the Auro-3D 11.1 mix. The bulk of the extra mixing time will be limited to optimizing all phase and level relationships between the two output formats. Naturally, this process requires some exercise but each engineer (or postproduction studio) capable of delivering the 5.1 surround *and* Auro-3D 11.1 mix for almost the same budget will have a big advantage against competitors without this experience." (see Figs. 3.35 and 3.36 as examples of Auro-3D 13.1 loudspeaker layout as implemented in a professional recording studio).

Fig. 3.35 Rear part of Auro-3D loudspeaker setup at Galaxy Studios, the first ever mixing studio for 3D audio with an AMS-Neve DFC 3D console installed in 2011 (from Van Daele and Van Baelen 2012a, b)

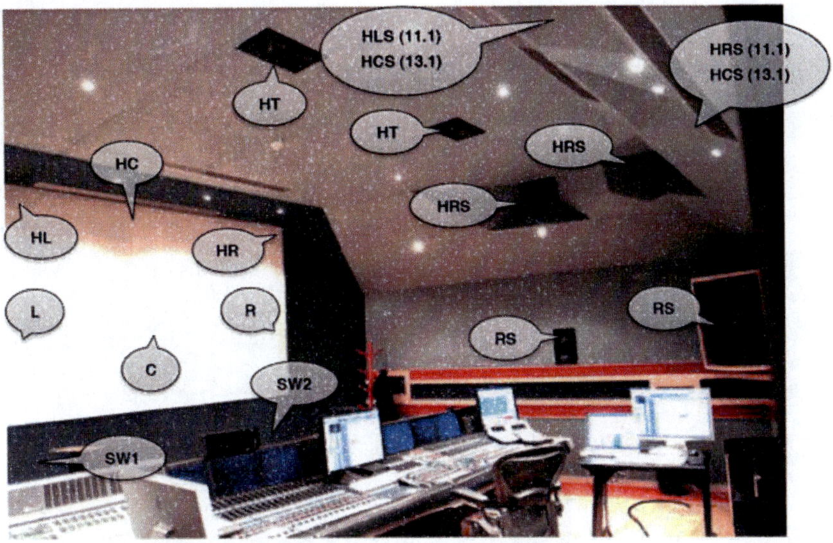

Fig. 3.36 Front part of Auro-3D loudspeaker setup at Galaxy Studios, The Netherlands; photo is showing Galaxy Studios' Auro-3D compatible pre-dubbing stage with up to 18.2 playback channels and AMS-Neve DFC with Encore II automation capable of Auro-3D compatible mixing for all surround formats (5.1, 6.1, 7.1) and Auro-3D formats (8.0 up to 13.1) (from Van Daele and Van Baelen 2012a, b)

3.6.4 Auro-3D Stem Layout

"Since most mixing consoles have stems with a bus-width limited to eight channels, a workaround is needed to create the possibility to mix Auro-3D content. This workaround is based on three items:

1. The Auro-3D Encoder plug-in that mixes the height and lower layers in a dynamic, artistically controlled way.
2. A stem layout that provides easy routing in Auro-3D as well as compatibility with surround sound formats (As an example, the AMS-Neve DFC console with Encore II already has all the necessary features onboard, including Auro-3D panning.)
3. For mixing consoles without the Auro-3D panning system, a special tool can be used to do this panning in-the-box.

The amount of channels per stem depends on the kind of content being mixed and the mixing engineer's artistic vision for the spatial experience. … " from Van Daele and Van Baelen (2012a, b), Van Daele (2016).

As a practical solution to this, the 'Creative Tools Suite' was developed by Auro Technologies, which provides the Auro-3D tools required for mixing and encoding, together with Auro-Matic Pro, all integrated e.g. into Digidesign's 'Pro Tools' DAW-Software (see Fig. 3.37).

Fig. 3.37 The 'Creative Tools Suite' (in Digidesign 'ProTools') with Auro-Panner, Auro-Matic, Mix engine and encoding engine software-tools (from Van Daele 2016)

Apart from this solution most Digital Audio Workstations (DAWs) such as Pro Tools, Nuendo, etc. are capable of working with (groups of) buses that have a sufficient amount of channels to do in-the-box mixing in the Auro-3D format.

Supported platforms for 'Creative Tools Suite' AU/VST3 on macOS:

- Logic Pro 10.7 and higher
- DaVinci Resolve 17 and higher (AU only)
- Cubase 12 and higher
- Nuendo 11 and higher
- Reaper 6 and higher.

The Windows version is expected soon.

From track channels, signals are captured with panner and upmixer plugins, allowing the user to define positions in 3D space, balance those sounds and send them through virtual busses. They are called 'virtual' because they are vector based and do not have pre-defined channels. Those busses are then sent to a mixing engine, which is running in a 'daemon' (a background processing in a multitasking computer operating system). The 'Mixing Engine' takes both the audio and the metadata through the virtual busses, creates a 3D mix and sends it back to the ProTools System for monitoring purposes or to record the Stem Mix. The Mixing Engine is also capable of encoding the content with the Auro-3D Codec, to create an object based data-stream, which is one file with all the object-based information directly from the mix session.

3.6.5 *Encoding and Authoring for Auro-3D*

Auro-Codec

The Auro-Codec is called a 'near lossless codec' because it delivers the audible sound quality of a lossless codec but the least significant bits, which are not used in a final master, are used for carrying metadata so that the Auro-Decoder can retrieve the original signals with the same audible sound quality from that PCM channel (see Fig. 3.38).

How does it work?

Each bit is representing 6 dB dynamic range, so a 24bit signal can have more than 140 dB dynamic range. One needs to add the background noise of the listening room on top of the dynamic range which means that a speaker system needs to be able to deliver 160 dB sound pressure level in order to reproduce the dynamic range of a 24-bit signal. Such high sound pressure levels would immediately damage human ears with unrecoverable serious hearing damage. It is like Tomlinson Holmen said: "you can only experience once in your life the full dynamic range of a 24-bit signal and I do hope you will never have that experience because life afterwards would not be the same again when your hearing system is totally damaged".

Although 24-bit is very useful during the workflow, it is not when it comes to create a delivery master in which the dynamic range is very rarely above 120 dBdynamic range (20 bit). Additionally, consumer playback systems do not have the ability to reproduce 24-bit dynamic range, most of them have a limitation between 18 and 20-bit maximum because the noise floor of the amplifiers will be higher. This is the 'hidden bottom'—as Van Baelen calls it—that enables us to combine multiple PCM channels in a single PCM stream, using the same bandwidth as a single PCM channel and using mathematical solutions so that the decoder can reconstruct the original PCM signal.

Even dynamically controlled down-mixing is possible, which means that the height channels can be dynamically mixed into the surround master channels and

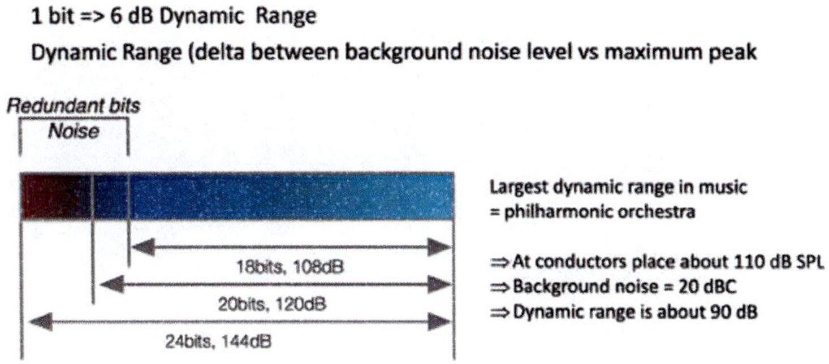

Fig. 3.38 The 'Auro-Codec', making use of LSBs of the 24bit PCM audio data

that the decoder can reconstruct the original height channels as to how they were originally sounding before the down-mix. This whole process happens in the digital domain and each ear-level (surround) channel will be encoded with its related height channel. The top layer (Voice-Of-God) is encoded in all the 'corner channels' of the ear-level layer (i.e. L, R, LS, RS).

In principle, the decoder needs only *one* (1) sample-cycle for each channel to be decoded. There is no psycho-acoustical optimization—as we know it from lossy codecs like mp3 or Dolby Atmos—which do suffer from audible artifacts, while Auro-3D maintains the (data-wise) uncompressed PCM audio quality throughout the whole chain.

The Auro-Codec is also used for Hi-Res streaming with a standard internet connection. On 28 October 2021 there was a big event in Japan (together with the famous Japanese broadcaster WOWOW and NTT) during which a live concert was streamed in Auro-3D over a standard internet connection all over Japan and to Europe in the same quality which you would get from a Blu-ray disc (this means same Auro-Codec with 2 K picture quality). This is considered a major milestone as it was the first-ever live performance that got streamed in high-quality immersive audio with HD-picture over standard IP (see Fig. 3.39).

Reiji Asakura, the popular Audio–Video journalist in Japan, was present at that event and later wrote in an article on Yahoo.com: "Many things are changing to 3D nowadays, and the trend toward 3D will continue to accelerate in the future. This experiment is groundbreaking because it can bring the sound experience in concert

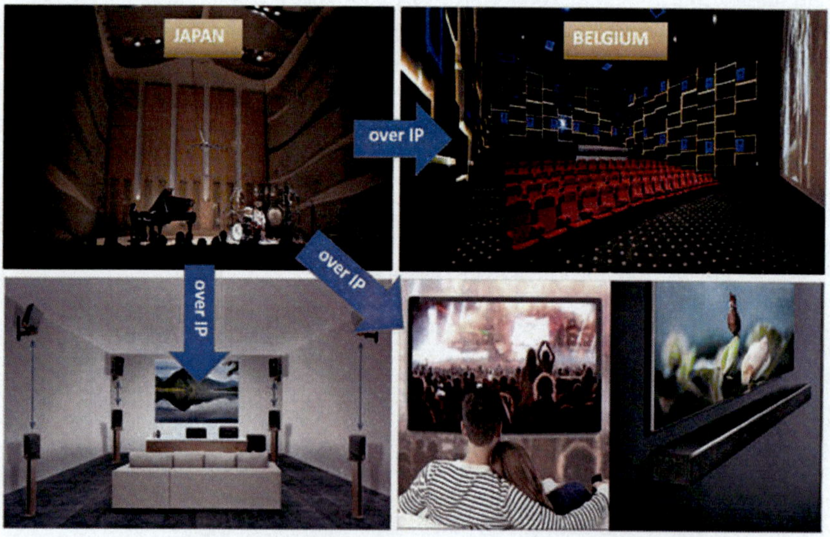

Fig. 3.39 Photos from the first-ever live performance that got streamed in high-quality immersive audio with HD-picture over standard IP in October 2021 (courtesy and © Van Baelen)

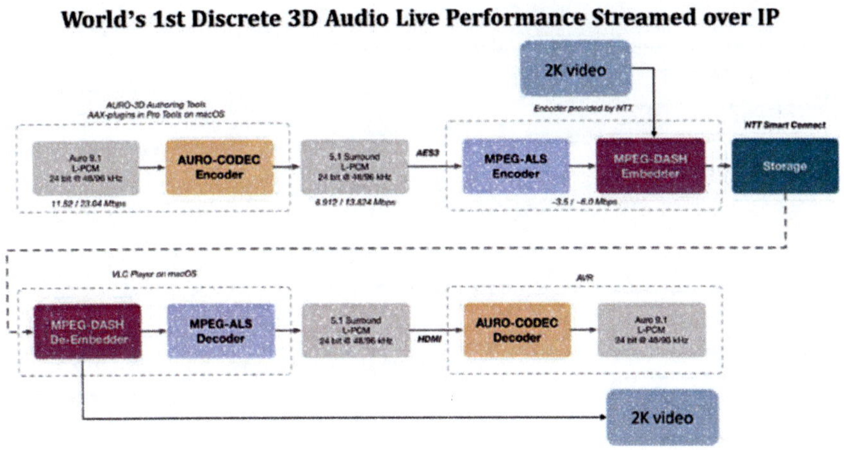

Fig. 3.40 Signal-flow diagram from the first-ever live performance that got streamed in high-quality immersive audio with HD-picture over standard IP in October 2021 (courtesy and © Van Baelen)

hall itself into the home, and *I think it is a revolution in the way that Edison's 'Mary's Sheep' was recorded on a gramophone*" .

The bandwidth used for this Auro 9.1 discrete stream was about three Mbit/s, which can be decoded by any AV-Receiver with Auro-3D (Fig. 3.40).

From (Van Daele and Van Baelen 2012a, b) and (Van Daele 2016):

"… The last stage before the release of Auro-3D audio material is the creation of the Auro-3D Encoded 5.1 PCM Master. This stage consists of the following three main steps:

1. Encoding of the Cinematic 11.1 Auro-3D and 5.1 surround sound versions into one 5.1 Auro-encoded PCM-file for Digital Cinema (X-Curve based mix for DCP – 'Digital Cinema Package') (see Fig. 3.41).

"On a standard Auro 11.1 system, the content will be on a DCP. Theatre projectors contain a component called the IMB (Integrated Media Block), which is responsible for decrypting and decoding the audio and video content to send out to what is called the 'B-Chain' audio post-processing (i.e. Crossovers, EQ, amps, speakers, etc.). In Digital Cinema, everything is heavily protected with encryption and keys, thus the decrypted content needs to be watermarked before leaving the system. The Auro-Codec decoder is located inside the media blocks as well as being present in the audio post-processor, which is also capable of up-mixing, if required. It is also able to decode what is called 'alternative content': Auro-encoded content sourced from a satellite link source, for example.

2. Bouncing of the consumer 9.1 Auro-3D and 5.1 surround sound versions into one 5.1 Auro-encoded PCM-file for BD (Blue-ray Disk) and DVD (Digital Versatile Disk) (Linear mix)

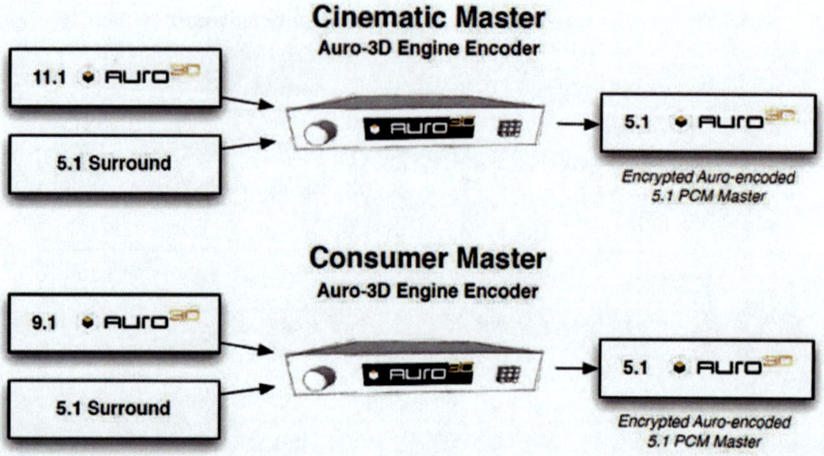

Fig. 3.41 Auro-3D encoding with the Auro-3D engine hardware encoder or plug-in for cinematic (above) and consumer (below) masters (from Van Daele and Van Baelen 2012a, b)

In home hi-fi, several brands of AV-receivers such as Denon, Marantz, Trinnov, Lyngdorf and Datasat have an Auro-3D engine integrated. What concerns the automotive industry, Auro Technologies have a partnership with Continental and it is expected that the Auro-Matic upmixer will be installed in several new car models, equipped with a 9.1 Auro-3D loudspeaker system.

3. Encryption of the files (copy protection)

The main advantage of the Auro-encoded PCM -files is that the same DCP can be used by theaters equipped with an Auro-3D Decoder and those without, effectively providing the highest compatibility between both groups as well as being future-proof.

Authoring

The authoring process is now simplified as there is only one audio-stream to be taken into account. Moreover, since this is a 5.1 PCM-stream, no further encoding process is necessary anymore. However, in case the audio-stream takes more bandwidth than originally budgeted, the Auro-encoded PCM-stream can still be further compressed using the currently available *lossless* audio codecs. This will further reduce the stream by up to 40%.

Single Inventory Distribution

As already mentioned earlier, one of the strongest points of the Auro-3D codec is the fact that it guarantees maximum compatibility with all existing standards. The final carrier format for distribution is *one* single standard multichannel (5.1) PCM-stream, in which the other formats (11.1 or 9.1) are encoded (see Fig. 3.42).

Fig. 3.42 Auro-encoded 5.1 PCM stream with and without decoding (from Van Daele and Van Baelen 2012a, b)

The same master can thus be sent to cinema theaters with or without an Auro-3D setup. Theaters not yet equipped for Auro-3D will then simply play the audio-stream as a standard 5.1 mix, while those that have the Auro-3D decoder and playback system will be able to play back the full 11.1 listening format. For playback at the homes from a BD, the same way of working is possible: one single master can be played back as 5.1 surround sound for those that do not own an Auro-3D decoder, while the consumers that do own an Auro-3D equipped home theater system will be able to enjoy the full 9.1 Auro-3D version of the movie ..."

Apparently, Auro Technologies also has made its entrance to the mobile phones and gaming market with the Auro-Matic upmixer being installed into the operating systems of the respective mobile phones. In the gaming and VR industry (Virtual Reality) they have established a partnership with the Canadian company AudioKinetic and developers are able to conduct real time panning of objects within the gaming or VR application, using head-tracking information (via Oculus Rift, HTC Vive, etc.) (see Van Daele 2016).

3.6.6 Covering the Auro-3D System with Only Eight Products

Auro-Codec

A unique, near lossless, multichannel audio codec. Allowing for 'one file distribution': multiple masters (surround sound, Auro-3D etc.) that can be played back exactly as intended by the creators from the same single file. This technology revolutionizes distribution, bringing content to all markets using existing standards and delivery formats, like Blu-ray or downloads, without the need for any extra bandwidth (see Fig. 3.43).

Auro-CX

A Next Generation Audio (NGA) codec that delivers high-quality immersive and interactive audio, including all Auro-3D formats, at the same quality as the highly

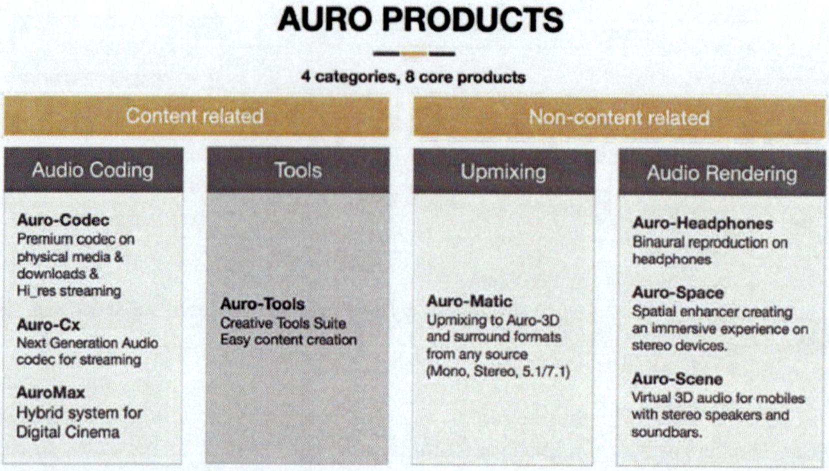

Fig. 3.43 Auro-3D related products

acclaimed Auro-Codec, but at scalable bitrates (from lossless to low-bitrate lossy). This technology is specifically developed from the ground up for streaming applications as well as digital downloads. Auro-Cx is created with the latest requirements for broadcast and OTT transmissions in mind, providing exciting interactive features enabled by object-based audio, such as Dialogue Enhancement.

Auro-Scene

Virtual speaker technology for soundbars and stereo speaker systems is integrated into the latest Auro-3D Engine v3. Providing an engaging and vivid experience and tuned to deliver optimal results alongside our acclaimed Auro-Codec and Auro-Matic technologies. Capable of playback from any source (e.g. stereo, 5.1/7.1) whatever the requirements may be: across movies, tv, music and gaming, this new virtual speaker solution is able to transform any room creating a virtual 3D space around the listener.

Auro-3D Creative Tools Suite

A complete package for the creation of native 3D audio content, it enables the creation of immersive and interactive content as AAX/AU/VST3 plug-ins for Pro Tools on MacOS or as VST3 plugin for Windows. Auro-3D Creative Tools Suite is now accessible in LogicPro, DaVinci Resolve, Cubase, Nuendo and Reaper. Mixing in Auro-3D with object based audio and creating multiple derived deliverables 'on the fly' is possible thanks to the advanced mixing engine. Each plug-in is designed to be easily integrated into any existing content-creation workflow, whilst solving many technical limitations present in DAWs. This makes Auro-3D authoring tools one of the most flexible and most efficient solutions on the market.

Auro-Headphones

An algorithm that creates a binaural version of the Auro-3D immersive sound experience. This technology is available for use on any kind of headphones or earbuds. Available on a wide variety of devices such as smartphones, tablets, notebooks, smart headphones, PCs, Consoles and AVRs. With Auro-Matic for Headphones, any content can be converted in to a binaural 3D experience on mobile phones (see Fig. 3.44).

Auro-Space

Sound enhancement technology: Auro-Space provides an instantaneous immersive improvement of sound quality for daily content playback over all devices from stereo speakers up to 13.1; including low- and mid-end soundbars, Smart Speakers and car stereo systems. Auro-Space is fully compatible with existing audio standards, such as CD-audio, mp3, AAC.

AuroMax

The ultimate step in the Digital Cinema sound systems combining the very best immersive sound with channel and object-based technology to create even more precision for sound localization with a minimum of 20 individually amplified speakers or speaker arrays installed. It consists of proprietary rendering technologies, whilst maintaining compatibility with the announced interoperable SMPTE-standard for immersive sound.

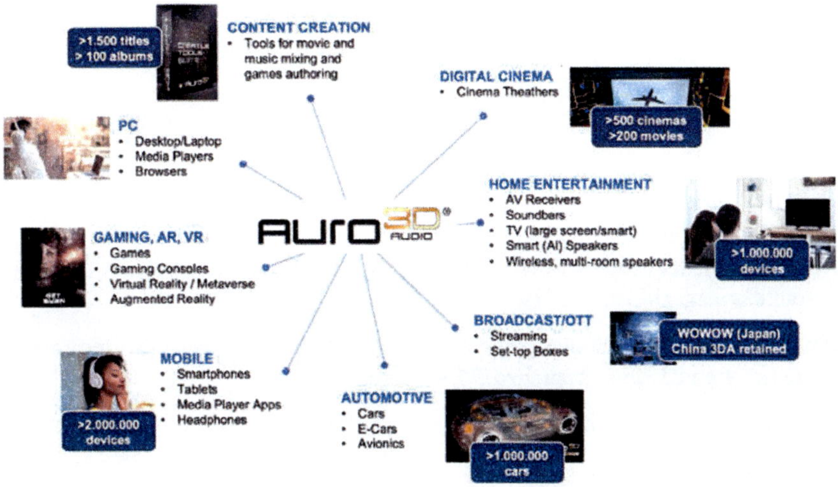

Fig. 3.44 The Auro-3D system, established in various technological and industry segments, as of 2021 (courtesy Auro Technologies NV)

Fig. 3.45 Dashboard in Porsche automobile, showing consumer control over the amount of Auro-Matic impact

Auro-Matic

Is an intelligent and highly efficient audio up-mix algorithm which can take any content (mono, stereo, surround sound content) and create an immersive sound experience for all Auro-3D speaker setups. Movie and music content is enhanced with a new 'heightened' immersive sound experience (see Fig. 3.45).

3.7 Practical Experience: Ronald Prent on 'Recording and Mixing in Auro 3D'

The following interview conducted by Nigel Jopson with sound engineer Ronald Prent (of Valhalla Studios New York) has originally been published in the British Resolution magazine in 2016 (see Prent 2016).

What are some of the challenges of recording and mixing in Auro-3D immersive audio in comparison to standard 5.1 surround?

In classical-style recordings, it's important to ensure that the additional height mics have a good time/distance relationship to the other mics to prevent phase anomalies in the recording itself, especially later in fold-down. In non-classical recordings, there aren't too many differences except recognizing that room mics may later be spread out among the channels and may require some artificial manipulation in the mix process to resolve potential time or phase issues. For mixing, it's even more fun than standard surround! There's more space to put stuff, and it's not really more difficult once you've got an appropriate technical workflow set up. Then it's really a creative extension of 5.1.

How much of the material you've mixed in Auro-3D was recorded with 3D in mind versus re-purposing stereo or 5.1 material?

I would say about 80% of what I've worked on was from recordings that were not originally made with 3D in mind, but that's primarily non-classical. In the world of film, orchestral or acoustic music, I think a much higher percentage of material being released in Auro was recorded with a plan for 9.1. For example, Morten Lindberg of 2 L in Norway was recording his productions in 9.1 for many years now.

How easy or difficult is it to repurpose an album from stereo to 3D?

It's usually not too difficult. It's trickiest when there aren't really enough 'voices' in the original to fill the extended channels, but plugin tools like Auro-Matic and Penteo exist to allow you to upmix elements to fill the soundfield. I also use a combination of mono, stereo, 5.1, and quad reverbs to create artificial spaces where appropriate. Also, you don't always have to have something happening in all the channels all the time. Sometimes silence is an effective musical tool.

How are artists responding to working in 9.1? Do they find it difficult to get their heads around it or does it feel like a natural progression?

The artists I've worked with, including Ozark Henry, Tiësto, and Prash Mistry, were very enthusiastic and seem to have found it both easy and exciting to make the transition into 3D, immediately thinking in terms of how their music can take advantage of the 3D soundfield. Ozark Henry has aptly described the joys of working with this expanded sonic canvas in his recent TedXTalk and Google Talk.

From a recording and mixing standpoint, do you find that Auro-3D lends itself to a particular genre of music, or have you been able to adapt diverse styles to the format?

In my experience, it works well for all genres as long as you adapt your approach according to the music itself. Auro-3D can be as simple and intimate as you like. For example, a singer-songwriter track with only voice and acoustic guitar can be a simple document of the space and performance that, to the listener, feels just like being there in the room. Or you can create a completely artificial experience—in electronic music, for example—where there is no basis in reality, you can fly things around to create energy and excitement, taking the listener on a sonic rollercoaster ride. It's in the hands of the artist and engineer to use the 3D soundfield in a way that best expresses the intention of the music.

When repurposing material that wasn't created with 3D in mind, and therefore not recorded as such, do you find there are any rules with regard to what you assign to the upper channels?

Not really, what is possible is generally determined by the music itself. In electronic-based music where the environment is entirely artificial, all bets are off. You can more or less do anything you want. In 'reality-based' material, where there is a kind of mental image created—a rock band, jazz ensemble, or live concert event—it makes sense to create a more traditional, stable image with perhaps some interesting 'accents' in places. In short, the 3D mix has to make sense with the material and the arrangement, but when the intention of the music is to be unusual and progressive, there's no reason not to go there.

What kinds of consoles have you been mixing on for Auro-3D so far? Were they designed for this?

So far, I've done my 9.1 work on the Avid (formerly Euphonix) System 5 and the API Vision, both at Wisseloord Studios. Both consoles were modified for working in immersive audio formats. The Euphonix had some DSP and software mods implemented, and the API has some extended hardware for a wider buss architecture and expanded monitor section.

Any particular pitfalls that engineers should watch out for while working in Auro-3D?

The electrical time alignment between the channels is extremely important because even if discrete stereo and 5.1 mixes are included in the delivered media, the 9.1 also needs to successfully down-mix for encoding and streaming purposes. Be sure to listen to how it all folds down. It may seem self-evident but it's worth mentioning that, when working in 3D, having a well-matched, well-aligned monitor setup that you can trust is very important. You can't catch potential problems if you can't hear them. Also, when splitting up direct or spot mics and room mics between the upper and lower layers, this can introduce timing issues that can result in phase problems, especially when more than one set of room mics is being combined. You can compensate for this by carefully manipulating the elements with regard to time in the digital domain during postproduction. It just takes some practice and a bit of courage, and again, a listening environment that you can trust.

Any tips or tricks?

Don't sit in the sweet spot all the time when mixing. Walk around to really feel what kind of image or environment you've created for the listener. And have fun, but be sure to approach the mix to suit the music, not just to entertain yourself with the technology.

3.8 Practical Experience: Darcy Proper on 'Mastering in Auro-3D'

The following interview conducted by Nigel Jopson with sound engineer Darcy Proper (of Valhalla Studios New York) has originally been published in the British Resolution magazine in 2016 (see Proper 2016).

How would you compare working in Auro-3D with working in stereo and 5.1 surround?

As one might suspect, mastering in 3D literally adds another dimension to watch out for—listening for balances not only from left to tight—but front to back, and top to bottom, being careful not to introduce any timing issues in my signal flow that could cause problems. As always in mastering, I'm very much dependent on the quality of the mixes I receive. Just like in stereo and 5.1, if there are inherent timing or phase issues between elements within the mix, they can't be solved in mastering—and with more channels, of course, the risk of these kinds of occurrences increases. But, in spite of the seemingly complicated aspects of 3D, the goal in mastering remains the same as with every other musical format: to help the listener connect to the emotion of the music by providing a well-balanced, coherent master.

Have you some advice for EQing for Auro-3D?

Keep in mind that the whole can be different than the expected sum of its parts: meaning that the 'perfect EQ' for L&R when soloed, plus the 'perfect EQ' for LS/RS soloed, plus the 'perfect EQ' for the upper channel pairs when soloed, doesn't necessarily create the 'perfect' result when listening to the whole. Experienced mix engineers know this in their work, too. It's important to evaluate and adjust, respecting the individual details, but always with the big picture in mind.

What about compressing or limiting Auro-3D masters?

With regard to compression and limiting, I find that I tend to be very conservative in 3D, using only what is absolutely necessary musically. Oversquashed material really becomes an assault on the senses when barking at you from so many channels. It can be exciting for a moment, for effect, but an entire album in such a state tends to make the listener try to detach and escape, rather than to engage and enjoy. The biggest challenge for me at the moment stems from the fact that, if I want to work using my usual preferred analogue signal processing, I don't have enough channels available in my console or outboard gear to accommodate all ten channels at once! Therefore, I end up mastering in two passes, lower 5.1 and upper four height channels. Since I can't hear all the channels processed together, it's a bit a process of trial and error: record the two passes then play them back and see if my plan actually worked. Usually it does, but sometimes the combined results disappoints and requires adjustment and reprinting which can be time-consuming. Needless to say, ensuring accurate timing between the upper and lower streams is of utmost importance.

Is there a future for immersive audio in music?

Overall, I find Auro-3D immersive sound fascinating, thanks to the sonic possibilities it opens up for artist and listener alike. You can be as conventional or outrageous as you like. And it will only get more exciting as more artists compose with 3D in mind. On the technical side, the necessary tools will definitely become more readily available and more affordable allowing more productions to take advantage of the format."

References

22.2 System https://en.wikipedia.org/wiki/22.2_surround_sound. Accessed 19 June 2022

Auro Technologies NV (2015) Auro-3D home theater setup—installation guidelines. Rev 6. http://www.auro-3D.com. Accessed 28 Oct 2015

Dabringhaus W (2000) 2+2+2—kompatible Nutzung des 5.1 Übertragungsweges für ein System dreidimensionaler Klangwiedergabe klassischer Musik mit drei stereophonen Kanälen. In: Proceedings to the 21. Tonmeistertagung des VDT

Genelec (2021) Genelec ist zu Besuch in Morten Linbergs atemberaubendem Immersive Audio Studio. https://www.youtube.com/watch?v=QnjaD10201U. Accessed 25 Jun 2022

Hamasaki K (2011) The 22.2 multichannel sound and its reproduction at home and personal environment. Paper presented at the 43rd international conference of the Audio Engineering Society, Pohang (Korea)

Hamasaki K, Van Baelen W (2015) Natural sound recording of an orchestra with three-dimensional sound. Paper presented at the 138th convention of the Audio Engineering Society, Warsaw

Hamasaki K, Hiyama K, Okumura R (2005) The 22.2 multichannel sound system and its application. Paper 6406 presented at the 118th convention of the Audio Engineering Society, Barcelona

ISO 2969 (1987) International Standard. Cinematography—B-chain electroacoustic response of motion-picture control rooms and indoor theatres—Specifications and measurements

ITU Recommendation ITU-R BS.775-3 (2012) Multichannel stereophonic sound system with and without accompanying picture. Int Telecommun Union

Matsui K (2015) 22.2 Multichannel sound reproduction system for home use. Broadcast Technology 59. https://www.nhk.or.jp/strl/english/publica/bt/59/3.html Accessed: 29 June 2022

Oode S, Sawaya I, Ando A, Hamaski K, Ozawa K (2011) Vertical loudspeaker arrangement for reproducting spatially uniform sound. Paper presented at the 131st convention of the Audio Engineeing Society, New York

Prent R (2016) Recording and mixing in Auro 3D. Resolution 15(5):36–37

Proper D (2016) Mastering in Auro 3D. Resolution 15(5):37

Röhbein M, Langhammer J, Ramon Menzinger J (2019) AuroMax immersive audio: how the three-layer approach leads to a larger sweet-spot. Whitepaper of company BARCO, dated 25 July 2019

Sawaya I, Oode S, Ando A, Hamasaki K (2011) Size and shape of listening area reproduced by three-dimensional multichannel sound system with various numbers of loudspeakers. Paper presented at the 131st convention of the Audio Engineering Society, New York

SMPTE Standard ST 2098-2:2019 (2019) Immersive audio bitstream specification

SMPTE Standard 202M (1998) Standard for motion pictures—dubbing theaters, review rooms, and indoor theaters—B-chain electroacoustic response

THX 10.2 https://en.wikipedia.org/wiki/10.2_surround_sound. Accessed 19 June 2022

Van Daele B (2016) Auro-3D creative tools. Resolution 15(5):44

Van Daele B, Van Baelen W (2012a) Productions in Auro-3D: professional workflow and costs. White paper by Auro Technologies

Van Daele B, Van Baelen W (2012b) Productions in Auro-3D—professional workflow and costs. Rev. 0.6 http://www.auro-3D.com. Accessed 28 Oct 2015

Van Den Berghe G, Van Baelen W (2008) A method and encoder for combining digital data sets, a decoding method and decoder for such digital data sets and a record carrier for storing such digital data sets. WO Patent 2,008,043,858-A1

Chapter 4
The DOLBY® "Atmos™" System

Abstract After a brief introduction to digital cinema the progression from Dolby Surround 7.1 to Dolby Atmos, including height speakers, is explained. Details of cinema theatre speaker placement and orientation for the purpose of improved audio quality and timbre matching are unveiled. The concept of object-based mixing and metadata is explained, as well as the typical workflow, which includes dialogue, Foley, SFX and music mixing into traditional 'beds' (stems) and 'objects' content and conversion into the final Dolby Atmos or 5.1/7.1 mix. The basics of Dolby's RMU (Rendering and Mastering Unit) are explained, as well as the DCP (Digital Cinema Package). The core functionalities of the Dolby Atmos Monitor and Dolby Atmos Panner software-plugins are introduced briefly. Practical applications within AVID mixing consoles and Pro-Tools Systems are touched as well as upmixing tools from third-party developers. The chapter concludes with case-study experiences gained in using Dolby Atmos for pop music live sound reinforcement, as well as an in-detail practical description of necessary steps and options for music production in Dolby Atmos.

Keywords Dolby Atmos · Dolby Surround · Height speakers · Object-based mixing · Digital Cinema Package · Dolby Atmos Panner · Upmixing

4.1 The Introduction of Digital Cinema

" … The introduction of digital cinema provided the opportunity for the industry to evolve beyond the technical limitations in place with sound on film. With the creation of standards for digital cinema, 16 channels of audio were now available within a DCP to allow for greater creativity for content creators and a more enveloping and realistic auditory experience for cinemagoers. During the advent of digital cinema, the industry focused primarily on the development of technologies and standards relating to image and security. At the same time, the industry has enjoyed the ability to use existing 5.1-equipped dubbing theatres and cinemas for the creation and playback of

The section below is a combination of citings from various Dolby whitepapers (see Dolby 2012, 2014a, b) in occasionally slightly altered form, unless otherwise noted.

E. Pfanzagl-Cardone, *The Art and Science of 3D Audio Recording*,
https://doi.org/10.1007/978-3-031-23046-2_4

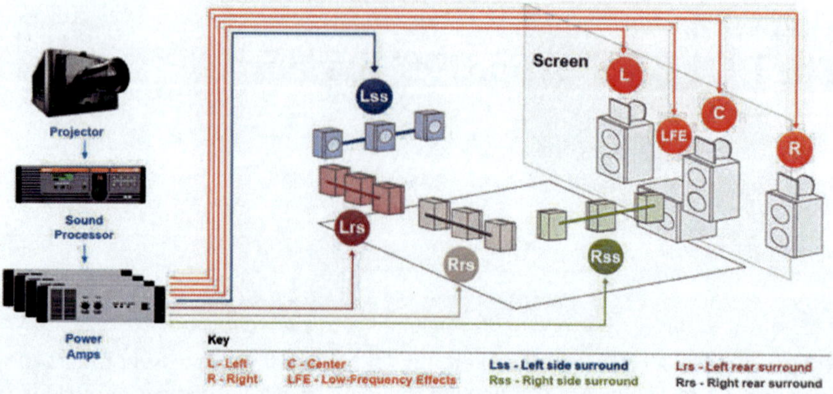

Fig. 4.1 Dolby 7.1 Cinema Surround (from Dolby Atmos 2014a)

soundtracks using effectively the same content for both digital cinema and 35 mm playback.

In 2010, the first step in enhancing digital cinema sound was undertaken with the introduction of Dolby Surround 7.1. This format continues the pattern of increasing the number of surround channels by splitting the existing Left Surround and Right Surround channels into four "zones," as shown in Fig. 4.1. The increased ability for sound designers and mixers to control the positioning of audio elements in the theatre, along with improved panning from screen to surrounds, has made the format a success in both the rapid adoption by content creators and the speed of conversion of theatres. With more than 400 titles and 4000 screens equipped in less than three years after its launch, the success of Dolby Surround 7.1 has indicated a desire within the motion picture industry to embrace new audio technologies.

Throughout the development of Dolby Surround 7.1, Dolby continued to investigate the future of cinema sound, working toward a new audio format. Dolby equipped dubbing theatres with various loudspeaker configurations to determine which loudspeaker locations were compelling to a content creator. Remixed movie content was taken into different auditoriums in various countries, that were equipped with appropriate loudspeaker locations, to determine what was effective in theatres of varying size and shape. Finally, these tests were demonstrated to global exhibitors to gain their feedback on what would work for their customers and what they would be willing and able to install.

This cycle of research, along with Dolby's cinema product and technology footprint, has allowed precise targeting of requirements for the next generation of digital cinema sound, in areas from sound design and editing to rerecording, mastering, packaging, distribution, and replay in theatres. For example, although many cinemas are equipped with inner left (Left Center, or Lc) and inner right (Right Center, or Rc) replay channels, these channels are rarely used because a dedicated five-screen channel mix must be created to support them. However, on larger screens, additional channels could provide both smoother pans and more accurate placement of sound to match the image. Similarly, while the use of surround arrays can arguably create a suitably ambient effect with appropriate content, the introduction of Dolby Surround 7.1 demonstrated that significant improvement in localization of sound results from increasing the number of surround zones within the auditorium.

In parallel to research into a new audio format, Dolby revisited critical areas of the theatrical replay environment, including the technology and standards by which dubbing theatres and cinemas are aligned and monitored. Introduction of a new audio format allows changes to be implemented without breaking compatibility, making it an ideal opportunity to revisit existing standards. In some areas, the current practice is ratified; in others, it is improved upon as technology evolves.

This exhaustive research, along with lessons learned from decades of introducing new cinema sound formats, culminated in Dolby's 2012 introduction of Dolby Atmos as the next generation of sound for cinema. The Dolby Atmos platform encompasses products, services, and technologies that build on existing workflows and technologies to deliver an audio experience well beyond the best available to date.

4.2 Dolby Atmos—An Overview

The Dolby Atmos system includes new authoring, distribution, and playback tools. It also offers a new *Cinema Processor* featuring a flexible rendering engine that optimizes the audio quality and surround effects of the Dolby Atmos soundtrack to the loudspeaker layout and characteristics of each room. In addition, Dolby Atmos has been designed from the ground up to maintain backward compatibility and minimize the impact on the current production and distribution workflows (Fig. 4.2).

Audience Immersion

Three critical elements significantly improve the audience experience over 5.1 and 7.1 systems:

Sounds originating overhead
Improved audio quality and timbre matching.
Greater spatial control and resolution

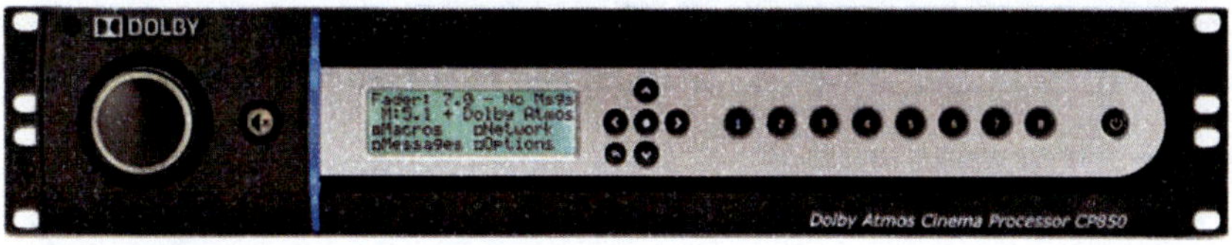

Fig. 4.2 Dolby Atmos Cinema Processor CP850 (from Johnson 2016)

Overhead Sound

In the real world, sounds originate from all directions, not from a single horizontal plane. An added sense of realism can be achieved if sound can be heard from overhead, from the 'upper hemisphere.'

The first example is of a static overhead sound, such as an insect chirping in a tree in a jungle scene. In this case, placing that sound overhead can create greater listener envelopment within the sound scene.

Another example is a helicopter elevating on the screen and flying off over the audience. The use of more discrete surround zones, as in Dolby Surround 7.1, helps achieve the perception of front to back movement, but adding overhead loudspeakers gives a much more convincing impression of the helicopter moving overhead. Finally, adding ceiling loudspeakers makes it possible to move sounds off the walls and into the room (Fig. 4.3).

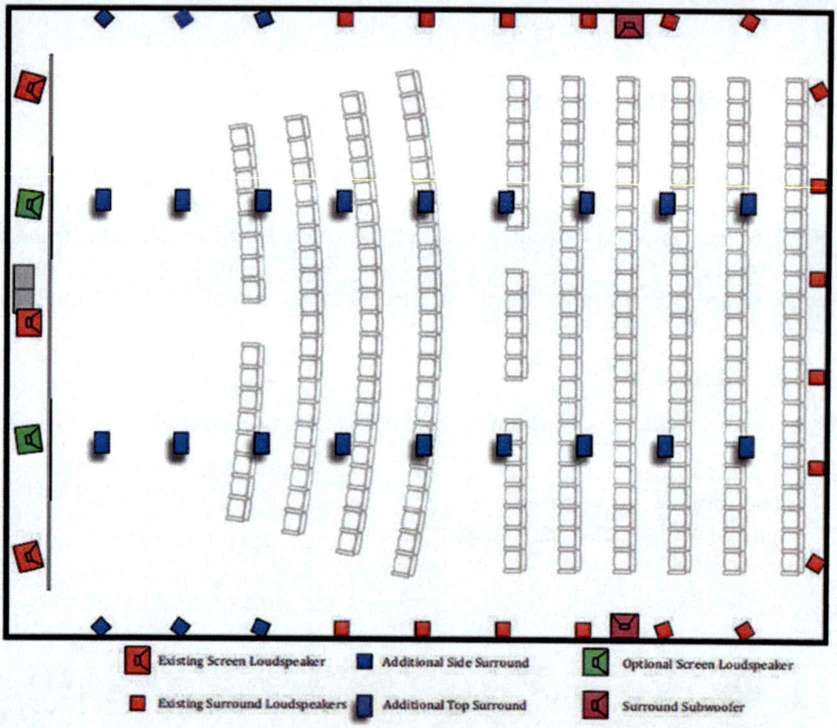

Fig. 4.3 Dolby Atmos—basic loudspeaker layout, top view (from Dolby Atmos 2014a)

4.3 Multichannel Speaker-Layout: Improved Audio Quality and Timbre Matching

In addition to the spatial benefits, the Dolby Atmos core audio quality is an improvement over existing state-of-the-art multichannel systems.

Traditionally, surround loudspeakers do not support the same full-range frequency response and level when compared to the screen channels. Also, the calibrated sound pressure level for surround channels in previous multichannel formats is lower than for the screen channels. As a result, any sound panned from the screen to the surrounds drops in level. Historically, this has created issues for mixers, reducing their ability to freely move full-range sounds from screen to room. As a result, theatre owners have not felt compelled to upgrade their surround channel configuration, creating a chicken-and-egg dilemma that has prevented the widespread adoption of higher-quality installations.

The timbral quality of some sounds, such as steam hissing out of a broken pipe, can suffer from being reproduced by an array of loudspeakers. The ability to direct specific sounds to a single loudspeaker gives the mixer the opportunity to eliminate the artifacts of array reproduction and deliver a more realistic experience to the audience.

Dolby Atmos improves the audio quality in different rooms through such benefits as improved room equalization and surround bass management so that the mixer can make use of all the loudspeakers (whether on- or offscreen) without concern about timbre matching (Fig. 4.4).

Top surround speakers (i.e. height speakers) should be installed in two arrays from the screen to the back wall, nominally in alignment with the (additional) Lc and Rc screen channels for a typical auditorium. Height speakers should always be placed

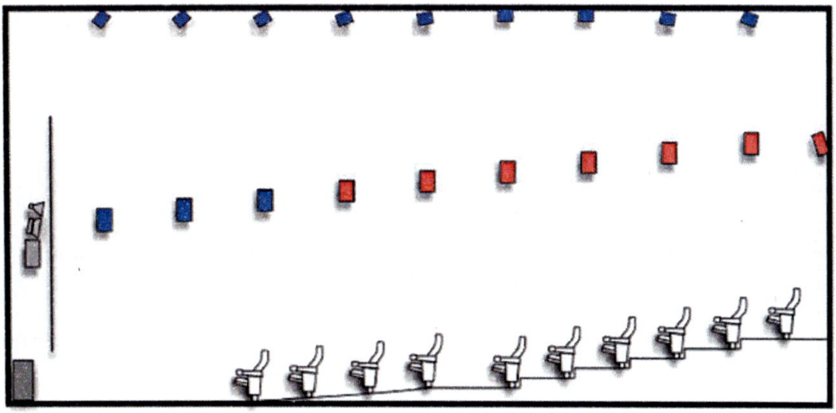

Fig. 4.4 Dolby Atmos—recommended surround- and top-speaker positions and orientation (from Dolby Atmos 2014a)

symmetrically with respect to the center of the screen. Height speakers should have the same design characteristics as the side and rear surround speakers for consistent matching of timbre.

The number and spacing of the height speakers should be based in the position of the side surround speakers. The lateral position of the arrays should be chosen to optimize spatial immersion and uniformity across the listening areas. As stated earlier, placing the height speakers in alignment with the Lc and Rc screen channels will generally produce good results. For rooms in which seating area is significantly wider than the screen, or the height speakers are mounted significantly higher than the top of the screen, it is desirable to have the overhead arrays more widely spaced. The minimum width for spacing of height speakers conforms to the spacing of the Lc and Rc screen speakers. The maximum width between height speakers should be determined based on elevation angle, as follows:

Let E be the elevation angle of the nearest side-surround speaker measured from the reference position, a point two-thirds back in the auditorium in the middle of the seating area. The elevation angle of the corresponding height-speaker array should be greater than or equal to 45° plus half of angle E, as shown in Fig. 4.5. For example, if E is 20°, then the elevation angle of the height-speaker array should be greater than or equal to 55° (Table 4.1).

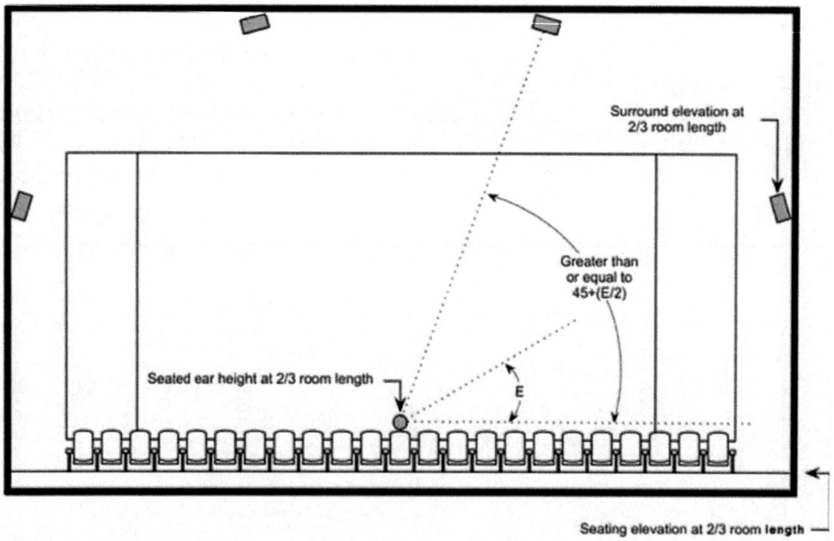

Fig. 4.5 Dolby Atmos—guideline for surround- and top-speaker positions and angling (from Dolby Atmos 2012)

Table 4.1 Dolby Atmos channel abbreviations (Table 1-2, pg 7 modified from "Authoring for Dolby® Atmos™ Cinema Sound Manual", Issue 1, Part No9111800, Software v1.0)

Abbreviation	Channel
L	Left
R	Right
C	Center
S	Surround
LFE	Low-Frequency Effects
Ls	Left Surround
Rs	Right Surround
Lrs	Left rear surround
Rrs	Right rear surround
Lss	Left side surround
Rss	Right side surround
Lts	Left top surround
Rts	Right top surround

4.3.1 Top-Speaker Aiming

In order to provide optimum coverage, the height speakers should be angled laterally (across the auditorium) to a position halfway between the top surround speaker's lateral position and the center line of the auditorium.

Height speakers should be angled longitudinally (along the length of the auditorium) in the same manner as the side-surrounds. Taking 0° as aiming vertically downward, the following rules apply:

- no speaker angel should exceed 45°
- speakers adjacent to the front and rear of the seating area should not exceed 30°
- speakers over the central listening area should be left at 0° (Fig. 4.6).

More information on the positioning and angling of the various speakers in a Dolby Atmos configuration can be found in (Dolby Atmos 2012).

4.3.2 Spatial Control and Resolution

For many years, cinema benefited from discrete screen channels in the form of Left, Center, Right, and occasionally inner left (Lc) and inner right (Rc) channels. These discrete sources have sufficient frequency response and power handling to allow sounds to be accurately placed in different areas of the screen, and to permit timbre matching as sounds are moved or panned between locations.

In a 5.1-channel configuration, the surround zones comprise an array of loudspeakers, all of which carry the same audio information within each Left Surround

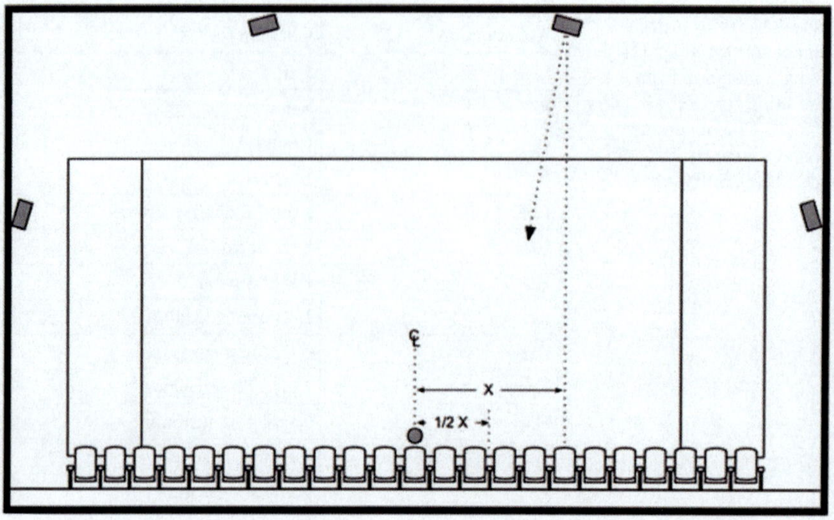

Fig. 4.6 Dolby Atmos—guideline for surround- and top-speaker positions and aiming (from Dolby Atmos 2012)

or Right Surround zone. Such arrays are particularly effective with ambient or diffuse surround effects.

However, in everyday life many sounds originate from randomly placed point sources. Consider the example of being in a restaurant. In addition to ambient music apparently being played from all around, subtle but discrete sounds originate from specific points: a person chatting from one point, the clatter of a knife on a plate from another. Being able to place such sounds discretely around the auditorium can add a heightened sense of realism.

A less subtle example is the sound of a gunshot fired from somewhere off screen. Being able to pinpoint this sound opens new possibilities. The increased resolution of the Dolby Surround 7.1 configuration helps add realism to such effects, but the ability to individually address surround loudspeakers in addition to the 7.1 arrays takes realism to a new level.

A fundamental role of cinema sound is to support the story on the screen. Dolby Atmos supports multiple screen channels, resulting in increased definition and improved audio/visual coherence for onscreen sounds or dialogue. The ability to precisely position sources anywhere in the surround zones also improves the audio/visual transition from screen to room. If a character on the screen looks inside the room toward a sound source, the mixer has the ability to precisely position the sound so that it matches the character's line of sight, and the effect will be consistent throughout the audience.

In contrast, in a traditional 5.1 or Dolby Surround 7.1 mix, the effect would be dependent on a viewer's seating position. Increased surround resolution creates new opportunities to use sound in a room-centric way. This approach is an important innovation, quite distinct from the traditional approach in which content is created assuming a single listener at the 'sweet spot' (that is, the ideal listening position).

What immersive audio formats have in common is the addition of an array of height speakers: and the concept of object-based mixing using metadata for the 3D pan information. The „Atmos " system by Dolby was one of the first 3D audio technologies to launch and—as it seems—is the most common format of them all at the moment.

Dolby Atmos can drive up to 64 discreet speaker channels, however, for domestic applications and premixing it can be scaled right down as far as 5.1.2. The '5' refers to the number of traditional speakers, '1' the number of subwoofers and '2' the number of overhead speakers. Different to many previous Dolby Surround technologies, the surround channels are all full range. A very typical smaller home or premix setup is a 7.1.4 configuration, which takes your traditional 7.1 speaker layout (left, center, right, left side, right side, left back and right back) and adds a quad array of height channels (left top front, right top front, left top rear and right top rear) to give that immersive experience.

4.4 Objects and Metadata

To accurately place sounds around the auditorium, the sound designer or mixer needs more control. Providing this control involves changing how content is designed, mixed, and played back through the use of audio objects and positional data. Audio objects can be considered as groups of sound elements that share the same physical location in the auditorium.

Objects can be static, or they can move. They are controlled by metadata that, among other things, details the position of the sound at a given point in time. When objects are monitored or played back in a theatre, they are rendered according to the positional metadata using the loudspeakers that are present, rather than necessarily being output to a physical channel.

Thinking about audio objects is a shift in mentality compared with how audio is currently prepared, but it aligns well with how audio workstations function. A track in a session can be an audio object, and standard panning data is analogous to positional metadata. In this way, content placed on the screen might pan in effectively the same way as with channel-based content, but content placed in the surrounds can be rendered to an individual loudspeaker if desired.

Although the use of audio objects provides desired control for discrete effects, other aspects of a movie soundtrack also work effectively in a channel-based environment. For example, many ambient effects or reverberations actually benefit from being fed to arrays of loudspeakers. Although these could be treated as objects with sufficient width to fill an array, it is beneficial to retain some channel-based functionality.

Dolby Atmos therefore supports 'beds' in addition to audio objects. Beds are effectively channel-based submixes or stems. These can be retained as separate bed

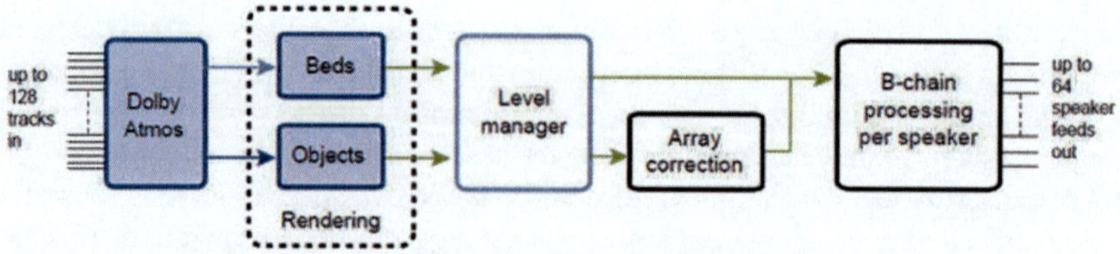

Fig. 4.7 Dolby Atmos Rendering—block diagram (from Dolby Atmos 2014a)

stems through the mixing process; they are combined into a single bed as part of the print-master process. These beds can be created in different channel-based configurations, such as 5.1, 7.1, or even future formats such as 9.1 (including arrays of overhead loudspeakers) (Fig. 4.7).

There are two distinct audio track types when mixing in Dolby Atmos. First you have the main 9.1 bed, on top of which sit your objects. You would treat this similarly to your 7.1 mix of old, but with the addition of left and right overhead arrays. As Dolby Atmos allows up to 128 tracks to be packaged,—in addition to the 9.1 bed—you can have 118 simultaneous mono objects, each of which can be placed individually within a virtual 3D space as automation metadata. This 3D model is then rendered down to produce a mix specifically for the individual monitoring environment, based on the number and position of speakers in your room and the metadata of the objects.

"… The renderer takes these audio tracks and processes the content according to the signal type. Bed channels are mapped to individual loudspeakers (in the case of the screen channels) or loudspeaker arrays. Objects are positioned within the room, and rendered in real time based on the physical location of the loudspeakers. The Dolby Atmos cinema processor assigns delays and equalization to channels, objects, and loudspeakers for optimal playback quality and consistency. The Dolby Atmos cinema processor supports rendering of these beds and objects for up to 64 loudspeaker outputs.

The rendering algorithm also takes into account the power handling and frequency response of the system loudspeakers. Additionally, the support for bass management of the surround loudspeakers through the installation of optional rear subwoofers allows each surround loudspeaker to achieve improved power handling and potentially use smaller cabinets.

Finally, the addition of side surround loudspeakers closer to the screen than has been the practice for previous formats ensures that objects can smoothly transition from screen to surround. It is important to note that these additional side surround loudspeakers are not used to replay content destined for a surround array (for instance, in Dolby Surround 7.1 rendered output, or in a 5.1 bed as part of a Dolby Atmos mix), because this would compromise the experience of using a sidewall array.

"… This metadata gives the mix engineer control of the position within the X, Y and Z axes, as well as the size of the object and the speaker zones to isolate an object to a particular array of speakers, for example the overheads.

When automating the position there are three *Elevation Snap* modes available, which automatically control the Z axis by altering the shape of the room. Ceiling Elevation is flat in the back 80% of the room, but curves down towards the screen in the front 20%. *Sphere* gives the room a domed ceiling and *Wedge* gives the room a peaked ceiling.

4.5 Workflow Integration—In the Dubbing Theatre

"... Dolby Atmos technology integrates seamlessly into existing postproduction workflows: the hybrid model of beds and objects allows most sound design, editing, premixing, and final mixing to be performed in the same manner as they have been in previous formats.

Plug-in applications for digital audio workstations, along with software updates for most of the major large-format film mixing consoles, allow existing panning techniques within sound design and editing to remain unchanged. In this way, it is possible to lay down both beds and objects within the workstation in 5.1-equipped editing rooms.

Object audio and metadata are recorded in the session in preparation for the premix and final-mix stages in the dubbing theatre.

Metadata is integrated into the dubbing theatre console surface, allowing the channel strips' faders, panning, and audio processing to work with both beds or stems and audio objects. The metadata can be edited using either the console surface or the workstation user interface, and the sound is monitored using a Dolby Rendering and Mastering Unit (RMU).

The bed and object audio data and associated metadata are recorded during the mastering session to create a 'print master', which includes a Dolby Atmos mix and any other rendered deliverables (such as a Dolby Surround 7.1 or 5.1 theatrical mix). This Dolby Atmos print-master file is wrapped using industry-standard Material Exchange Format (MXF) wrapping techniques, and delivered to the digital cinema packaging facility using standard DCP techniques that allow file validation prior to packaging (Fig. 4.8).

4.5.1 Dolby Certification

'... If you are mixing in a Dolby Atmos-certified theater, you will have a Dolby Rendering and Mastering Unit (RMU). It is the 'brains' behind the platform and provides the rendering engine for the mix stage. Each RMU is programmed by Dolby with the specific configuration information of the room it is rendering for. This includes the amount of surround speakers and subs, their location, corrective EQ and level control in order to have a calibrated monitoring environment. It is also where your *Print Master* is created for embedding into your *Digital Cinema Package*

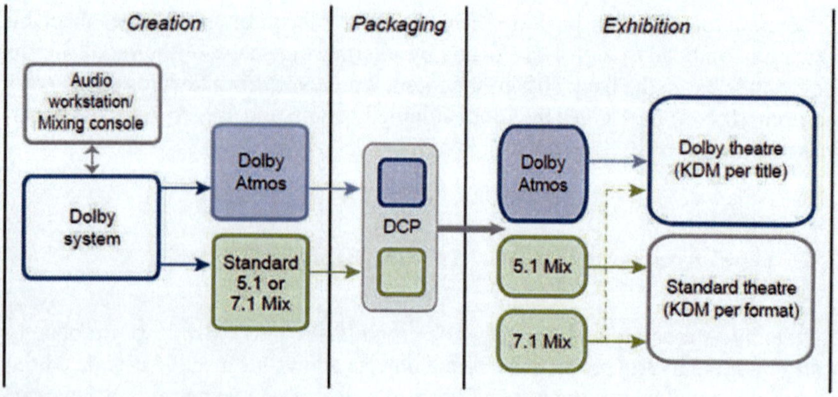

Fig. 4.8 Dolby Cinema audio workflow diagram (from Dolby Atmos 2014a)

(DCP). Physically, your 9.1 bed and 118 objects are routed from your DAW or mixing console into the RMU via two MADI connections. The metadata is provided over a network connection from the chosen host. This could be a traditional console—such as a Harrison MPC, AMS Neve DFC or Avid System 5—but equally could be done directly within your DAW. Fairlight 3DAW and Avid Pro Tools both have Dolby Atmos integration, with Steinberg providing their implementation in Nuendo.

4.5.2 *Packaging*

Dolby Atmos print-master files contain a Dolby Atmos mix. The main audio mix can be rendered by the RMU in the dubbing theatre, or created by a separate mix pass if desired. The main audio mix forms the standard main audio track file within the DCP, and the Dolby Atmos mix forms an additional track file. Such a track file is supported by existing industry standards, and is ignored by Digital Cinema Initiatives (DCI)–compliant servers that cannot use it.

4.5.3 *Distribution*

The Dolby Atmos packaging scheme allows delivery of a single SMPTE-standard DCP to the cinema, which contains both main audio and Dolby Atmos track files. A single key delivery message (KDM) targeted to the cinema media block enables controlled playback of the content, and a DCI-compliant server with appropriate software can play the composition.

4.5.4 *In the Cinema*

A DCP containing a Dolby Atmos track file is recognized by all servers (with appropriate software) as a valid package, and ingested accordingly. In theatres with Dolby Atmos installed, the Dolby Atmos track file is ingested into the server and, during playback, is streamed to the Dolby Atmos cinema processor for rendering. Having both Dolby Surround 7.1 (or 5.1) and Dolby Atmos audio streams available, the Dolby Atmos cinema processor can switch between them if necessary.

4.6 Audio Postproduction and Mastering

Consider the workflow in audio postproduction. There are many steps, some of which occur in parallel, that lead to the creation of a final mix. Three main categories of sound are used in a movie mix: dialogue, music, and effects.

Effects consist of groups of sounds such as ambient noise, vehicles, or chirping birds—everything that is not dialogue or music. Sound effects can be recorded or synthesized by the sound designer or can originate from effects libraries.

A subgroup of effects known as Foley, such as footsteps and door slams, are performed by Foley actors.

Dolby sound consultants work globally on all film soundtracks using Dolby technologies, and continue to provide services in all aspects of the audio postproduction workflow. The following sections outline initial integration of Dolby Atmos into a feature film.

4.6.1 *Production Sound*

Sound is recorded on set, and hundreds of sound files are created. Spotting sessions determine which files, including dialogue or Foley content, are of acceptable quality.

4.6.2 *Editing and Premixing*

Dialogue

Production dialogue that is not considered usable is rerecorded in ADR (automated dialogue replacement or additional dialogue recording) sessions. The dialogue editor uses both production dialogue and ADR, and the dialogue mixer creates dialogue premixes containing mono dialogue tracks and several channel-based beds of 'loop group', such as crowd noise. At this point, dialogue that would benefit from being placed or panned throughout the auditorium is marked as an object and panned accordingly.

Foley and Effects

The Foley editor uses production and recorded effects to create several channel-based beds of Foley. Any Foley that would benefit from being placed precisely in the auditorium is marked and panned as an object.

The effects editor uses designed and library sound effects to create what could be hundreds of sound effects elements and beds of ambiences. The effects mixer uses these sessions, along with the Foley content, to create effects premixes of both individual tracks and channel-based beds. Again, any suitable effects are identified and positioned as objects. Effects may be further split into groups such as atmospheres, crowds, and movements, such as rustling cloth.

Music

Music is mixed by a scoring mixer and passed to the music editor and music mixer for creation of music premixes, which can again consist of tracks and channel-based beds. The use of full-range surround loudspeakers allows mixers to move music offscreen to the surround zones and maintain the same timbre and fidelity as the screen loudspeakers.

Mixers have also found that assigning tracks as objects and moving the objects offscreen will change the perception of the music size and can enhance the audience's sense of envelopment. Additionally, moving music offscreen frees up the screen loudspeakers for audio effects and can help clarify dialogue tracks.

4.6.3 Final Mixing

All of the music, dialogue, and effects are brought together in the dubbing theatre during the final mix, and the rerecording mixers use the premixes along with the individual sound objects and positional data to create stems as a way of grouping (for example, dialogue, music, effects, Foley, and background). In addition to forming the final mix, the music and effects stems are used as a basis for creating dubbed foreign-language versions of the movie.

Each stem consists of a channel-based bed and several audio objects with metadata. Stems combine to form the final mix.

Using object panning information from both the audio workstation and the mixing console, the RMU renders the audio to the loudspeaker locations in the dubbing theatre. This rendering allows the mixers to hear how the channel-based beds and audio objects combine, and also provides the ability to render to different configurations. In this way, the mixers retain complete control of how the movie plays back in each of the scalable environments supported by Dolby Atmos (Fig. 4.9).

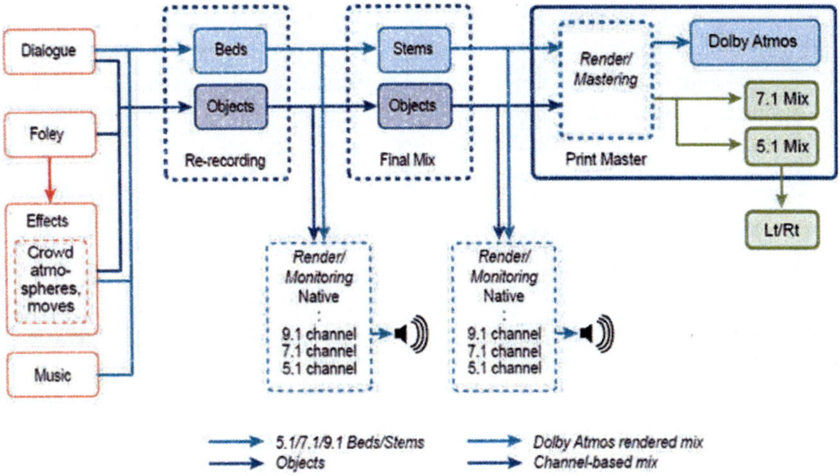

Fig. 4.9 Dolby Atmos—audio post-production workflow diagram (from Dolby Atmos 2014a)

4.6.4 Mastering

During the mastering session, the stems, objects, and metadata are brought together in a Dolby Atmos package that is signed off in the dubbing theatre and is carried through to exhibition in the cinema.

The RMU can render the necessary channel-based mixes, thereby eliminating the need for additional workflow steps in generating existing channel-based deliverables. The audio files are packaged using industry-standard MXF wrapping techniques to minimize the risk of changes, and delivered to the digital cinema packaging facility.

As has been standard practice for several decades, the dubbing theatre is equipped and calibrated by Dolby sound consultants in exactly the same manner as the playback theatres to ensure complete confidence that what is created in the studio will translate predictably to the cinema.

In addition to rendering channel-based theatrical deliverables, the Dolby Atmos master file can be used to generate other deliverables, such as consumer multichannel or stereo mixes.

4.6.5 Digital Cinema Packaging and Distribution Audio File Delivery

The Dolby Atmos audio files delivered to the packaging facility can be imported into an appropriate digital cinema packaging system, such as the Dolby Secure Content Creator (SCC2000), to create a DCP. The audio track files may be locked together to help prevent synchronization errors with the Dolby Atmos track file that was signed off in the dubbing theatre.

The packaging system can also respond to data in the print-master file, such as first frame and last frame of action, to ensure accurate synchronization of sound to picture as was signed off in the dubbing theatre.

4.6.6 Track File Encryption

Upon creation of the DCP, the main audio MXF file (with appropriate additional tracks appended) is encrypted using SMPTE specifications in accordance with existing practice. The Dolby Atmos MXF file is packaged as an auxiliary track file, and is optionally encrypted using a symmetric content key per the SMPTE specification" (Fig. 4.10).

4.7 Practical Experience: John Johnson on the Process of Dolby Atmos Mixing

" … You can no longer get Dolby certification for a new 5.1/7.1 theater because Dolby is actively pushing its new technology. It is also fair to say that not everyone has a room suitable for a full-blown Dolby Atmos-certified theater due to either the size of room, a lack of ceiling height, not meeting certain acoustic properties (reverb time, for example) or even just the potentially prohibitive cost of the additional

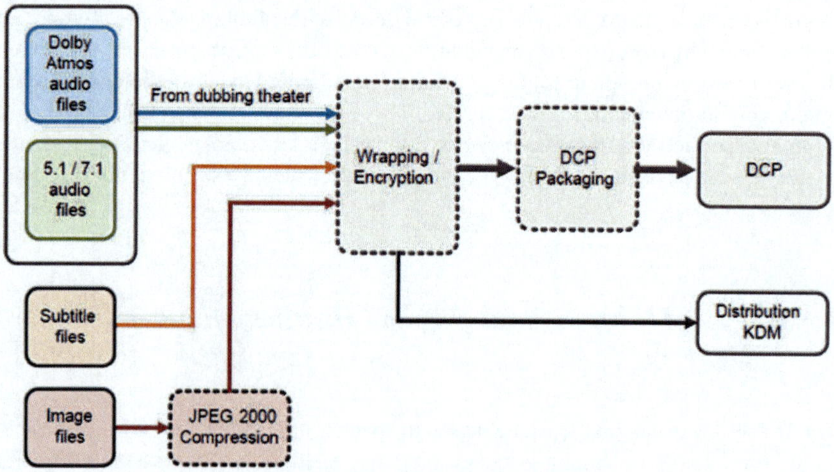

Fig. 4.10 Dolby Atmos—Digital Cinema Packaging Workflow (from Dolby Atmos 2014a)

speakers. Some facilities have several 5.1/7.1 equipped rooms, but only one Dolby Atmos-certified mix stage.

In these cases, people still want to be able to premix for Dolby Atmos prior to taking the mix into a certified room to final mix. Thankfully, there is a workflow available for just such cases.

4.7.1 The Local Renderer

Provided free of charge (subject to application) by Dolby, the local renderer is a software-based rendering option that can handle up to 16 speaker channels in a variety of configurations. This gives access to the same tools in a premix or track-lay room, enabling you to place the audio within a virtual 3D space without necessarily monitoring it. For example, I can build my session and assign my sounds as objects, then record complex pan moves around the overheads whilst only monitoring a 5.1 downmix. Equally, I could have a small Dolby Atmos setup such as a 7.1.4, which would allow me to correctly monitor my 9.1 bed as well as a basic quad array of overheads. When I then take my project into a certified theater, all of my object metadata translates across seamlessly.

When working with the *Local Renderer*, the Dolby Atmos plugin suite provides a series of AAX plugins for your Pro-Tools session as well as a standalone bit of software called '*Dolby Atmos Monitor*', which can also be utilized with the RMU (Fig. 4.11).

Dolby Atmos Monitor offers you an overview of your objects' audio activity and their position within the 3D space. The rendering runs in the background with discreet bed and object audio passing to it from the Pro Tools mixer via a set of plug-ins and then rendered content returned back to Pro-Tools via another set, allowing you to route it out to your monitoring. Finally, there is the Dolby Atmos *Panner* Plug-in, which you put on each of your object tracks in order to generate the positional metadata, which is again sent to the renderer. It does not allow you to create DCP, however this metadata is carried in your Pro-Tools session automation into the main mix stage, at which point it can be authored into the RMU print master and generate a DCP (Fig. 4.12).

4.7.2 Monitor Control

If you opt for the local renderer route, monitor control is an important consideration. There are very few monitor controllers on the market that can handle anything above 5.1, however, there are options. Many people using the local renderer with Pro-Tools already have an Avid ICON series console with an XMON at the heart of their monitoring. Colin Broad of CB Electronics has designed a product called XPand, which can bolt on the additional monitoring channels to XMON's existing

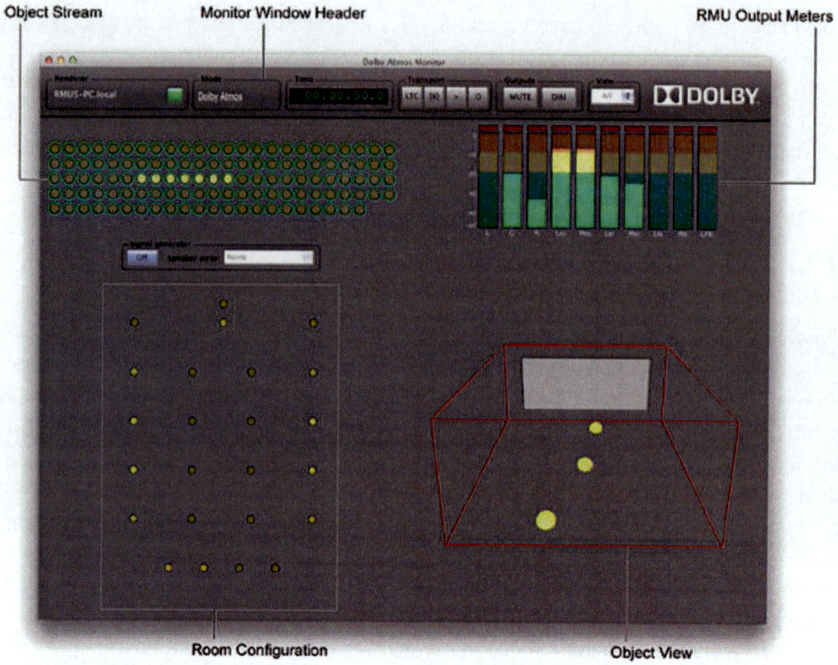

Fig. 4.11 Dolby Atmos—monitor application (from Dolby Atmos 2014b)

7.1 capabilities. This means that you potentially do not have to change too much of your existing infrastructure, but are instead just adding four speakers, an Xpand box and some additional cabling.

You may require some extra analogue outputs, but that depends on your current Pro Tools I/O, however, you are very likely to require an HDX2 system when working in Dolby Atmos due to a lack of voices in Pro Tools. The 256 voices of a single HDX card are eaten up very quickly in a Dolby Atmos project. When working with the RMU and HDX2, two MADI I/Os are essential in order to provide the required 128 channels of audio (9.1 bed plus 118 objects). In many cases a Dolby Atmos theater is likely to have at least two Pro Tools HDX rigs, an HDX2 for playback and an HDX1 running as a recorder or dubber, which will usually be playing back the reference picture as well. When running multiple Pro-Tools rigs (some theaters run with dozens of systems) audio routing can become quite complex. With a traditional DSP console, you would simply pass the audio tracks/stems from each system into the console, which would then mix them and route them on to the RMU, but when 'mixing in the box' and using a controller such as the Avid S6 you have to work a little differently.

Avid S6 also has an 'S6-Joystick Module' available which can be mapped to the X, Y, Z and Size controls within the Dolby Atmos *Panner*.

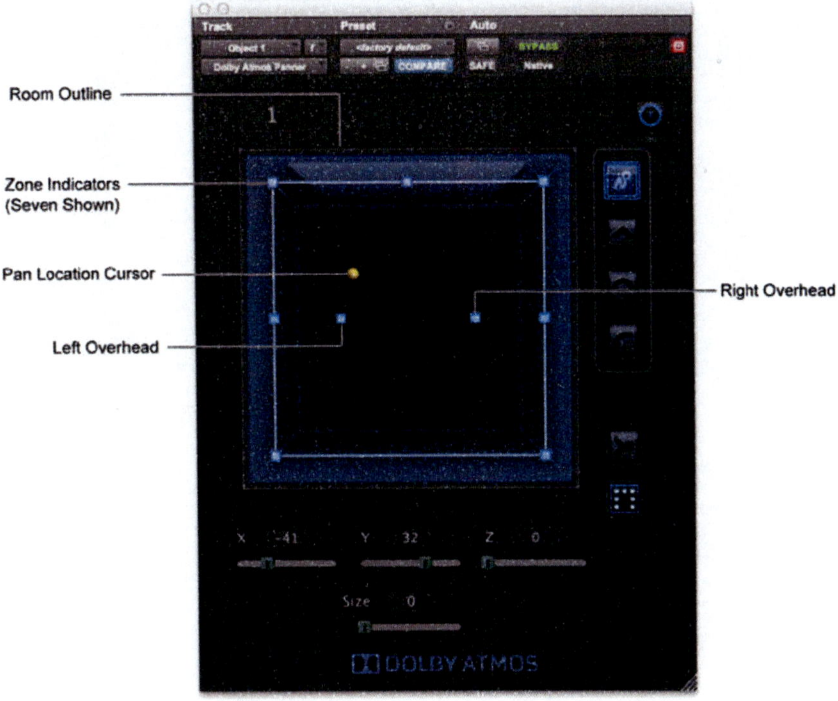

Fig. 4.12 Dolby Atmos—panner plug-in (from Dolby Atmos 2014b)

Additionally, Dolby has made an iPad app and there are also some standalone MIDI/ USB panner options from companies including JLCooper and Schapiro Audio.

In larger theaters and fully certified rooms, it is very common to see large arrays of horn-loaded JBL, Meyer Sound or custom Exigy speakers, however, in the smaller installations there is a large Genelec presence. This is in no small part due to a combination of a wide range of mounting hardware (crucial for ceiling speakers) and a strong offering of speakers in a variety of sizes, which maintain a consistent tone across the range. This means you can drop down a few sizes for your overhead speakers, often buying crucial inches of ceiling clearance.

4.7.3 Upmixing

With wider beds comes more of a requirement for upmixing.

NUGEN Audio's HALO Upmix plugin can handle that and is really intuitive, quick to set up and provides very good sounding results. When all your source material is often only stereo or 5.1 content, the ability to upmix that to your 9.1 bed

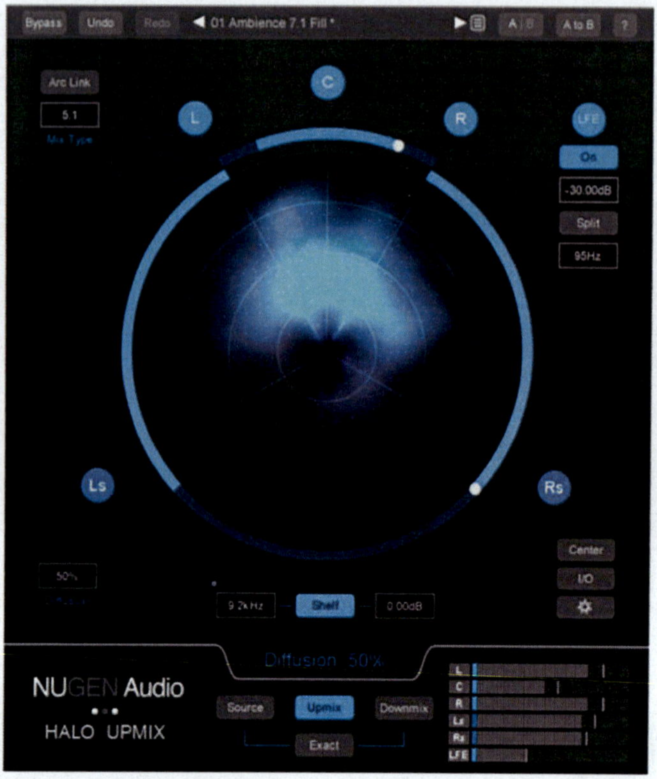

Fig. 4.13 Nugen Audio—'Halo' Upmix-PlugIn

without having to manually pan it anywhere or do too much is very convenient. It saves a lot of time, but still gives the operator the control should he/she need to get in and tweak the balance how much to put into the overhead channels or how wide to pull out the image around (Fig. 4.13).

4.7.4 The Future of Object-Based Audio

The other side of Dolby Atmos is in the music domain: Dolby has evolved its technology to enable Dolby Atmos to be rolled out to DJs and music venues. The music club 'Ministry of Sound' in London was the first venue to install it. It currently has a 60-speaker, 22-channel 'Dolby Atmos' setup in their main room 'The Box'.

The tools that Dolby has produced for the music side are very similar to the post-production tools for film and television, but obviously tailored more to DJs being able to take their tracks and place their stems and instruments all around you.

Object-based audio is progressing into live broadcast as well, though not exclusively for the purpose of 3D Audio. One of the main tricks is actually all about the end user having audio streams to choose from. An example would be a live sports game where you have an unbiased commentary by default, but an option for a home or away-biased commentary. Similarly, there can be options for additional commentary languages. To add to these options, additional control of the crowd mics to opt for more of the home/away fans or a mic closer to the pitch is possible. It is using the same concept of objects and metadata, but this time the objects are user-selectable, with rules dictated at the time of authoring as to how the audio is treated when the user's selection is mixed down in their TV or tuner.

Most home cinema AV receiver manufacturers now produce a few boxes in their range which can decode Dolby Atmos and DTS:X, and there is a large amount of Blu-Ray Disk titles available with immersive audio soundtracks. 3D immersive audio also has real importance when you consider developments in the video domain, including 360° video and virtual reality.

Without immersive audio, I feel that VR is an incomplete experience. Youtube already has full VR implementation, employing Ambisonics as its 3D audio solution. These technologies are now becoming readily available to the consumer. Dolby Atmos is in homes on formats including Blu-Ray, Amazon Prime and Netflix, where previously it was exclusive to only the largest screen at a handful of cinema multiplexes. ... " (from Johnson 2016).

4.8 Dolby Atmos in Live-Sound Reinforcement and Mixing

While all elements of a traditional stereo P.A. are typically time-aligned, that's not necessarily best practice with a Dolby Atmos system, reports Jed Harmsen, Dolby's vice president of cinema and content solutions. "If everything is perfectly time-aligned and you pan, you're not going to get the sensation of panning. You want to ensure the overall production is cohesive, coherent and consistent. But we want to ensure that when the creatives choose to pan elements or add static objects, you notice that effect and it's not getting washed out by the time alignment."

There is a learning curve when first mixing in Dolby Atmos, as Kevin Madigan, Santana's FOH mixer since January 2019, observes: "You don't have a left–right bus to send to anymore. That changes things quite a lot."

The flipside is that assigning elements of the mix as objects offers the mixer a set of tools to be even more creative or to overcome challenges. Madigan assigned audio groups rather than individual inputs as objects. "The [DiGiCo] SD5's capabilities were certainly put to work," he says. "With everything going on, I was up to 41 stereo groups. It's the most complicated DiGiCo session I've ever written."

Santana's percussionists at stage right and left can occupy a lot of real estate in a stereo mix, but Atmos enabled Madigan to widen the soundstage. "Being able to push those wide and along the side walls is a cool and unique thing," he says.

To achieve even greater separation, he says, he could send individual elements to separate speakers. "That makes things easier to deal with instead of trying to do it with master bus compression and EQ and carving out space." Plus, because the high-frequency content of the percussion can be easily localized by the listener, "a little movement in those elements is very obvious," says Madigan.

Similarly, he says, he could bring guitar solos into the room. "I wasn't doing fader pushes to bring guitar solos out. It was just a movement change that brought it forward and out of the mix."

As for getting creative, Dolby Atmos enables a mixer to give the audience an experience a stereo rig simply cannot deliver. "Being able to do a rotation with a B2 and Leslie solo is really cool," he reports.

Yet Madigan's overall philosophy was to create a generally static mix. "You could do dramatic things that were interesting for the audience, but Carlos and the band are the focus. You don't want to do anything that distracts from that."

Dolby's software panner, which is also used in post-production applications, generates the dynamic spatial positioning metadata for each object in the renderer. Looking down the road, that control, currently in "advanced beta" for live use, will likely become more tightly integrated into third-party products.

In the meantime, Dolby is learning from its interactions with mixers how best to optimize the software for live musical and theatrical performances. "We want to make sure that our solution set covers all the bases, while still being flexible and easing the creation process. We want to be able to cover all the bases with a single solution set, but make sure it's still approachable and not too difficult to implement," says Harmsen (from Harvey 2020).

4.9 Practical Experience: Iker Olabe on Music Production in Dolby Atmos

4.9.1 Context

Immersive audio technology called Dolby Atmos was created initially for the film industry. This might explain some of the difficulties in Dolby Atmos music mixing. Early 1900s methods for reproducing movies and music generally used mono single- or multi-point source technology. The 1930s saw the advent of stereo in film and music (see Blumlein 1931; Keller 1981), marking the start of a protracted stereo era for popular music. Dolby surround for film, the first commercial two-dimensional system widely used in theaters and mixing stages, was introduced by Ray Dolby in the 1970s. In 2.0, 5.1 and 7.1 systems, panning can be allocated to a specific loudspeaker. These systems are all referred to as channel-based systems. Stereo is a one-dimensional system, which should be remembered (you receive sound from a single axis) Dolby Atmos is a 3D dimensional technology, preceded by Dolby Surround, which is a bidimensional systems.

4.9.2 Introduction and Software

The goal of Dolby Atmos is to replicate the cinematic 360° audio experience in your living space or on a mobile device. With the use of Atmos, content producers such as sound engineers, cinematographers, and broadcasters may precisely place particular sounds in the soundscape to ensure that you hear the action as it was intended to be heard. We will introduce the Dolby Atmos technology specifics used in music production environments in this chapter. The sound of a pianissimo harp playing at the extreme back of a concert hall or a loud fortissimo gong can both be accurately reproduced via Atmos technology. Due to the 3-axis reproduction technology, Atmos will produce spatially correct sound; in essence, it will fool the brain into believing that the listener is in the heart of the action. Because Dolby Atmos is all about the details, it employs a 7.1.2 or 9.1-bed channel (composed of stationary sounds like background ambience, instrumental accompaniment, FX returns, etc.) to provide that genuine sound. To generate an immersive soundscape, 128 tracks and up to 118 simultaneous sound objects are used. This subject can occasionally be somewhat perplexing due to the abundance of jargon. The phrases immersive audio, 360, and spatial audio, among others, all relate to the same thing, despite the fact that Apple has dubbed their own immersive audio technology after the term spatial audio, which has entered their lexicon.

4.9.3 Dolby Atmos Renderer (DAR)

It is a program used to transform your Dolby Atmos mixes into deliverable formats and monitor, record, and render them. Input and master are the DAR's two primary sources. The portion where signals from your DAW will be watched is referred to as input. All inputs are disarmed when the Master option is used to playback your recorded mixes. Offline rendering and re-rendering of mixes is also done in this part. The output format for distribution, as will be seen later, is typically ADM WAV—Audio Definition Model Wave Format. These mixes can be recorded in real-time. With up to 128 CH (@ 24bit/48 kHz) and substantial metadata required for the Atmos Mix, this is an interleaved .wav file (Fig. 4.14).

4.9.4 Loudspeaker Monitoring and Routing

The DAR will keep an eye on your Atmos mixes. There are monitoring setups available where both speaker configuration and routing are specified. The data supplied includes both an audio stream and metadata, enabling the DAR to create downmixes to 5.1, stereo, etc. in real-time (Fig. 4.15).

Fig. 4.14 Input window overview of the Dolby Atmos Renderer

Four monitoring schemes are shown in Fig. 4.16. To check for format compatibility when mixing, you may switch from 7.1.4 to stereo or 5.1. From the routing pane, as illustrated in Fig. 4.16, you may configure the audio outputs on your interface. When operating in loudspeaker monitoring mode, headphone output is also routed through this window.

This routing would be done in headphone-only mode from the options panel, as seen in Fig. 4.17.

4.9.5 Dolby Atmos Music with Headphones

Dolby Atmos was initially only compatible with a very small number of studios, and the majority, if not all, of the projects were for movies. We should really explore tailoring our Atmos music mixes for headphone listeners more than the various speaker layouts, since we know that more than 75% of our listeners stream their music through headphones. With headphones, we can mix in Dolby Atmos, which is a terrific strategy for musical projects. Binaural rendering was incorporated when Dolby created the Atmos Renderer. This makes it possible to listen to a 7.1.4 Atmos mix through headphones that was converted to an Atmos binaural format. Using the given metadata stream from the audio objects, the DAR performs this in real-time. It is advised to experiment with object positions and binaural render modes to find the optimum values for each component of the mixes in order to increase their accuracy and effectiveness.

One of the general guidelines mentioned for Dolby Atmos music deliveries is as follows:

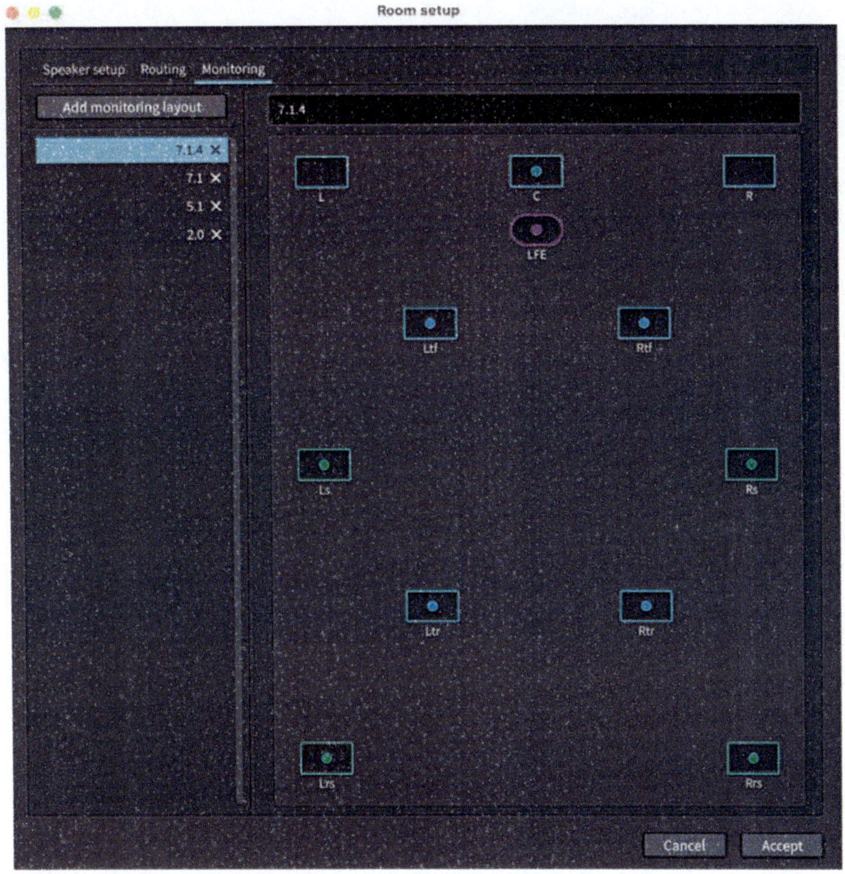

Fig. 4.15 Personalized monitoring layouts

"*All deliverables must have been reviewed and authorized for home listening in a room with at least a 7.1.4-ch speaker setup.*" (see Dolby Knowledgebase 2022, Module 9.3).

Dolby Atmos music is therefore compatible with headphones; however quality checking in a 7.1.4 monitoring system is advised.

4.9.6 Binaural Render Mode

For its headphone experience, Dolby Atmos features special binaural settings. To enhance spatialization, beds and objects can be placed close together, in the middle, farther apart, or off to one side. The last setting will totally disable binaural processing.

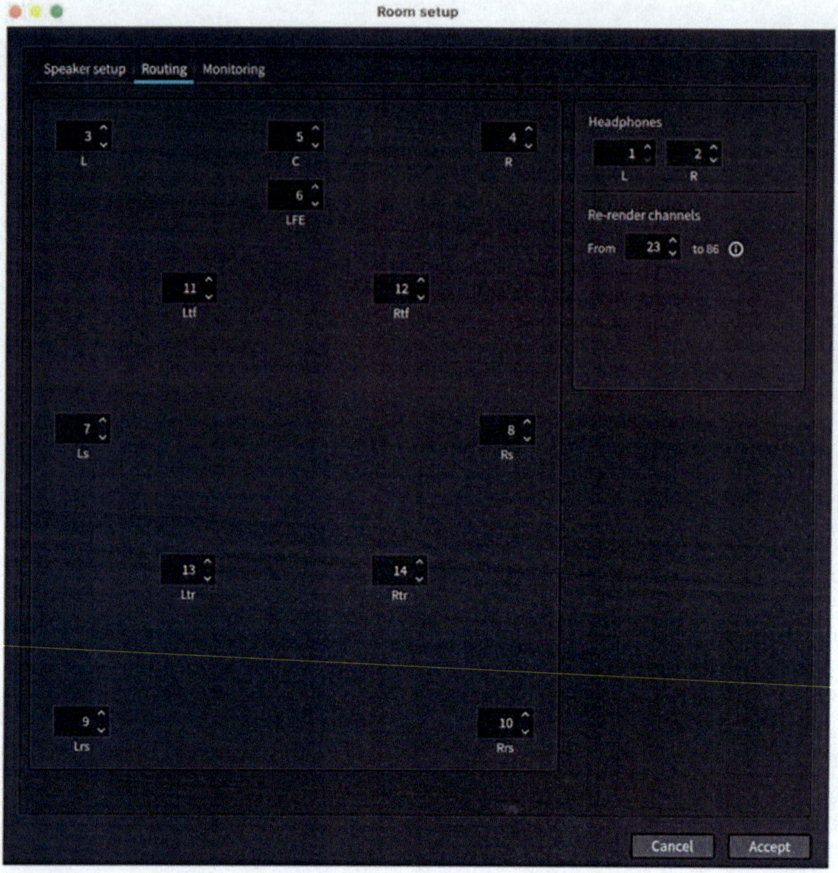

Fig. 4.16 7.1.4 speaker layout output routing

A sound from a monitoring system 1.5 m away would be replayed at mid position. We might walk far to get a wider perspective or near if we wanted to bring our music "closer" (Fig. 4.18).

"*The distribution of objects evenly between close, middle, and distance is a good practice*" (see Dolby Knowledgebase 2022).

4.9.7 Dolby Atmos Personalized Rendering and PHRTF

Dolby recently (March 2022) announced a Beta version of the Dolby Atmos Personalized Rendering software, which uses PHRTF (i.e. 'Personalized Head Related Transfer Function') technology to deliver a tailored immersive mixing experience

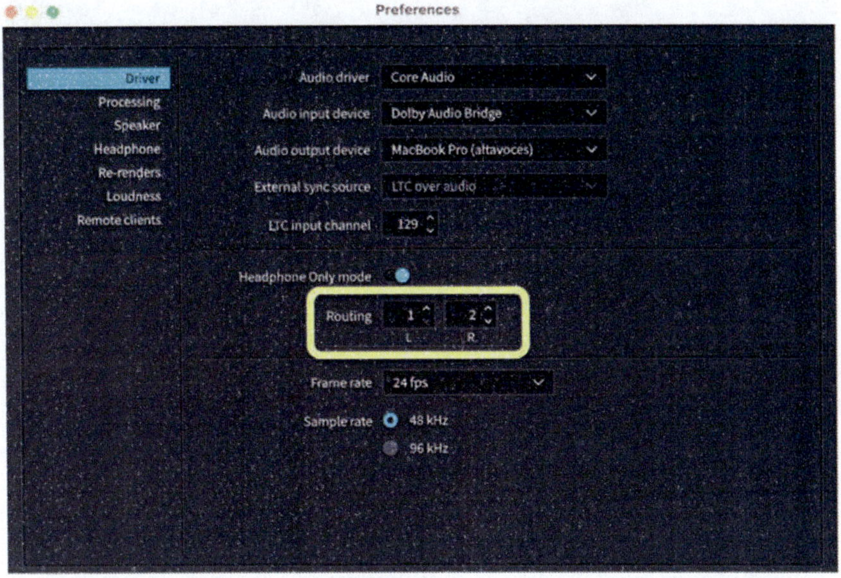

Fig. 4.17 Headphone only mode, output routing

Fig. 4.18 Showing configuration for the binaural render mode

customized to one's anatomy instead of having a 'one size fits all' binaural renderer (Fig. 4.19).

For years, binaural playback has relied on default HRTFs that ignore the diversity of individuals, and this has created wild inconsistency in the way immersive audio is experienced over headphones" (from Dolby Knowledgebase 2022).

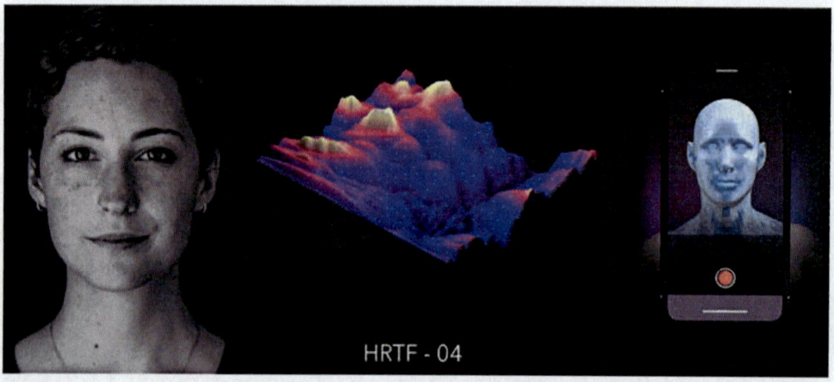

Fig. 4.19 Dolby Atmos personalized rendering beta app (graphics © Dolby Laboratories, Inc)

Fig. 4.20 Selection of the DAR input and output configuration preferences

4.9.8 Renderer Sources

Input and *master* are the DAR's two primary sources.

- The portion where you will record and monitor signals received from your DAW is referred to as 'input'.
- The master option is chosen to generate your finished product and playback your recorded mixes. We can export our ADM WAV for delivery, reference MP4s, or re-renders in formats other than Atmos from the master position.

Note: *A crucial need to check before recording a master file is that the input configuration in the DAR must match the output configuration. To simplify this process Dolby includes a shortcut in their program which says 'copy settings from input to master' or 'copy settings from master to input' (in case the opposite was wanted)* (Fig. 4.20).

4.9.9 Dolby Atmos Music Panner (DAMP)

The positioning of audio objects within a 3D space is done with this AAX, AU, or VST3 plugin in your DAW. Ensure that the Dolby Atmos Renderer is receiving the plugin in the proper manner. DAMP features a sequencer to sync automation of objects with DAW tempo as an example.

You might prefer to use a 3D panner built into some DAWs rather than DAMP, such as Nuendo and Pro-Tools. DAW's like Reaper do not have an integrated 3D object panner till this date (Figs. 4.21 and 4.22).

Fig. 4.21 Dolby Atmos music panner

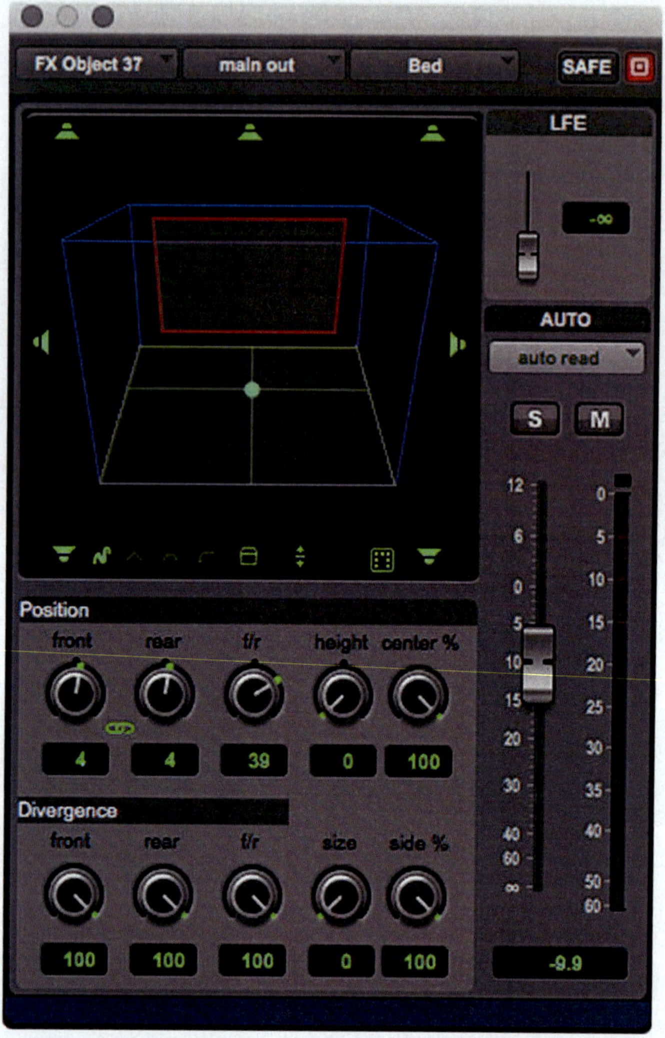

Fig. 4.22 Pro-tools ultimate panner

4.9.9.1 LTC-Generator

To sync your DAW to the DAR, the Dolby LTC generator 'grabs' the timecode from your DAW and allows you to internally pass this signal to the Dolby Atmos Renderer. This will be especially useful for synchronizing the DAR's REC/PLAY command with your DAW (Fig. 4.23).

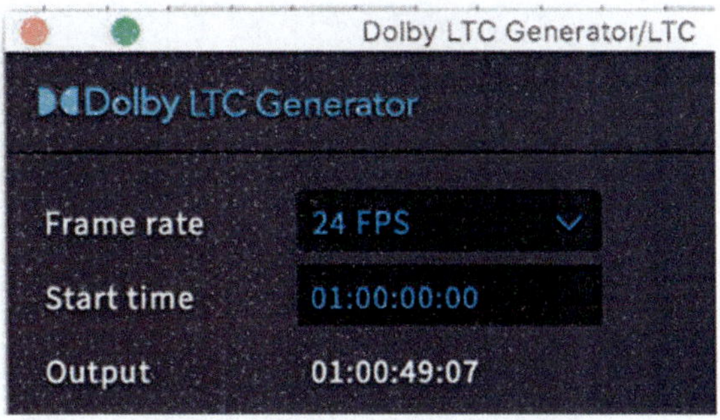

Fig. 4.23 Dolby LTC generator plugin

4.9.9.2 Sampling Rate and Channel Count

Two fixed sampling rates—48 and 96 kHz—are supported by Dolby Atmos sessions and renderers. The frame rate and sampling rate of the DAW session and renderer session will match. 128 input channels are supported by the Dolby Atmos renderer at 48 kHz: 118 objects plus 10 beds at most. 64 input channels are available at 96 kHz, which is essentially 'a cut in half' in terms of objects: a maximum of 10 beds and 54 objects.

"*Starting with 48 or 96 kHz (or higher) sample-rate source content will yield better results than up-sampled content*." (from Dolby Professional Support 2021).

4.9.9.3 Layout (Beds and Objects)

Mono/stereo/surround channels would be routed to buses, auxes, and finally our master channel in our DAW in typical channel-based mixing setups. The master bus would be the source of our last bounce. All of the master's outputs—left, center, right, left surround, etc.—are panning signals to loudspeakers. In Dolby Atmos the approach or work-practice is different: with a few exceptions, Dolby Atmos uses what we refer to as 'object-based mixing'.

Beds

The term 'beds' refers to the actual physical mapping of channels, or what is known as a 'channel-based system', which has been used up to this point with 5.1 and 7.1 systems. For instance, with a 7.1.2 bed, you would be able to route these channels to an additional two Z overhead channels in addition to a 7.1 XY layer with LCR, surrounds, and LFE. Please be aware that the LFE will only accept signals sent from beds. Beds are typically used for sound effects returns, instruments with fixed spatial positions, ambient pads, and other 5.1 or 7.1 surround sound sources without the necessity for precise spatial placement or room movement.

For the time being, these are locked at 7.1.2 since Dolby Atmos does not accept 7.1.4 or 5.1.4 buses as beds; rather, bed outputs are to be the top 10 outputs.

Fig. 4.24 The top row shows the default 7.1.2 bed created by the DAR ("Dolby Atmos Renderer"); (rem.: only the top 30 of all 128 objects are shown)

One fixed bed is automatically established in the Dolby Atmos renderer's (DAR) initial configuration (a maximum of 10 beds can be created) (Fig. 4.24).

Objects

Objects are standalone sources that can be positioned anywhere in space.

These provide a sophisticated metadata feed containing the XYZ coordinates and trajectory across time in addition to an audio stream. The Dolby Atmos Renderer application (hence referred to as DAR) must be used to continuously monitor objects.

The more isolated the sound, the stronger the location impact; they are typically utilized for sound sources that move around the room. This has altered various recording trends, including the usage of microphones with 'narrower' polar-patterns (like super- and hyper-cardioid) in addition to the introduction of new recording techniques. The size of object channels can also be changed, which will alter how the audio signal is perceived. The input matrix for beds and objects is shown in the image below. This is a 48 kHz session because we can see a 128-input panel. The default 7.1.2 bed is plainly selected in the top row. The blue circles indicate that these inputs are receiving metadata from the DAW 3D panner, and there are objects routed from inputs 11 to 24. Input 18 indicates that it is also receiving an audio stream in green (Fig. 4.25).

3D visualization

The DAR displays the locations of your things inside the 'room' in real-time. This provides extremely accurate visual feedback regarding the location of audio elements in space. It can be fairly complicated when the population of the mix grows, but you

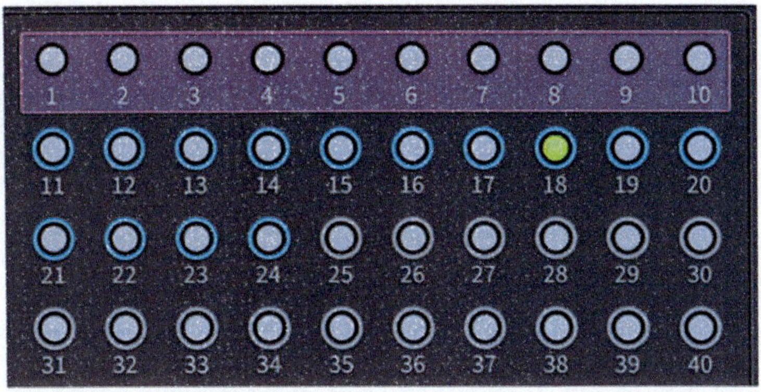

Fig. 4.25 The bed and objects input matrix (rem.: only the top 40 of all 128 objects are shown)

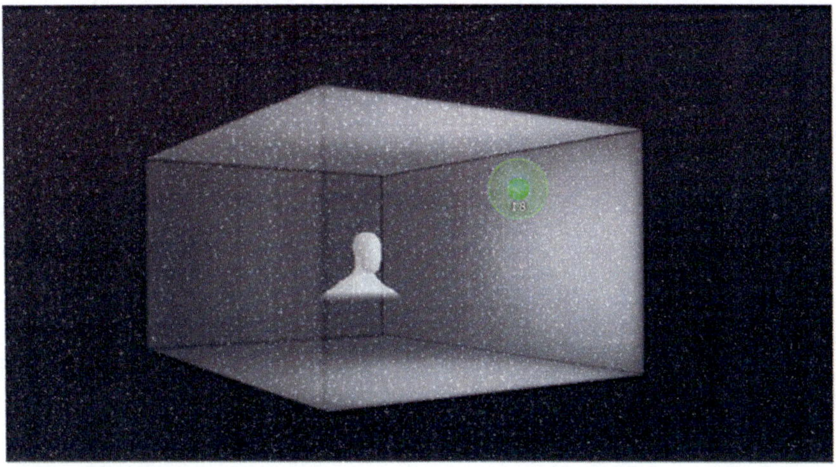

Fig. 4.26 3D visualization view in the DAR showing one audio object

can label the objects to display their name or number. One item is depicted in Fig. 4.26, identified by its number (18) (Fig. 4.27).

4.9.9.4 Dolby Atmos DAW Routing

Up to 128 channels can be sent to the DAR from your DAW (48 kHz). Your DAW is supposed to send LTC to the DAR via channel 129. The renderer's first 10 inputs are by default allocated to a 7.1.2 Bed. The I/O settings of the DAW also specifies these channels. A narrower Bed-channel width can be utilized instead of the Renderer's default single 7.1.2 Bed-channel (Figs. 4.28 and 4.29).

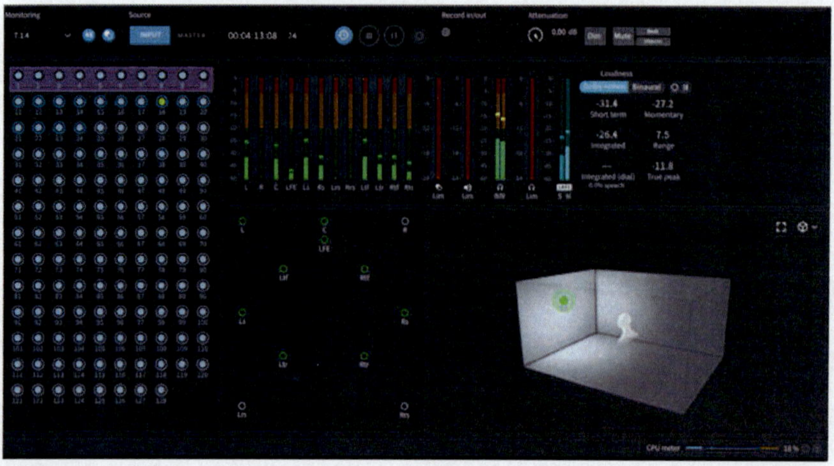

Fig. 4.27 Overview of the input section of the DAR with a single object playing through it in real-time

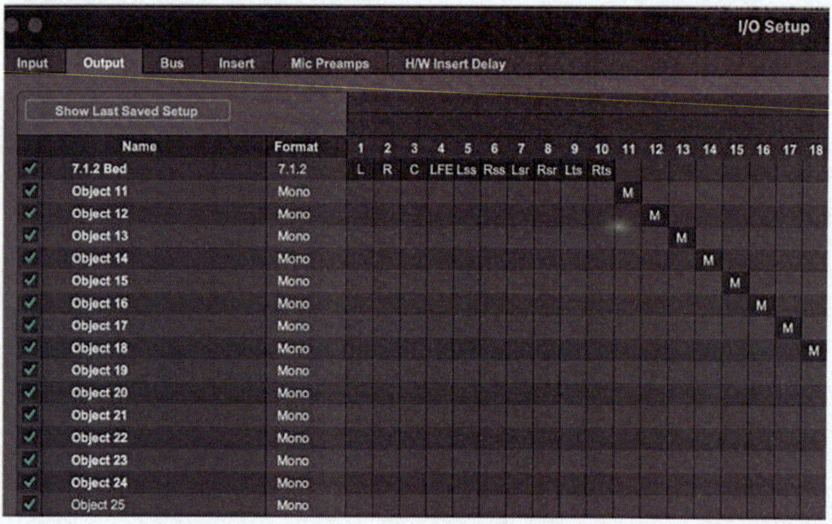

Fig. 4.28 Output matrix in ProTools DAW

4.9.9.5 Audio Bridge

The Dolby Atmos Renderer is frequently used in music production on the same computer as our DAW-software. Dolby includes a virtual audio driver called the Dolby Audio Bridge with 130 virtual channels of inputs and outputs to communicate between both apps as a way to allow internal routing between our DAW and DAR.

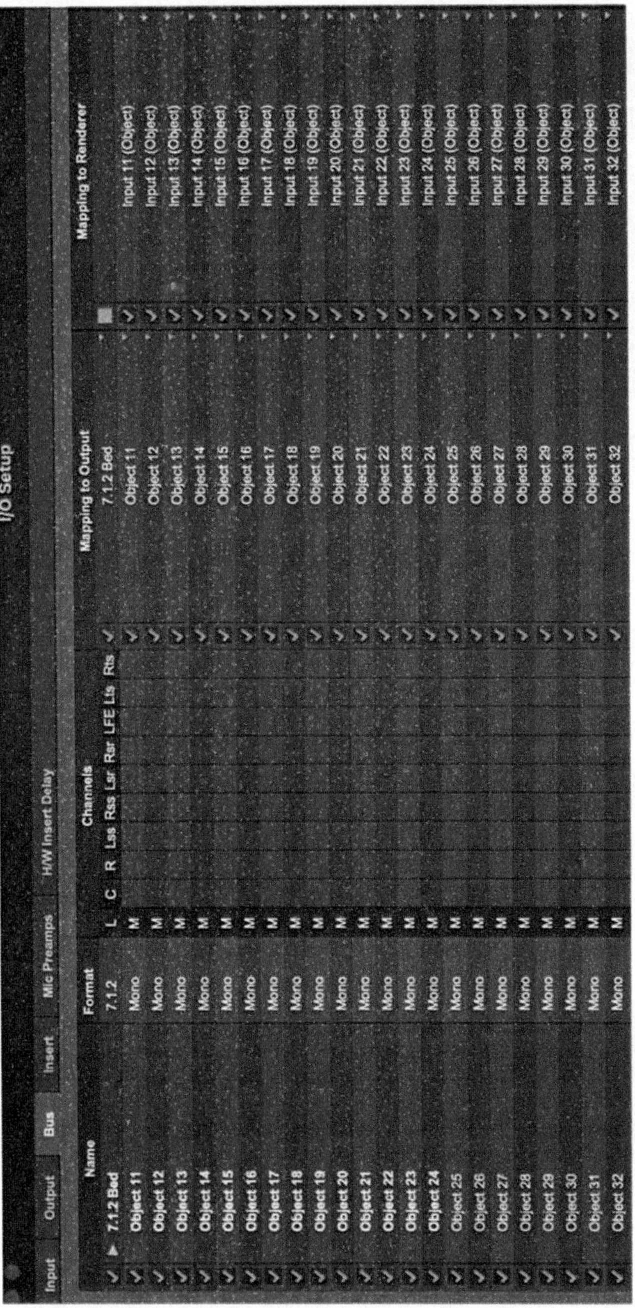

Fig. 4.29 Output "mapping to Renderer" in ProTools DAW

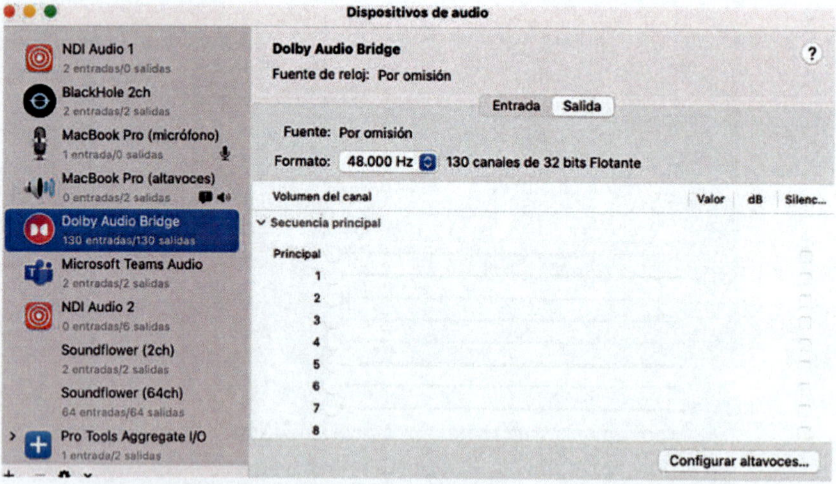

Fig. 4.30 Screenshot showing the Dolby Audio Bridge in the Mac OS config window

Your DAW must have chosen DAR as its primary input and the Dolby Audio bridge as its primary audio interface. The standard hardware interface for monitoring should be assigned to the DAR's main output. Only Macs can use this function (Figs. 4.30, 4.31 and 4.32).

4.9.9.6 Basic Setups for Dolby Atmos Productions

Dolby Atmos Production Suite's Internal Rendering

In music productions with fewer than 100 tracks, this configuration is the most typical. It allows for versatility by allowing the DAW and Dolby Atmos renderer to run simultaneously on the same computer. The program used is "Dolby Atmos Production Suite" (DAPS), which is the Mac-only version of the software. "Dolby Atmos Mastering Suite" is a comparable Windows and Mac version that has some features in common with the mastering license and others that are unique. Internal rendering offers advantages and disadvantages. The following are some of the primary benefits:

- An Atmos system can be set up using just one computer and audio interface
- More affordable choice
- Configuration setting is extremely flexible and recallable with ease

One of the biggest drawbacks is that it will put a strain on your computer's processing power, especially during lengthy sessions with active pan automation. Also, there is no speaker- equalization feature built into the renderer, so you will need hardware DSP or a strong hardware interface like the DAD AX32 or AX64 with SPQ signal processing cards to help. When monitoring in bigger venues, the

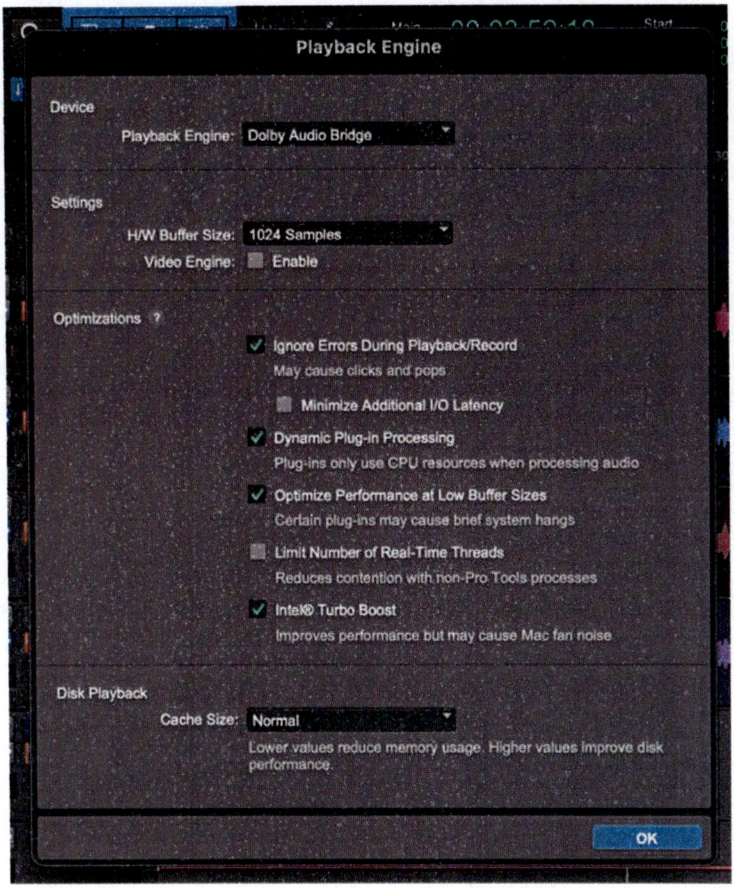

Fig. 4.31 ProTools DAW showing playback engine/Dolby Audio Bridge

lack of a speaker array mode could be a problem. Let's note that the majority of the drawbacks mentioned above would be a problem mostly for film or music productions that require a lot of work because there would be a general system overhead. For mixing in Atmos while away from the studio with a laptop and headphones or for bedroom producers, internal rendering with DAPS offers a very dynamic and extremely productive process.

Figure 4.33 shows a block diagram of a Macintosh computer with the DAW and DAPS installed, (from left to right). The virtual driver "Dolby Audio Bridge" and the Dolby Atmos renderer (DAR) are two of the programs in the package. The DAW's primary playback driver, seen in Fig. 4.30, is the Dolby Audio Bridge. Now, 128 outputs from the DAW's I/O configuration can be directed to the DAR+ Ch 129 for LTC transmission. In order to monitor the signals through an Atmos 7.1.4 loudspeaker system and two channels of Atmos binaural monitoring for headphones, the DAD

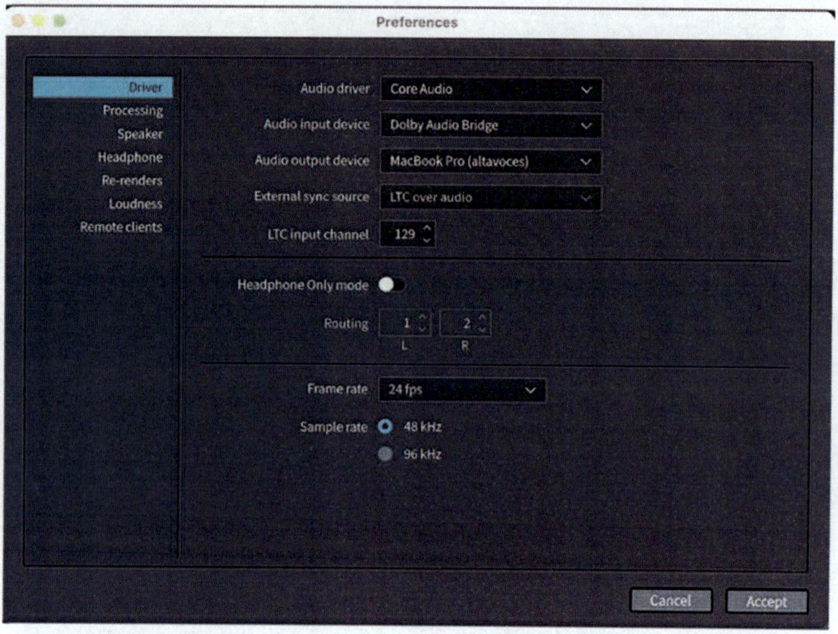

Fig. 4.32 DAR showing input device selected as/Dolby Audio Bridge

AX32 interface (DVS Dante virtual soundcard) is used as the main hardware output of the DAR.

External Rendering

This configuration also adds a second computer that is only used to run the Dolby Atmos Renderer with either a DAMs 5 or DAPs 6 license, depending on requirements. Due to the numerous tracks handled by the MADI, Dante, or Ravenna protocols, a hardware interface will be required for each computer. The management of the CPU is one of this system's primary advantages. Since the Dolby Atmos Renderer will have its own dedicated computer and audio interface, HDX systems (such as those used by ProTools Ultimate) take advantage of their own technology to ensure that AAX plugins and tracks operate without interruption. Larger sessions can be run without any bugs or technical problems thanks to external rendering. If a DAW has MADI, Dante, or Ravenna hardware interfaces, we can use those with all other DAWs (Figs. 4.34, 4.35 and 4.36).

Dolby Atmos Mastering Suite: In addition to the aforementioned rendering systems, the Dolby RMU (Rendering and Mastering Unit) for Home Entertainment (Nearfield monitoring) and the Dolby RMU for Theatrical (not for sale, only available through a special licensing agreement and the presence of a Dolby Engineer) are also used. The RMU performs all processing by creating the Dolby Atmos file and producing the

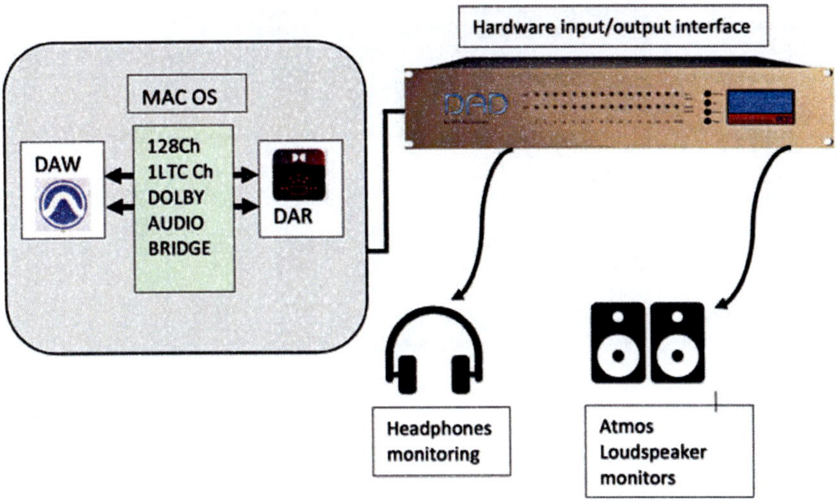

Fig. 4.33 Internal rendering config using the Dolby Audio Bridge

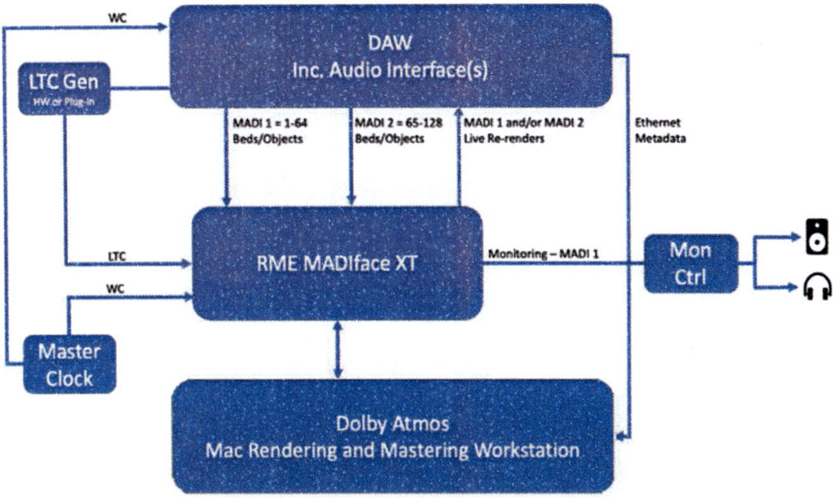

Fig. 4.34 External rendering with MADI using a Mac computer (from Dolby Technologies Inc.)

object-received information. This version additionally offers speaker combinations and room calibration (Fig. 4.37).

Integrated Rendering

Some DAWs integrate Atmos rendering in their programs to make the mixing and setting process easier. As there won't be any other apps to boot or route, there will

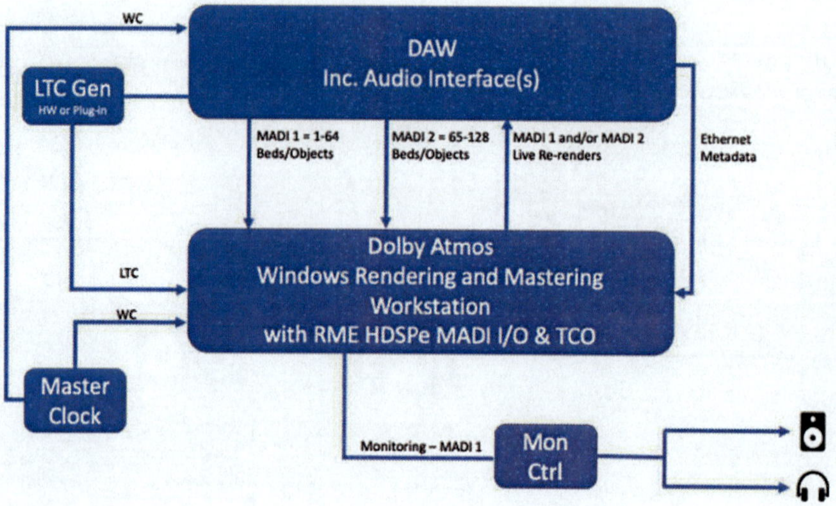

Fig. 4.35 External rendering with MADI using Windows computer (from Dolby Technologies Inc.)

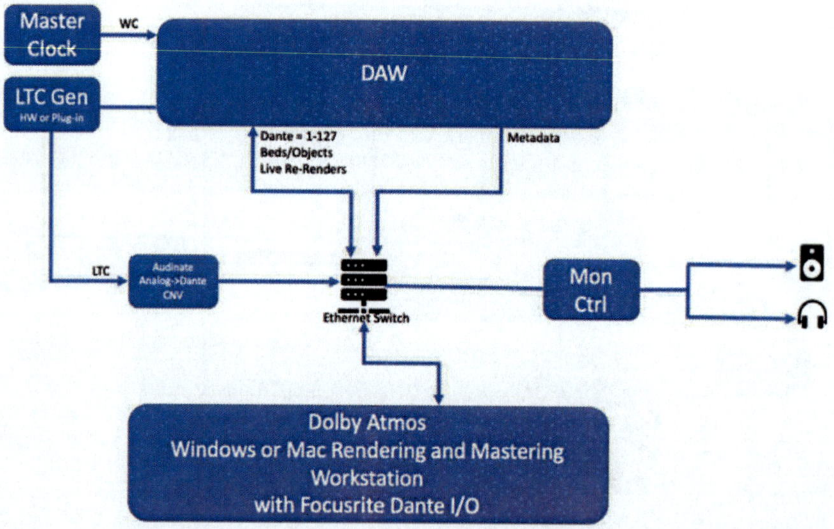

Fig. 4.36 External rendering with Dante using Windows or Mac computer (from Dolby Technologies Inc.)

Fig. 4.37 Dolby/Dell RMU for theatrical productions (RSPE audio solutions)

be a less steep learning curve. Although the outputs can be routed to a 7.1.4 monitor system (or smaller), this workflow was specifically designed for bedroom producers using headphones. Nuendo, Cubase, Da Vinci, and Logic are some of the DAWs with inbuilt Atmos rendering (V10.7 onwards).

By having an integrated 'Object Panner' and eliminating the requirement for LTC sync, Logic streamlines the process. The real-time conversion of your project's sampling rate by Logic to your final 48 kHz ADM BWF is a fantastic feature. It's crucial to keep in mind that Atmos mixes with Logic are only available with one bed.

A monitoring feature (see Fig. 4.38) of your Atmos mix was updated in Apple Spatial Audio version 10.7.4. The integrated renderer does not have any calibration tools, loudness measurement, or bass management on the monitoring side. Having said that, plugins (which monitor loudness) and hardware DSP can address all of these drawbacks.

4.9.9.7 Dolby Atmos Content Creation Signal-Flow

See Fig. 4.39.

4.9.9.8 Dolby Atmos Single Format

Dolby Atmos is a significant advancement for both the movie and music industries. The delivery of separate 5.1 Surround, 7.1 Surround, and 11.1 mixes for theatrical presentations previously implied a significant amount of work. These days are over—thanks to Dolby Atmos, which enables the fulfillment of numerous formats with a single delivery. The Dolby Atmos playback system accomplishes this in real time. 'Alexa' speakers can play Atmos content, and Dolby Atmos will handle Speaker Virtualization for soundbars in real-time (Fig. 4.40).

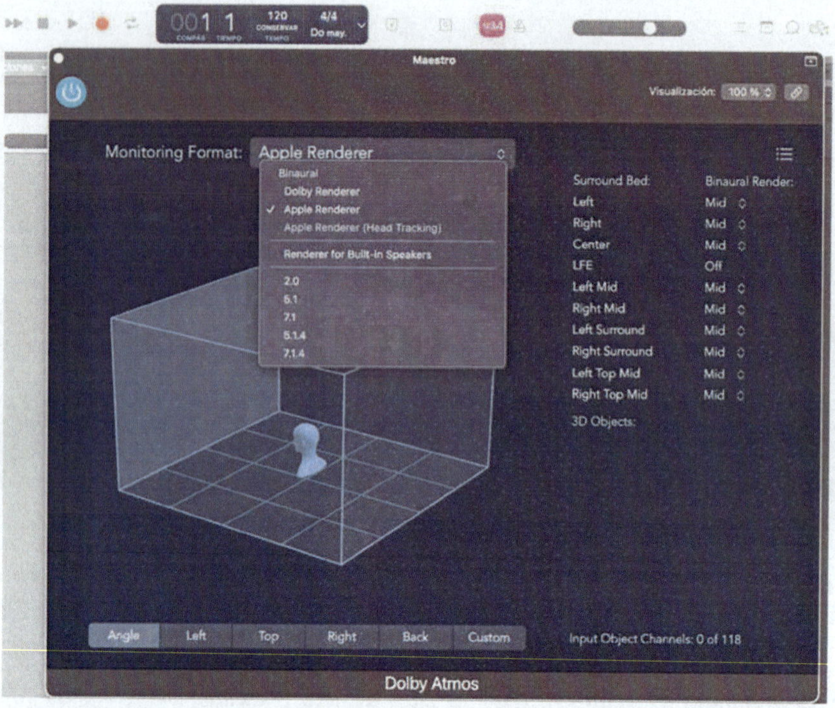

Fig. 4.38 Logic (Version 10.7.4) and its integrated Dolby Atmos and Apple Spatial Audio renderer

Re-Renders:

Channel-based surround and binaural formats can be exported using re-renders. For instance, if your client requests a reference for the 7.1 surround version audio, you can send him an auditioning re-render. Binaural, stereo, 5.1 surround, AmbiX, etc. can all be used in the same way. There are two types of re-renders:

- **Live re-renders**

  Performed in real-time from the DAR’s input window, these can be mapped back to your DAW, for example to monitor the loudness levels of a certain mix with your favorite plugin.
- **Offline re-renders**

This method of exporting masters for non-Atmos projects and rendering offline from the DAR’s master window is frequently used for reference versions. Using spatial coding and real-time object-based mixing, Dolby Atmos combines the listening experience across a wide range of playback platforms. Any sound engineer would love to achieve this (Fig. 4.41).

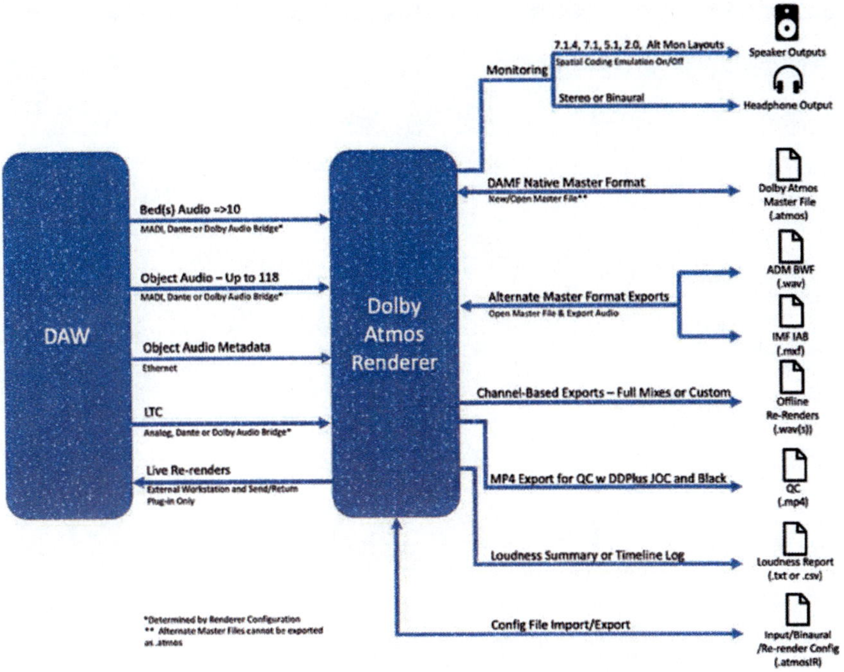

Fig. 4.39 The Dolby Atmos content-creation signal-flow (from Dolby Knowledgebase 2022, Module 1.4)

Fig. 4.40 Diagram showing orientation and reflections of HL and HR speakers in Samsung Dolby Atmos soundbar (from Samsung Soundbar 2022)

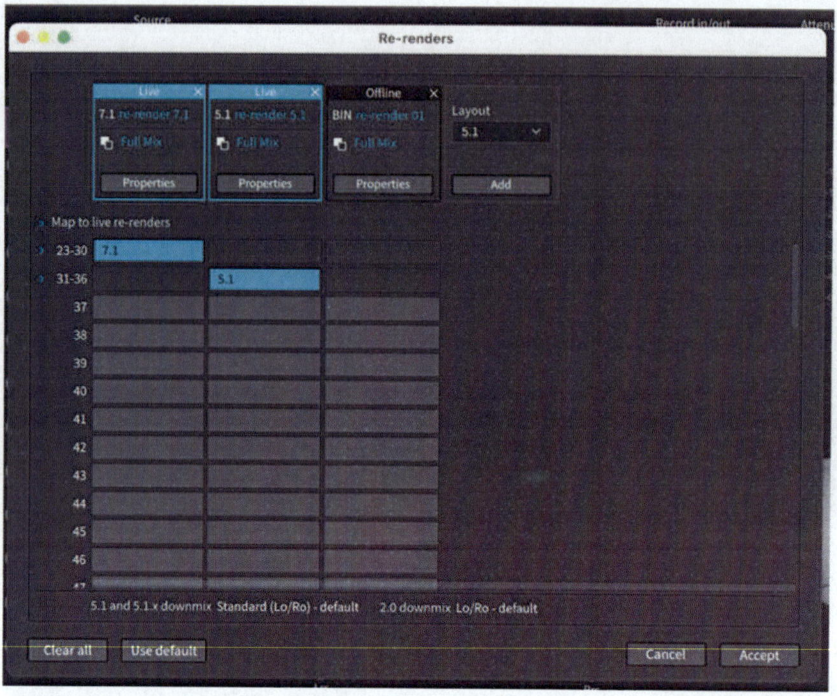

Fig. 4.41 Re-render window panel showing live and offline re-renders

4.9.9.9 Formats and File Delivery

- **DAMF** Dolby Atmos Master File

Native master format: When in the destination, it needs encoding. The customizable structure in a DAMF looks like this (see Dolby Knowledgebase 2022):

- **Atmos**: This top-level file, which is in XML format, offers crucial details about the presentation that is part of the master file set.
- **Atmos.Metadata**: All of the 3D positional coordinates for each object's audio in the .audio file are contained in this file.
- **Audio**: All bed signals and objects' audio data are contained in this file. It is a Core Audio file that is encoded in interleaved PCM (CAF—Core Audio Format).
- **ADM BWF**: Stands for Audio definition Model Broadcast Wave File. ADM is a stream of metadata that contains extensive data necessary to locate objects. (Defined in EBU Technical Paper 3364). It is the delivery type used most frequently by music streaming services. A unique XML file with metadata and a PCM stream with up to 128 single tracks in an object-based format are also features of the ADM BWF format.

- **IMF IAB**: Stands for 'Interoperable Master Format Immersive Audio Bitstream'. This SMPTE standard, which dates to 2013, is essentially a multichannel. MXF is for audio and video, and it only supports 48 kHz. It supports Dolby Vision and operates at 29.97NDF. Netflix uses the IMF IAB standard delivery format, which is Dolby Atmos compatible.
- **DD + JOC (Dolby Digital Plus Joint Object Coding)**: An option to export MP4 files is provided by the Dolby Atmos Renderer. When exporting, DD + JOC ('Dolby Digital Plus Joint Object Coding') 12 for audio and H.264 for video encoding are both used. One of the most popular encoding types used by services like Amazon Music and others to stream Atmos mixes is this one. The outdated encoding method known as DD + JOC, which ignores all of your mix's Binaural Render mode parameters, was created for film.
- **AC4 IMS: Immersive AC4 encoding** This is the more recent codec created especially for binaural music. In contrast to DD + JOC, it takes the settings for Binaural Render mode into account.

Encoding and Delivery Exceptions

The process should be fairly simple. A DAMF (Dolby Atmos Master File) should typically be entered into our DAR, followed by an ADM BWF export and upload to our aggregator. Our aggregator should have a Dolby (Hybrik) cloud encoding server, and it should be able to deliver DD + JOC, AC4 IMS, True HD from there. They will 'fetch' one encoded version or another depending on the platform. For Dolby Atmos binaural, Tidal and Amazon employ AC4 IMS, whereas Apple Music simply uses DD + JOC. Why Dolby Atmos Binaural masters sound so differently in Apple Music is understandable. Apple also chooses to render the DD + JOC files using its own Spatial Audio technology, which produces 'strange' outcomes for our listeners in addition to not ingesting metadata for Binaural render mode. It goes without saying that this is a special circumstance to be aware of which is rather unfortunate for artists, engineers/producers and listeners, and will hopefully be changed in the near future ....

References

Blumlein AD (1931) Improvements in and relating to sound-transmission, sound-recording and sound-reproducing systems. British Patent 394,325, 14 Dec 1931 (reprinted in: Anthology of Stereophonic Techniques. Audio Eng Soc, 1986, pp 32–40)

Dolby (2012) Dolby Atmos—cinema technical guidelines. White paper by Dolby Laboratories Inc. https://www.dolby.com/us/en/technologies/cinema/dolby-atmos.html. Accessed 27 Feb 2018

Dolby (2014a) Dolby Atmos—next-generation audio for cinemas. White paper by Dolby Laboratories, Inc. https://www.dolby.com/us/en/technologies/cinema/dolby-atmos.html. Accessed 27 Feb 2018

Dolby (2014b) Authoring for Dolby Atmos—cinema sound manual (Issue 3 for Software 1.4). White paper by Dolby Laboratories Inc. https://www.dolby.com/us/en/technologies/cinema/dolby-atmos.html. Accessed 27 Feb 2018

Dolby Knowledgebase (2022) Dolby-Atmos-Music-Training. https://learning.dolby.com/hc/en-us/sections/4406037447828-Dolby-Atmos-Music-Training Accessed 4 Jun 2022

Dolby Professional Support (2021) What sample rate does Dolby support. https://professionalsupport.dolby.com/s/article/What-sample-rates-does-Dolby-Atmos-support?language=en_US. Accessed 4 Jun 2022

Harvey S (2020) Dolby Atmos moves into live sound. Prosoundnetwork.com. https://www.prosoundnetwork.com/live/dolby-atmos-moves-into-live-sound?utm_source=Selligent&utm_medium=email&utm_campaign=16253&utm_content=PSN+4%2F6%2F20+&utm_term=914538&m_i=tQdFaVmQAhNukHD40nNKTMIfUt2OYcSMbPYVy2biSC32RqF0SC42Qx0UIhlgnanOQNbGEnc8C3RM1kcjpovPjDJlKrPf7Lttt1&M_BT=418331185039. Accessed 20 Apr 2020

Johnson J (2016) Atmos mixing for pro tools. Resolution 15(5):41–43

Keller AC (1981) Early Hi-Fi and stereo recording at bell laboratories (1931–1932). J Audio Eng Soc 29:274–280

Samsung Soundbar (2022) como-usar-dolby-atmos-con-tu-soundbar-samsung/ https://www.samsung.com/es/support/tv-audio-video/como-usar-dolby-atmos-con-tu-soundbar-samsung/. Accessed 5 Jun 2022

Chapter 5
HOA—Higher Order Ambisonics (Eigenmike®)

Abstract This chapter starts with an analysis of the reproduction accuracy limits of higher order Ambisonics (HOA) microphones up to 25th order. A short comparison between HOA and other mic-techniques from stereo to 3D audio is made. First order Ambisonics (FOA) is identified as delivering just sufficient resolution for up to 5.1 surround, but inadequate quality for 3D audio purposes. The results of a subjective listening comparison of four commercially available FOA (SoundField MKV, Sennheiser 'Ambeo', Core Sound 'TetraMic', Zoom 'H2n') and one HOA microphone (mh Acoustics Eigenmike® M32) are presented. After that, experiences of the practical application of Ambisonics in multichannel broadcast, as well as 360° web-based VR are evaluated by a seasoned specialist in on-location recording and tv- and cinema sound mixing, including the various aspects from the sound aesthetics involved up to a detailed workflow analysis.

Keywords 3D audio · Immersive audio · Ambisonics · Higher-Order Ambisonics (HOA) · First-Order Ambisonics (FOA) · Head-Related Transfer Function (HRTF) · Virtual Reality (VR)

5.1 Theoretical Background

In the chapter on surround microphones in (Pfanzagl-Cardone 2020) we have taken a look at the first order Ambisonics 'SoundField' microphone; in (Nicol 2018) we find the following analysis on the characteristics of HOA microphones:

> ...The tetrahedral microphone arrangement of the 'SoundField' Ambisonic microphone allows the capturing of a full 3D sound field. However, since it is composed only of four capsules, its spatial resolution is low, which means that the discrimination between the sound components is not accurate ....

The number of microphones of the spherical array imposes the maximal order that can be extracted. On example is the 'Eigenmike®',which—if composed of 32 capsules—allows HOA encoding up to the fourth order. For sound reproduction, a second step of appropriate decoding is needed to correctly map the spatial information contained in the HOA components to the loudspeaker array, in order to compute the loudspeaker input signals.

E. Pfanzagl-Cardone, *The Art and Science of 3D Audio Recording*,
https://doi.org/10.1007/978-3-031-23046-2_5

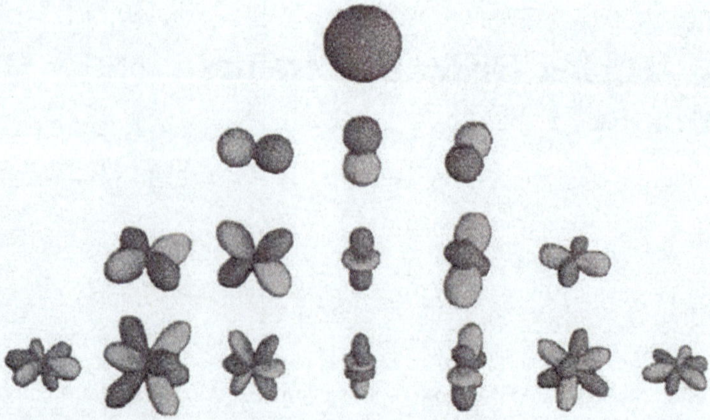

Fig. 5.1 Visualization of spherical harmonics up to order three

In its original definition, Ambisonics is based on the Spherical harmonics expansion limited to zeroth and first order components. HOA (Higher Order Ambisonics) generalizes this concept by including components of order m greater than 1, as shown in Bramford (1995) and Daniel (2001). … The HOA components convey spatial information as a function of the azimuth and elevation angles (see Fig. 5.1).

Each order m is composed of $(2\,m + 1)$ components with various directivities. It should be noted that some components are characterized by a null response in the horizontal plane. The consequence is that they do not contribute to any spatial horizontal information. By contrast, the directivity of the remaining components is symmetrical to the horizontal plane. These latter components are referred to as the '2D Ambisonics components', in the sense that, if the sound field reproduction is restricted to the horizontal plane (i.e. the loudspeaker setup is limited to the horizontal plane), only these components must be considered. On the contrary, if a full 3D reproduction is expected, all the components are used and the reproduction setup requires both horizontal and elevated loudspeakers to render spatial height information.

The component of zeroth order corresponds to the spatial equivalent of the DC component and is characterized by no spatial variation. In other words the zeroth order component, W, is the monophonic recording of a sound field by a pressure microphone. The three first order components are characterized by a figure-of-eight variation (i.e. cosine or sine function). As the order increases, the spatial variation is faster and faster as a function of the angle, as illustrated in Fig. 5.1. A first benefit of including components of higher order is therefore to enhance spatial accuracy and the spatial definition (resolution) of the sound field representation. This is due to the increase of the high frequency cutoff of the associated spatial spectrum. The resulting effect on the reproduced sound field is complex: both the size of the listening area and the bandwidth of the 'time spectrum' are affected. Indeed, first order Ambisonics reproduction is penalized by a phenomenon of 'sweet spot': the sound field is correctly reproduced only at the close vicinity of the center of the loudspeaker setup. In addition, for a given reproduction area, low frequencies, which are linked to large wavelengths and therefore to slow spatial variations, are better reconstructed than high frequencies. Adding Ambisonics components of order higher than M = 1 increases both the size of the listening area and the high-frequency cutoff of the time spectrum. Thus, small movements of the listener are then allowed. In Fig. 5.2, the sound field reproduced by Ambisonics systems of various orders is illustrated in the case of a plane wave. It is observed that a low-frequency plane wave ($f = 250$ Hz) is well reconstructed over a wide area by a fourth order system. If the frequency increases up to 1 kHz, the area of accurate reproduction shrinks considerably. An upgrade to a nineteenth order system is needed to achieve a listening area the size of which is equivalent to that obtained by the fourth order system at $f = 250$ Hz. Thus, if the sound field reproduced is observed over a fixed area, the high-frequency cutoff decreases as a function of the maximal Ambisonics order M. In the same way, if the sound field is observed at a fixed frequency the size of accurate reproduction decreases as a function of the maximal Ambisonics order M.

In (Ward and Abhayapala 2001), a rule of thumb was proposed to estimate the reproduction order as a function of the wave number k and the radius r of the reproduction sphere, to achieve a maximum threshold of the truncation error equal to 4%. The order M is obtained as:

$$\mathrm{M} = kr \tag{5.1}$$

rounded up to the nearest integer. For instance, if we consider a radius of the reproduction sphere equal to 8.5 cm, which is close to the average radius of a human head, first order Ambisonics achieves valid reconstruction of the sound field (i.e. truncation error lower than 4%) only up to 637 Hz.

To increase the frequency cutoff up to 16 kHz for the same area, it is needed to include HOA components up to the M = 25th order" (from Nicol 2018).

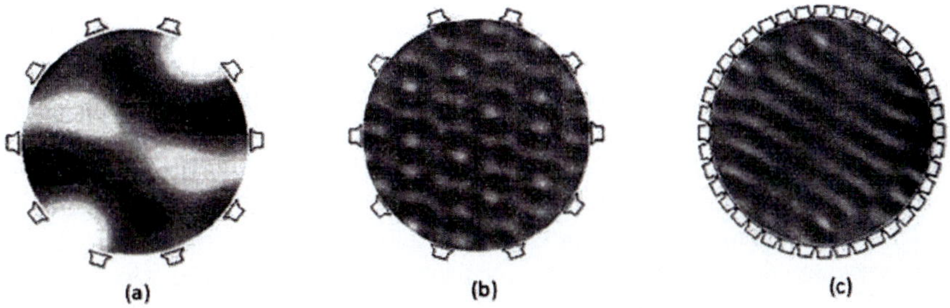

Fig. 5.2 Benefits of adding higher order Ambisonics component: **a** the 250 Hz plane wave is well reconstructed over a 4th order system; **b** increasing the frequency to 1 kHz yields a poorer reconstruction using the same 4th order system; **c** to reconstruct the 1 kHz plane wave with the same accuracy as in (**a**), the system must be increased to a 19th order (from Nicol 2018)

5.2 A Comparison Between HOA and Other Mic-Techniques for 3D Audio

Based on well founded, professional recording experience we find a critical comparison of the sonic restrictions or drawbacks between Ambisonics and other stereo as well as surround microphone techniques in (Wittek and Theile 2017):

> "... 3D Audio can give distinctly better spatial perceptions than 5.1. Not only is the elevation of sound sources reproduced, but noticeable improvements can also be achieved with regard to envelopment, naturalness, and accuracy of tone color. The listening area can also be greater; listeners can move more freely within the playback room without hearing the image collapse into the nearest loudspeaker.

5.2.1 Why is Stereo Different?

It is crucial to differentiate between 'sound field reconstruction' and stereophonic techniques because they differ fundamentally in the principle by which sources are perceived, as found by Theile (see Theile 1980, 1991). In contrast to the common theory of 'summing localization', Theile assumes that loudspeaker signals are perceived independently, and that their level and time differences thus determine the location of phantom sources just as in natural hearing. It is essential that this superposition of only two loudspeakers does not lead to audible comb filtering, as the physical properties of the sound field would suggest. A stereophonic system can very easily create phantom sources in various directions, with good angular resolution and without sound-color artifacts. This makes it superior to imperfect sound field reconstruction principles such as wavefield synthesis with excessive loudspeaker spacing, or Ambisonics of too low an order, both of which create artifacts (Wittek et al. 2007).

When recording and reproducing stereophonically, closely-spaced microphone pairs are used, which create time and/or level differences between the microphone signals. These signals are routed discretely to the loudspeakers. The inter-channel differences lead to the creation of phantom sources (Wittek and Theile 2002). Stereophonic systems with more than two channels, such as 5.1 or 9.1 Surround, may be considered as systems consisting of multiple individual loudspeaker pairs with time and/or level differences that create phantom sources (Theile and Wittek 2012).

There is a fundamental difference between a first order Ambisonics microphone and a stereophonic array for 5.1, even though the microphone arrays may look similar. An Ambisonics array aims for physical reconstruction of the original sound field, but cannot achieve it because of the early truncation of the order of the reproduced spherical harmonics. A stereophonic array aims to capture time and/or level differences in individual microphone pairs, but often cannot achieve that because of excessive crosstalk between the pairs. Hence both approaches have their own artifacts, as well as methods for overcoming them (see Wittek et al. 2007; Theile and Wittek 2011).

5.2.2 *What is an Ambience Microphone?*

Often the sound source to be recorded is a speaking voice, an instrument or the like. These sources can easily be recorded with a single microphone, and reproduced either by one loudspeaker or panned between two loudspeakers. If multiple individual sources have to be captured, e.g. a pop band with four instruments, multiple individual microphones can be used. However, if the sound source is spatially extended, if the room sound is to be captured as well, or if there simply are too many sound sources, this method fails. In that case a so-called 'main microphone' or 'room microphone' pair/setup serves for the stereophonic pickup of these sources in an efficient way, because these arrangements of two microphones (or the five microphones of a stereophonic array for 5.1 surround) are designed so that the recorded scene is properly reproduced between the loudspeakers (Wittek and Theile 2002). Typical 'main microphone' techniques are A/B, ORTF and X/Y (for two-channel stereo), and OCT, IRT Cross/ORTF Surround or a Decca Tree (for 5.1 surround).

An 'ambience microphone' arrangement is a 'main microphone' arrangement as well. The only difference is that the sound source is 360° around the listener instead of only in front (as in concert recording). Hence an ambience microphone has no 'front' direction, but an equally-distributed image of phantom sources throughout the entire space spanned by the loudspeakers. Often the Center channel is omitted in the design of an ambience microphone, because it would destroy this equality of energy distribution.

5.2.3 *One Recording Method for All 3D Formats?*

There are various 3D Audio playback systems, so the recording techniques that work best for each of them will naturally be different. For sound field synthesis systems, multichannel microphone arrays can be a solution, while for 3D stereo, stereophonic miking techniques are the norm. For binaural reproduction in the simplest case, a dummy head can be used.

But all these systems share one requirement when recording complex, spatially-extended sound sources such as ambient sound: stereophonic techniques must be used, because they alone offer both high-quality sound and high channel efficiency (even two channels may be enough). It is impossible or inefficient to reproduce in high quality the sound of a large chorus, for example, or the complex, ambient sound of a city street, by compiling single point sources recorded with separate microphones.

In the same way, multichannel microphone arrays for sound field synthesis, such as higher-order Ambisonics ('HOA') or wavefield synthesis, fall short in practice because their channel efficiency or sonic quality are too low. If on the other hand the number of channels is reduced, e.g. with first order Ambisonics, the spatial quality becomes burdened with compromise.

For binaural playback, the dummy head technique is clearly the simplest solution—but it does not, in itself, produce results compatible with virtual reality glasses, in which the binaural signals must respond to the user's head motions. That would be possible only through the 'binauralization' (Nicol et al. 2014) of a stereophonic array—a technique that is already well established.

5.2.4 *Is First-Order Ambisonics Adequate for 3D?*

There is a common assumption that Ambisonics would be the method of choice for 3D and VR. The professional recording engineer would do well to examine the situation more closely.

Ambisonics, which has existed for a long time by now, is a technology for representing and reproducing the sound field at a given point. But just as with wavefield synthesis, it functions only at a certain spatial resolution or 'order'. For this reason, we generally distinguish today between 'first-order' Ambisonics and 'higher-order' Ambisonics ('HOA').

First order Ambisonics cannot achieve error-free audio reproduction, since the mathematics on which it is based are valid only for a listening space the size of a tennis ball. Thus, the laws of stereophony apply here—a microphone for first order Ambisonics is nothing other than a coincident microphone with the well-known advantages (simplicity; small number of recording channels; flexibility) and disadvantages (very wide, imprecise phantom sound sources; deficient spatial quality) of that approach in general.

Creation of an Ambisonics studio microphone with high spatial resolution is an unsolved problem so far. Existing Ambisonics studio microphones are all first order, so their resolution is just adequate for 5.1 surround but too low for 3D Audio. This becomes evident in their low inter-channel signal separation as well as the insufficient quality of their reproduced spatiality.

The original first order Ambisonics microphone was the SoundField microphone. The Tetramic or the Sennheiser Ambeo microphone have been built in a similar way. The Schoeps 'Double M/S System' (Wittek et al. 2006) works in similar fashion, but without the height channel.

Ambisonics is very well suited as a storage format for all kinds of spatial signals, but again, only if the order is high enough. A storage format with only four channels (first-order Ambisonics calls them W, X, Y, Z) makes a soup out of any 3D recording, since the mixdown to four channels destroys the signal separation of the 3D setup.

Ambisonics offers a simple, flexible storage and recording format for interactive 360° videos, e.g. on YouTube. In order to rotate the perspective, only the values of the Ambisonics variables need be adjusted. Together with the previously mentioned small first order Ambisonics microphones, 360° videos are very easily made using small, portable cameras.

For virtual reality the situation is different, however. The acoustical background signal of a scene is generally produced by 'binauralizing' the output of a virtual loudspeaker setup, e.g. a cube-shaped arrangement of eight virtual loudspeakers. The signals for this setup are static; turning one's head should not cause the room to spin. Instead, head tracking causes the corresponding HRTFs to be dynamically exchanged, just as with any other audio object in the VR scene.

As a result, most of the advantages of first order Ambisonics do not come into play in VR. On the contrary, its disadvantages (poor spatial quality, crosstalk among virtual loudspeaker signals) only become more prominent.

If practical conditions allow for a slightly larger microphone arrangement, an ORTF-3D setup would be optimal instead as an ambience microphone for VR.

5.2.5 *Criteria for Stereophonic Arrays Used in Ambience Capture for 3D Audio*

Stereophonic arrays are thus the approach of choice for ambience recording in all 3D formats. The requirements for 3D are the same as in two- and five-channel stereophony (Theile and Wittek 2011):

- Signal separation among all channels in order to avoid comb filtering: No one signal should be present at significant levels in more than two channels.
- Level and/or arrival time differences between adjacent channels to achieve the desired imaging characteristics
- Decorrelation of diffuse-field sound for optimal envelopment and sound quality.

5.2.5.1 2-Channel Stereophony

These demands are still easy to fulfil in two-channel stereophony; a suitable arrangement of two microphones and two independent channels can provide the desired imaging curve. Tools such as the "Image Assistant" (Wittek 2000) application (available as an iOS app or on the Web at www.ima.schoeps.de) have been developed for this purpose.

They take into account not only the creation of phantom image sources, but also the ever-important channel decorrelation. A classic, positive example is the ORTF technique, which has a 100° recording angle and delivers a stereo signal with good channel decorrelation.

5.2.5.2 5-Channel Stereophony

The above requirements are distinctly more difficult to meet with five channels, and there are numerous geometries that fail to meet them, e.g. a microphone that looks like an egg the size of a rugby ball, with five omni capsules that can deliver only a mono signal at low frequencies.

Five independent channels simply cannot be obtained with any coincident arrangement of first-order microphones. A coincident arrangement such as first order Ambisonics is thus a compromise for 5.1, though highly workable because of its advantages in compactness and post-production flexibility.

One optimal solution for ambient recordings in multichannel stereophony is the 'ORTF surround' system [remark of the editor: see Chap. 4 on surround microphones in Pfanzagl-Cardone 2020] in which four super-cardioids are arranged in a rectangle with 10 by 20 cm side lengths. Here the distances between microphones help with decorrelation, and thereby lend the sonic impression its spatial openness. The microphone signals are routed discretely to the L, R, LS and RS channels. The signal separation in terms of level is ca. 10 dB; thus, the sonic image during playback is stable even in off-axis listening positions.

5.2.5.3 8 or More Channels

With eight or nine channels, the arrangement of the microphones becomes very difficult if the above-mentioned requirements are to be met. The simplest method for maintaining signal separation is to set up eight or nine microphones far apart from one another. "[Rem.: another possibility would be to make use of the high directionality of a 'Blumlein-Pair' (or BPT 3-channel) based microphone setup, which—despite being a one-point coincident technique—also yields very high signal de-correlation; see Chap. 4 in Pfanzagl-Cardone 2020]. " … Thus, a large nine-channel 'Decca Tree' arrangement is very well suited for certain applications, although it has severe disadvantages that limit its practical usability. For one, the sheer size of the arrangement is greater than 2 m in width and height. And the signal separation in terms of level difference is nearly zero; every signal is more or less available in all loudspeakers. Thus, this array can represent a beautiful, diffuse spaciousness, but stable directional reproduction is not achieved beyond the 'sweet spot'. This can be helped by adding spot microphones." (from Wittek and Theile 2017).

Wittek then proceeds in promoting the ORTF-3D recording method as an optimal ambience arrangement for 8-channel recording (for more details on the ORTF-3D microphone array please see Chap. 8 in this book).

5.3 A Comparison Among Commercially Available Ambisonic and Other 3D Microphones

A very interesting comparison between several Ambisonic based, commercially available microphones was carried out by Enda Bates et al., about which they report in Bates et al. (2016, 2017):

" … This paper presents some further experiments devised to assess the performance of an expanded number of commercially available Ambisonic microphones.

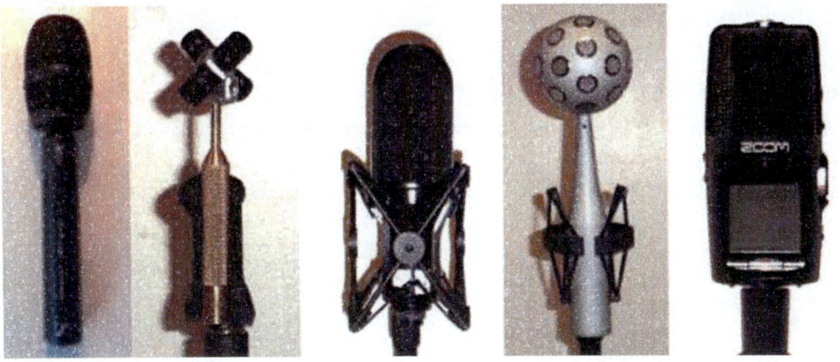

Fig. 5.3 The compared microphones from L to R: Sennheiser 'Ambeo', Core Sound 'TetraMic'; SoundField 'MKV', MH Acoustics 'Eigenmike'®, Zoom 'H2n' (graphic courtesy of Enda Bates)

The subjective timbral quality of five microphones (SoundField MKV, Core Sound TetraMic, Sennheiser Ambeo, mh acoustics Eigenmike®, and Zoom H2n) is assessed using listening tests and a recording of an acoustic quartet. Localization accuracy is assessed using an objective directional analysis and recordings from a spherical array of 16 loudspeakers. Intensity vectors are extracted from 25 critical frequency bands of the Bark scale and used to compute the angle to the source location. Significant differences were found between microphones, with the SoundField MKV and Eigenmike® producing the best results in terms of timbral quality and localization respectively (Fig. 5.3).

Below short descriptions of the various microphone systems under test, from the preprint:

5.3.1 *Sennheiser 'Ambeo'*

The Sennheiser Ambeo was released in 2016 and is based around a similar design to the SoundField MKV using four cardioid capsules mounted on the vertices of a tetrahedron. The microphone outputs a raw A-format signal via a breakout cable which is then calibrated and converted to B-format using the Sennheiser conversion VST plugin. This conversion plugin includes an optional 'Ambisonics Correction Filter'. As the default settings for the conversion plugin includes this filter, it was used for all the Ambeo recordings in this listening test.

5.3.2 SoundField 'MKV'

The SoundField MKV system consists of a rack-mounted control unit and microphone containing four sub-cardioid capsules mounted on the vertices of a tetrahedron. The MKV control unit contains four calibrated pre-amplifiers and also performs the conversion of the raw, A-format microphone signals into a line-level B-format output.

5.3.3 Core Sound 'TetraMic'

The Core Sound TetraMic was first released in 2007 and consists of four cardioid electret-condenser capsules, mounted so that the capsule grills align with the surfaces of a tetrahedron. Intended as a lower-cost alternative to the SoundField microphone, the TetraMic outputs a raw A-format signal via a breakout cable that also contains four Phantom Power Adaptors (PPAs). The calibration of these 4 unbalanced microphone level signals is supplied by Core Sound for each specific microphone and applied to the raw A-format signals using a customized version of the VVMic VST plugin developed by David McGriffy, which also handles the A-to-B-format conversion process.

5.3.4 mh Acoustics 'Eigenmike'®

The Eigenmike® is a spherical microphone array developed by mh Acoustics and is the first commercially-available microphone system that can produce HOA recordings. The system consists of an 8.4 cm rigid sphere upon which thirty-two 14 mm omni-directional electret microphones are mounted. The pre-amplifiers and 24-bit A/D converters are also contained within the spherical enclosure while the raw 32-channel signal (as well as power and other control signals) is transmitted to the Eigenmike® Microphone Interface Box (EMIB) via a Cat-5 cable. The EMIB produces a Firewire audio stream and appears as a 32-channel audio driver in the host computer. The EigenStudio application software can be used to record the raw microphone signals, apply the calibration for that specific microphone, and to generate both First and Higher Order Ambisonic signals.

5.3.5 Zoom 'H2n'

The Zoom H2n is a portable digital audio recorder containing five microphone capsules arranged in two, opposite facing stereo pairs. The output consists of Mid-Side stereo (which was chosen as the front facing side of the recorder) and 90° X/Y stereo, recorded separately as two stereo files, which can then be processed into a horizontal only B-format signal using a conversion matrix. Despite the lack of a vertical Z component and extremely low cost, the H2n was used for audio capture in Google's JUMP system for immersive virtual reality content, and the associated commercial camera system, the GoPro Odyssey.

For the second experiment, a conversion was applied automatically using the native Ambisonics recording mode released by Zoom as a firmware update in 2016.

The test of the first preprint (Bates et al 2016) used modified Multiple Stimuli with Hidden Reference and Anchor (MUSHRA) technique for the listening test, and did not yet include the Sennheiser 'Ambeo' microphone, which was only used in the experiments reported about in the second preprint.

The results of the first experiment in terms of mean azimuth and elevation error as well as diffuseness are being presented in Figs. 5.4 and 5.5:

> while the Eigenmike® has the significantly lowest angular error, the SoundField MKV seems to be bit better in terms of diffuseness (although not significantly, see overlapping 95% confidence intervals). The number of subjects was 19 for the first test. (for further results see also Figs. 5.6 and 5.7).

The conclusions from the second study are summarized below:

" ... The results from these two experiments largely match the findings of part 1 of this study, with the SoundField MKV once again producing the best results in terms of overall timbral quality, and with comparable results to the other microphones in

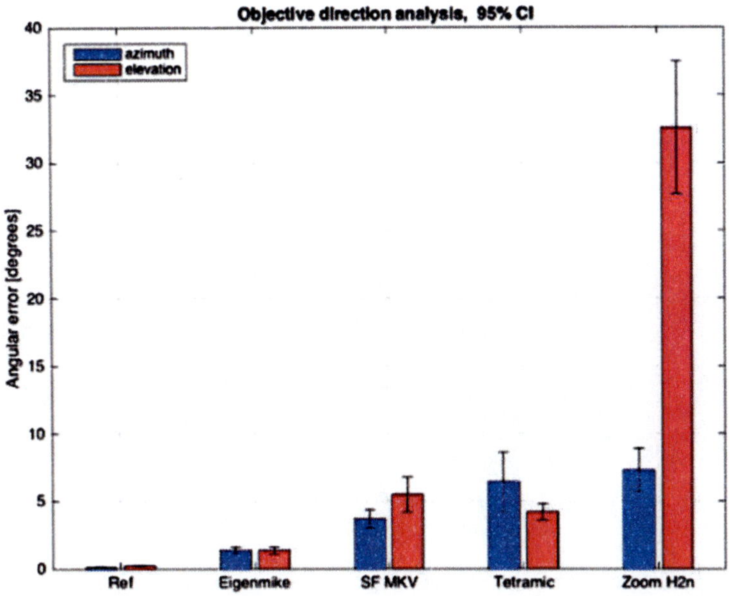

Fig. 5.4 Ambisonic microphones mean angular error (azimuth and elevation) with 95% confidence intervals (from Bates et al. 2016)

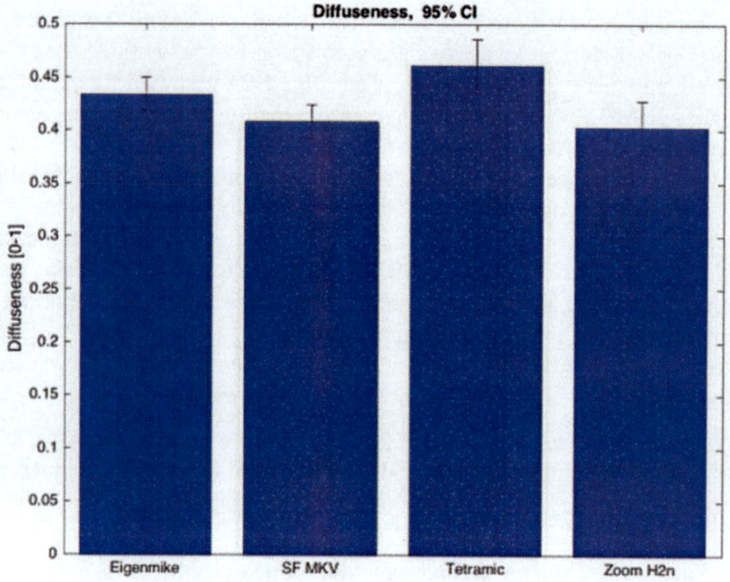

Fig. 5.5 Ambisonic microphones diffuseness with 95% confidence intervals (from Bates et al. 2016)

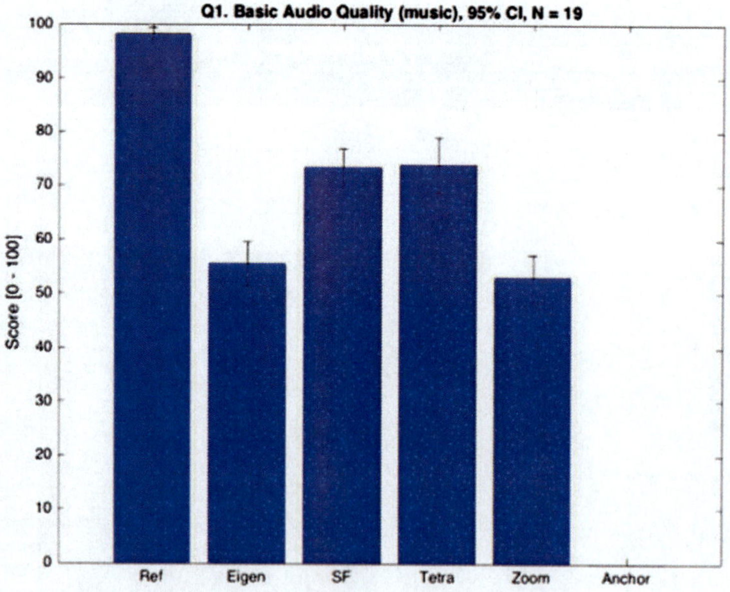

Fig. 5.6 Ambisonic microphones basic audio quality, with 95% confidence intervals (from Bates et al. 2016)

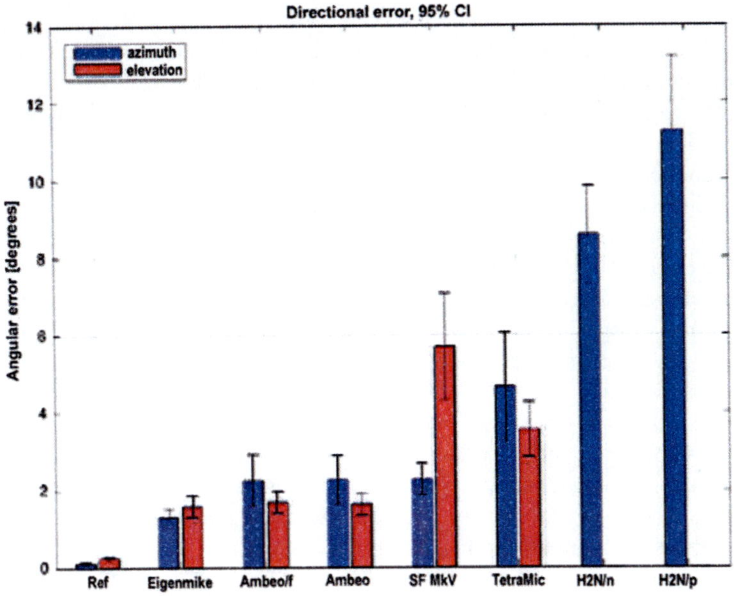

Fig. 5.7 Ambisonic microphones mean angular error (azimuth and elevation) for 16 sources, results from test 2 [with 95% confidence intervals; 21 test listeners] (from Bates et al. 2017)

terms of directionality, with the exception of the Eigenmike® and Ambeo. Although the findings of the directional analysis largely match previous results, with the same overall ranking of microphones (Bates et al. 2016), some differences were notable in the performance of the SoundField MKV in terms of elevation accuracy. While a similar trend was visible in both experiments, here a notable increase in variance in the results for the four lowest loudspeakers caused a reduction in elevation accuracy for this microphone. Supplementary experiments to determine if the microphone cradle or other changes in test conditions caused this increased variance revealed no significant changes in these results. The Ambeo performed very well in terms of directionality, on par with the Eigenmike®, and significantly better for elevation accuracy compared to both the SoundField MKV, and TetraMic. In terms of timbral quality however, the Ambeo was evaluated as less ideal and brighter than either the SoundField MKV or TetraMic. However, this is perhaps an unsurprising result given the noticeable high frequency boost applied in the correction filter used by default in the A-to-B-format conversion plugin. The worse performance of the TetraMic in terms of directionality, particularly elevation, is perhaps explained by the slightly greater inter-capsule spacing of this design compared to other models. While the use of speech signals and studio recordings in part 1 of this study revealed some issues with noise due to the low signal level outputted by the TetraMic, this was much less evident in this real-world recording in a reverberant acoustic environment. This suggests that given appropriate pre-amplifiers with sufficient gain, excellent results can be achieved with this microphone in terms of timbral quality, albeit with slightly reduced localization accuracy. As might be expected, the more elaborate designs of the Eigenmike® produced the best results overall in terms of directional accuracy, but with decreased performance in terms of timbre. As with part 1 of this study, this again suggests a trade-off between timbre quality and directionality, particularly when the number of individual capsules in the microphone is significantly increased. The Zoom H2n performed worse overall compared to all other microphones, however, its performance is still reasonable given its extremely low cost, and that it was not originally designed to produce B-format recordings." (from Bates et al. 2017).

It is therefore interesting to note that the FOA SoundField MKV microphone, despite being the oldest version of the Ambisonic microphones tested above, still seems to be—'overall quality' wise—at par with the HOA 'Eigenmike'® with the latter being superior in terms of localization accuracy, but less favorable in terms of 'timbre' or sound-color ...

For correctness it should be mentioned that the Sennheiser A-to-B conversion plugin was updated in the meantime and the high-frequency boost that was reported in the paper is no longer apparent. So that result in terms of the excessive brightness of the Ambeo mic compared to the others, while true at the time when the study was undertaken, is likely no longer the case.

5.4 The 'Eigenmike®', Octomic and ZM-1

[rem.: the following Sect. 5.4 is a copy of Sect. 4.3.7 from the author's previous publication (Pfanzagl-Cardone 2020)]

In (Meyer and Agnello 2003) a sphere-microphone with 24 miniature capsules and a diameter of only 7.5 cm (roughly 0.25 ft.) is described. With post-processing a microphone directivity of 3rd order can be achieved (Fig. 5.8).

With the Eigenmike® the number of microphones (single transducer elements) n defines the order which can be achieved:

$$n = (M + 1)^2 \tag{5.2}$$

with M being the order. (rem.: therefore with the 4 capsules of the SoundField-Microphone only first-order microphone characteristics can be realized).

With the 24 transducers of the Eigenmike®, above 1.5 kHz 3rd-order directional characteristics can be achieved, below this 2nd order and below 700 Hz only 1st order (see Eargle 2004).

The 'Eigenmike® m32', with its 32 capsules, is able to achieve 4th order Ambisonics. An arrangement similar to the Eigenmike® is registered as international patent WO 03/061636 A1 by Elko et al. (2003) "Audio system based on at least second order Eigenbeams".

In 2018 US-company Core Sound, which also produces the "TetraMic", has come up with the first commercially available 'Second Order Ambisonic' (SOA) microphone: the "OctoMic" is made up of 8 capsules with cardioid characteristics, which

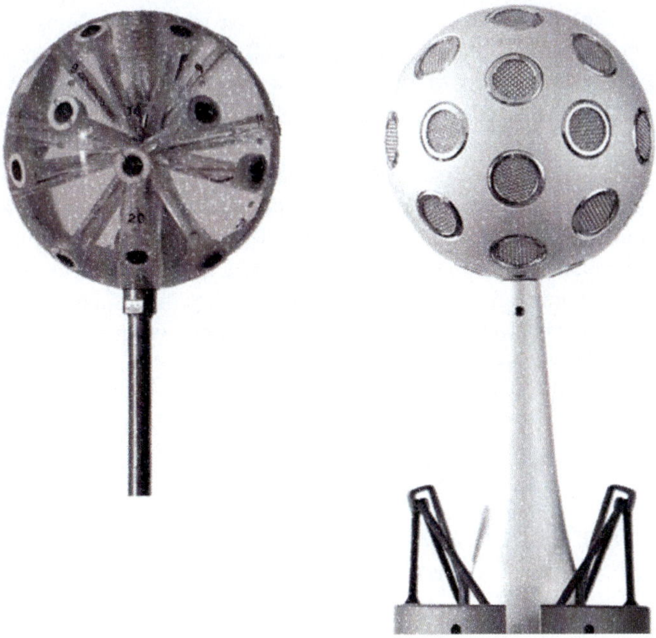

Fig. 5.8 Eigenmike® (left): older model with 24 capsules; (right): new version 'Eigenmike® m32' with 32 capsules (courtesy of mh acoustics LLC)

apparently allows not only for a broad range of 1st and 2nd order microphone patterns but also delivers more accurate spatial impression and localization due to better signal separation between the capsules. The enhanced microphone pattern flexibility of the OctoMic in terms of post-production is undoubtedly an advantage; however it still needs to be seen if some of the new 2nd order microphone patterns (e.g. with 3-dimensional '4-leaf flower' characteristics, looking a bit like a crossed figure-of-eight Blumlein-pair, but as *one unified* microphone pattern; for reference see the two left ones in the middle row of patterns in Fig. 5.1) will find a practical application as part of new microphone technique concepts.

Based on the estimate proposed in (Ward and Abhayapala 2001) and taking a look at the 'spatial resolution' or accuracy, which second order ambisonics is able to achieve in terms of soundfield reproduction, we can use Eq. 5.1 (with the wave number k substituted as in formula 1.5) and for a 'reproduction sphere' of 17 cm diameter (similar to the size of a human head) we will arrive at a frequency of approximately only 1300 Hz, up to which the OctoMic should be able to achieve a valid reconstruction of the soundfield (i.e. truncation error lower than 4%).

Around the same time, Polish company Zylia has released their 3rd order Ambisonics (TOA) microphone 'ZM-1', based on 19 omnidirectional capsules. Accompanying software enables beamforming and the creation of useful microphone patterns for 2D and 3D-Audio use, including binaural signal generation.

Repeating the calculation, which was made above for the OctoMic: based on the estimate proposed in (Ward and Abhayapala 2001) and an assumed 'reproduction sphere' (i.e. the 3D- analogy to the 'sweet zone' in 2D-surround sound) of 17 cm diameter, we arrive at a frequency of approximately 1900 Hz, up to which the ZM-1 should achieve a valid reconstruction of the soundfield, according to theory.

Zylia states a recording resolution of 48 kHz, 24bit which would result in a theoretical dynamic range of 24*6 dB = 144 dB (rem.: as each bit is the equivalent of an amplitude resolution/dynamic range of 6 dB). However, most likely due to the quality of the transducers and the electronics involved, in their technical specifications Zylia states a dynamic range of 105 dB and a signal-to-noise ratio of only 69 dB(A), which is similar to what well-aligned analog tape machines (without a noise reduction system) were able to deliver in previous times; all this results in an effective (usable) dynamic range equivalent to approximately 12bit, because the audio signal in the amplitude range of the lower 12 bits (of the total of 24 bits) will be drowned in noise.

5.5 Practical Experience: Dennis Baxter on Higher Order Ambisonics for Broadcast Applications

The text in the following section was taken from an article by on-location recording specialist and tv- and cinema sound-mixer Dennis Baxter (see Baxter 2016).

" … The quest of dimensional sound for radio and television began with the journey from mono to stereo, and as surround sound becomes commonplace there are even strong advocates for formats all the way up to 22.2. These ongoing advances in audio for electronic picture have been significant benchmarks in the enduring pursuit for sound that enhances the believability and reality of a two-dimensional picture.

The production of dimensional sound has been around for a while. But its nemesis is the expectation that the reproduction setup will have the same number of speakers and prescribed location as the production. Frankly, this channel and speaker based production has been the weak link since the introduction of surround sound. Even now with soundbars, you have to wonder if your mix is sounding like what you are crafting, much less an optimum sound experience.

Consider a concept where the audio elements reside in the audio stream and the consumer reproduction format renders the appropriate sound for everything from stereo to immersive. This completely different approach is to capture or generate the entire sound pressure field and convert it to Higher Order Ambisonics (HOA) and send the HOA signals to the playback device to be rendered. With the audio rendered at the playback device, the renderer matches the HOA sound field to the number of speakers and their location at the playback in such a way that the sound field created in playback resembles closely to that of the original sound pressure field.

Since HOA is based on the entire sound field in all dimensions, a particularly significant benefit with this audio spatial coding is the ability to create dimensional sound mixes with spatial proximity, horizontal and vertical localization and deliver the heightened mix to the listener. Significantly, with scene-based audio reproduction, the rendering of the HOA signals solves the problem of producing multiple accurate mix formats. By simply rendering the underlying sound field-representation, HOA ensures a consistent and accurate playback of the sound field across virtually all speaker configurations.

The sonic advantages of ambisonics reside with the capturing and/or creation of HOA. Ambisonics works on a principle of sampling and reproducing the entire sound field. Intuitively, as you increase the ambisonic order the results will be higher spatial resolution and greater detail in the capture and reproduction of the sound field.

However, nothing comes without a cost. Greater resolution requires more sound field coefficients to map more of the sound field with greater detail. Some quick and easy math: fourth order ambisonics requires 25 coefficients, fifth order requires 36, and sixth order requires 49 and so on.

Clearly the benefits are in the use of higher order ambisonics but this is channels and data intensive. Qualcomm, a technology giant which has been researching the benefits of HOA, has developed a "mezzanine coding" that reduces the channels up to 29th order HOA (i.e. 900 channels) to 6 channels + control track (Fig. 5.9).

Now consider a sound designer's production options: HOA provides the foundation for stereo, 5.1, 7.1, 7.1 + 4, 10.2 up to 22.2 and higher using only 7 channels in the data stream. The fact is that multi-formats will not go away. People want options for consumption and the challenge is how to produce and manage a wide range of audio experiences and formats as fast and cheaply as possible. 'Cheaply' means minimizing the amount of data being transferred. If the rendering is done in the consumer device, it inherently means the need to deliver more channels/data to the device. Data compression has significantly advanced to the point that companies

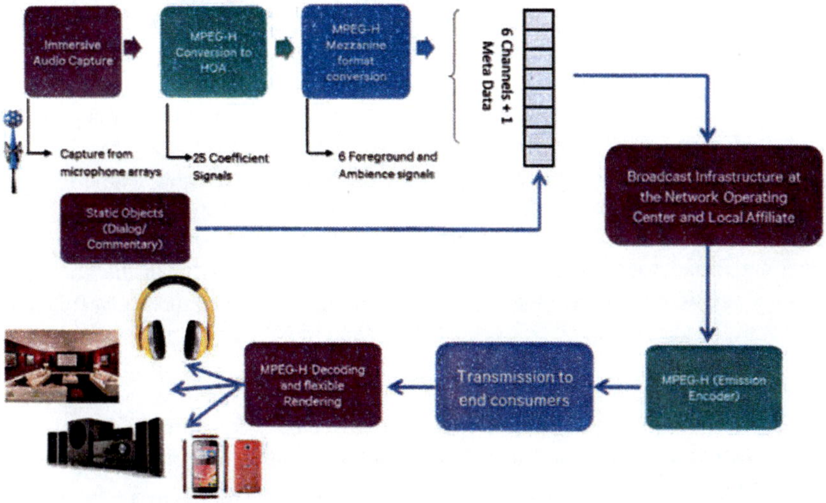

Fig. 5.9 HOA—Higher Order Ambisonics workflow for broadcast applications (from Baxter 2016)

like Fraunhofer have developed an audio codec ratified as the MPEG-H 3D audio standard that can deliver twice as much audio over the same bit stream as previous codecs. This enabled the content creator to transmit up 16 channels of audio to the device.

Now consider the upside of the production options using HOA. You have the ability to reproduce virtually all speaker formats over 7 channels plus, you have 9 additional channels for multiple languages, custom audio channels, and other audio elements that are unique to a particular mix. Additionally producing the foundation sound field separately from the voice and personalized elements facilitates maximum dynamic range along with loudness compliance while delivering consistent sound over the greatest number of playout options.

So where do we start? It all begins with sound design. Sound is essential to the believability and the sensation of reality and with the addition of height information, immersive sound easily completes the illusion. Immersive sound has been documented and researched to be a key component to the listening experience. Now the challenge for the professional audio community is to consider effective and practical production practices. Start with the basics by asking the question: what are you using the height channels for? In live sports, for example, typical sound design is presented from the perspective of the spectator or the participant. From the Point of View (POV) of the spectator in the stands, the natural sound above the audiences would probably be diffused ambiance and perhaps the PA. There could be a sound element or object, such as a bird, that may randomly appear in the sound field above the listener and/or there could be a sound element or object such as a drone or an airplane that moves through the sound field above the listener. In these examples, the sound above tends to be diffused with an occasional bit of definition.

Now considering using the height channels to define and enhance the perspective by placing sound events closer to the viewer—as if they were the participant. For sports, the sound of the POV of the participant is up close audio of the athlete, apparatus, or coach. In typical entertainment productions however, this design may not work for a singer in a concert-type live setting whose POV is the audience. When you consider that today's sports and live-entertainment sound design is generally produced with everything in front of the viewer/listener no matter how big the sound field. It may be time to consider alternatives to sound design and spatial imaging. That's when it gets interesting with immersive sound capabilities.

One proposal by Tsuyoshi Hirata, 22.2 sound designer from The Japanese National Broadcaster, NHK, is the use of back to front clarity in the aural definition along with bottom-to-top frequency banding resulting in more low frequency information at ear level and below the listener and more high frequency information above the listener. It does make common sense to use the front speakers and even the height speakers for more definition. The net result of focusing the attention of the viewer forward while using the rear height speakers for diffused spatial information makes perfect sense and may be a better replication of immersive audio over sound-bars.

Another consideration when producing in immersive sound is placing sound elements or objects in the sonic space above and around the listener using various degrees of elevation. In my listening tests I found that by separating sounds with slight degrees of elevation results in spatial unmasking and clarity in the mix, and there has been considerable research on spatial unmasking.

Typically, these sound elements are static in their dimensional space but I got a lengthy test of a dimensional panner resulting from a partnership with Jünger Audio, an innovative audio processing company, and the Fraunhofer Institute in Germany.

Dynamically moving a sound element in both the horizontal and vertical space is a powerful production tool that redefines the use of height sound. I had the pleasure of mixing a skateboard jumping event in the Technicolor Soundstage and moving the sound of the board up and down across the screen was a unique experience.

If you advance your vision of sound design there are creative uses for moving a sound object in space in a repetitive and accurate way. Think of the use of DVE, digital video effects and the ability to program a repetitive motion sequence, or several to augment the motion effect the visuals.

5.5.1 Capture or Create?

Natural capture, processing and mixing are components of a HOA sound foundation and I imagine a typical sports and live entertainment production would use all of these tools. To capture an immersive impression of an event, a well placed high density array microphones can go a along way to providing a stable immersive sonic foundation to build on. My problem with traditional closely correlated microphone arrays is the lack of detail beyond the useful sound capture zone ultimately requiring additional microphone for detail.

From what I have heard from the higher density array microphones is that they can improve the detail of the desired signal and reduce noise interference by steering the microphone beam—aiming the polar pattern at the desired sound. Interestingly beam steering can manipulate the sound at the capsules by reducing unwanted sound as well—once again the higher the order, the higher the resolution.

A significant production note: you can also use individual microphones at spatially separated arbitrary locations and use the capture from all of these microphones to derive the HOA coefficients. Additionally, content not created with a HOA foundation using mono and stereo stems, can be processed with a HOA encoder to generate 3D coefficients producing a dimensional sound scene when played back over speakers. Mixing mono, stereo and array microphones is a familiar and similar production workflow. This same workflow would be used for an immersive sound production, truly making Immersive and HOA production seamless for the mixer and audio producer.

There are cleanly significant advantages to HOA and Scene Based Delivery over current multi-channel audio workflows and practices. This technology fulfills the need to produce immersive content, distribute it in an efficient bit stream and have it played out on a wide variety of speaker configurations and formats in a creative, efficient and consumer-friendly way. … " (from Baxter 2016).

As to the question of 'practicality' of FOA and HOA it needs to be said that Higher Order Ambisonic microphone signals usually tend to have too much latency—due to processing—in the context of live broadcasting, so there is not much use of Ambisonics capture e.g. in live sports beyond some 1st order Ambisonics experimenting. Production in Ambisonics seems to have great potential, but currently there are not many advocates, at least in live broadcasting.

References

Bates E, Gorzel, M, Ferguson L, O'Dwyer H, Boland FM (2016) Comparing ambisonics microphones: Part 1. Paper presented at the Audio Eng Soc Conference on Sound Field Control, Gilford, 18–20 July 2016

Bates E, Doonery S, Gorzel M, O'Dwyer H, Ferguson L, Boland FM (2017) Comparing ambisonics microphones—Part 2. Paper 9730 presented at the 142nd Audio Eng Soc Convention, Berlin, 20–23 Mai 2017

Baxter D (2016) Dimensional sound. Resolution 15(5):41–48

Bramford JS (1995) An analysis of ambisonics systems of first and second order. Dissertation, University of Waterloo, Ontario, Canada

Daniel J (2001) Représentation de champs acoustiques, application á la transmission et á la reproduction de scènes sonores complexes dans un contexte multimédia. Dissertation, University of Paris VI, France

Eargle J (2004) The microphone book, 2nd edn. Focal Press

Elko G, Kubli R, Meyer J (2003) Audio system based on at least second-order eigenbeams. Int Patent WO 03(061636):A1

Meyer J, Agnello T (2003) Sperical microphone array for spatial sound recording. Paper 5975 to the 115th audio engineering society convention, New York

Nicol R (2018) Sound field. In: Geluso P, Roginska A (eds) Immersive sound. Focal Press, Routledge

Nicol R, Gros L, Colomes C, Noisternig M, Warusfel O et al. (2014) A roadmap for assessing the quality of experience of 3D audio binaural rendering. In: EAA joint symposium on auralization and ambisonics, Berlin, Apr 2014, pp 100–106

Pfanzagl-Cardone E (2020) The art and science of surround and stereo recording, Springer, Austria, part of Springer Nature. https://doi.org/10.1007/978-3-7091-4891-4_4

Theile G (1980) Über die Lokalisation im überlagerten Schallfeld. Dissertation, Technische Universität Berlin, Germany

Theile G (1991) On the naturalness of two-channel stereo sound. Paper presented at Audio Eng Soc 9th international conference, Detroit, 1–2 Feb 1991

Theile G, Wittek H (2011) Principles in surround recordings with height. Paper 8403 presented at the 130th Audio Eng Soc Convention

Theile G, Wittek H (2012) 3D audio natural recording. In: Proceedings to the 27.Tonmeistertagung des VDT, Cologne

Ward DB, Abhayapala TD (2001) Reproduction of a plane wave sound field using an array of loudspeakers. IEEE Trans Speech Audio Process 9(6):697–707

Wittek H (2000) Masters thesis, Institut für Rundfunktechnik (IRT)

Wittek H, Theile G (2002) The recording angle—based on Localisation Curves. Paper 5568 presented at the 112th Audio Eng Soc Convention, Munich

Wittek H, Theile G (2017) Development and application of a stereophonic multichannel recording technique for 3D Audio and VR. Paper presented at the 143rd Audio Eng Soc Convention, New York

Wittek H, Haut C, Keinath D (2006) Doppel-MS – eine Surround-Aufnahmetechnik unter der Lupe. In: Proceedings to the 24. Tonmeistertagung des VDT, Leipzig

Wittek H, Rumsey F, Theile G (2007) Perceptual enhancement of wavefield synthesis by stereophonic means. J Audio Eng Soc 55:723–251

Chapter 6
The Isosceles-Triangle, M.A.G.I.C Array and MMAD 3D (After Williams)

Abstract Based on several decades of experience in sound-engineering and in the demand for an optimized 3D audio recording and reproduction system, Michael Williams has arrived at designing univalent microphone/loudspeaker systems: each loudspeaker receives a signal from *one* microphone only. Based on his research on SRA (Stereophonic Recording Angle) for microphones with various directivities he divides the horizontal plane into segments with appropriate recording angle. In addition to the resulting 'surround sound array' (i.e. main layer) a 'height array' and zenith microphone are used for the upper hemisphere. Williams' 'Isosceles' triangle structures of microphone arrangements are presented in detail, which allow for correct localisation, based on time-of-arrival differences between microphones of the main and height layer. For his "Comfort 3D" array, a bottom layer of microphones is also added for an accurate, almost complete 'spherical' sound reproduction. After examining diverse versions of the Williams' M.A.G.I.C 3D microphone array we proceed to his last proposal, the MMAD 3D (3D Multichannel Microphone Array Design, now called "The Williams Tree"), which includes ETO (Electronic Time Offset) between the Height Array and the Main Array in order to steer the segment coverage angle in a desired direction, thereby obtaining perfect critical linking between the various parts of the system. Williams' most elaborate microphone arrays are comprised of up to 19 capsules and are compatible from 2-channel stereo via surround to various 3D audio arrangements.

Keywords Isosceles · Zenith microphone · 3D audio · MMAD · Mic arrays · Height information · Microphone technique · Spatial hearing

Before we go into more detail on the Isosceles principle as used by Williams it should be mentioned that 'Williams' shows designs using most first order directivity microphones, however many of his designs are fundamentally rooted in the use of microphone capsules with 'hypo-cardioid' (also known as 'infra-cardioid', 'sub-cardioid' or 'wide-angled cardioids') characteristics.

In Williams (2004) he explains why he prefers this characteristic for his designs:

"… The terms 'directivity factor' (DF) and 'directivity index' (DI) are the accepted way in which to define the total directional discrimination of a specific microphone directivity pattern.

E. Pfanzagl-Cardone, *The Art and Science of 3D Audio Recording*,
https://doi.org/10.1007/978-3-031-23046-2_6

Polar Diagram	Directivity Name	Attenuation at 180°	'A'	'B'	DF	DI
	Omni	0dB	1	0	1	0dB
	hypocardioid	6dB	0.75	0.25	0.583	-2.35dB
	hypocardioid 'Infra-cardioid or Wide Angled Cardioid'	10dB	0.66	0.34	0.474	-3.25dB
	hypocardioid	12dB	0.625	0.375	0.437	-3.6dB
	hypocardioid	15dB	0.59	0.41	0.404	-3.9dB
	"Cardioid"	infinity	0.5	0.5	0.333	-4.8dB
	hypercardioid	15dB	0.41	0.59	0.284	-5.4dB
	"Supercardioid"	11.5dB	0.366	0.634	0.268	-5.7dB
	hypercardioid	10dB	0.34	0.66	0.26	-5.8dB
	"Hypercardioid"	6dB	0.25	0.75	0.25	-6dB
	"Bidirectional" 'Figure-of-eight'	0dB	0	1	0.333	-4.8dB

Fig. 6.1 Microphone directivities and related directivity factors (DF) as well as directivity indexes (DI) (graphic is a reproduction of Table G from Williams 2004, p. 118)

Another term that is sometimes used to describe this same directional discrimination function is 'Random Energy Efficiency' (REE). The directivity factor is defined as a ratio, whereas the directivity index is the same energy ratio but expressed in decibels. The directivity factor is the ratio of the microphone's response (output level) to diffuse or reverberation sound coming from all directions around the microphone and with equal energy distribution, with respect to the response of a truly omnidirectional microphone of the same axial sensitivity and in the same acoustic environment. A cardioid microphone has a directivity factor of about 0.333, the directivity index is therefore 10 log(0.333) or −4.77 dB. Figure 6.1 shows the corresponding DF and DI for each of the specified directivity patterns.

To explain the meaning of the coefficients A and B in Fig. 6.1.

The mathematical model of directivity combines an omnidirectional directivity response with a bi-directional or figure-of-eight response, in varying proportions for each directivity pattern. The directivity response is given by the formula:

$$A + B\ \cos(\alpha) \tag{6.1}$$

where

$$A + B = 1 \tag{6.2}$$

The coefficient A gives the percentage of the omnidirectional response, and the coefficient B gives the percentage of the bi-directional response, the bi-directional response being represented by the term cos (α) where α is the angle of incidence of the sound wave. The representation of bi-directional directivity with the cos (α) in the mathematical model which is positive from 0° to 90° (and 270° to 360°) and negative from 90° to 270° corresponds to the inversion of polarity between the front and back of the real bi-directional microphone (i.e. between the front- and rear-lobe of the figure-of-eight pattern).

The case of one particular hypo-cardioid directivity is of particular interest. There is always an inevitable off-axis high frequency loss determined by the diameter of the diaphragm and the microphone body size. However with careful design, it is possible to integrate this change in directivity pattern into the overall directivity pattern of the microphone. If the microphone is designed with only pressure acoustic coupling in the high frequencies and pressure-gradient coupling for the rest of the audible frequency range, then the directivity pattern in the medium and low frequencies range can be made to correspond almost exactly with the natural directivity pattern at high frequencies—this is equivalent to designing the acoustic labyrinth as an acoustic low pass filter circuit. This is obviously different for each diaphragm diameter, but in the case of a small diaphragm condenser microphone, the high frequency directivity with pressure acoustic coupling corresponds to a hypo-cardioid directivity pattern with about 10 dB back attenuation (see Fig. 6.1). Schoeps have manufactured a microphone capsule (MK21) using this design approach. The result is a directivity pattern that in fact remains remarkably constant up to approx. 8 kHz. This directivity also has another advantage in that it has a better low frequency response compared to a cardioid microphone. Schoeps have given this directivity pattern the name 'Infra-cardioid' or 'Wide Angled Cardioid'…".

6.1 The 1st Layer—The M.A.G.I.C. Array

From (Williams 2012b):

> "… The main 1st layer of the 3D/Multiformat Microphone Array is the 7 or 8 channel Microphone Array Generating Interformat Compatibility (M.A.G.I.C. Array)…" as described in (Williams 2007) and (Williams 2008).

"… plus a 2nd layer array of vertically orientated figure of eight or super-cardioid microphones spaced at 52 cm…" as described in (Williams 2012a).

"… First of all it must be said that this type of array is a univalent microphone/loudspeaker system—each loudspeaker receives a signal from one and only one microphone. In addition, although the loudspeakers are positioned in the same order around the compass rose of reproduction, they do not necessarily have the same coverage segment angles as the microphones.

To understand the MAGIC Array concept it would seem appropriate to consider its development, stage by stage, starting with the front stereo pair. Most people will be familiar with the Stereophonic Recording Angle diagrams that I have published previously in AES preprints. In this case we will consider the SRA (Stereophonic Recording Angle) diagram for hypo-cardioid microphones with 10 dB back attenuation (see Fig. 6.1), a Schoeps MK21 or CCM 21 for instance…"

"… Figure 6.2 shows a particular solution for a SRA of +/− 45° using 90° between the axis of directivity of the microphones and 32.7 cm between the capsules. This means that the Stereophonic Recording Angle is the same as the angle between the microphones.

The next step is to apply these same dimensions to a quadraphonic square. Figure 6.3 shows the layout of a typical equal segment quad array—first published at the 91st AES Convention in New York in 1991 (see Williams 1991). This paper was updated to include up to eight channels in 2008" (see Williams 2008).

"…as shown in Fig. 6.3. From this figure we can see that the distance between hypo-cardioid capsules for a Quad Array must be 32.7 cm (see Figs. 6.3, 6.4 and 6.5)…"

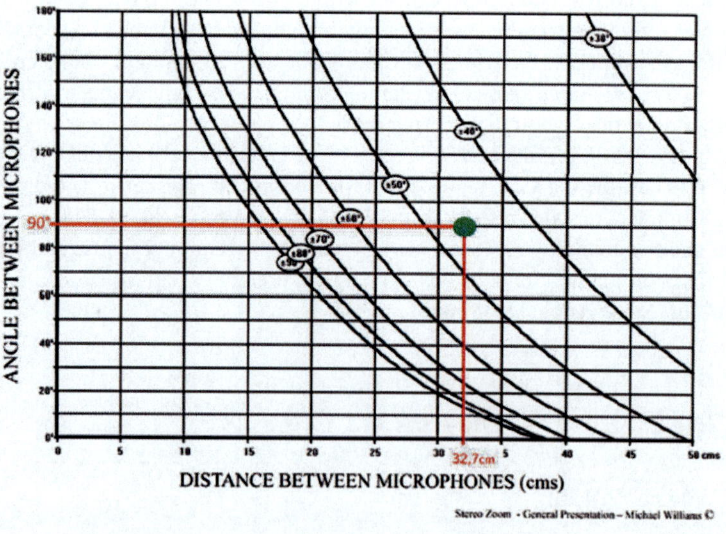

Fig. 6.2 Stereophonic recording angle diagram for hypo-cardioid microphones (from Williams 2012b)

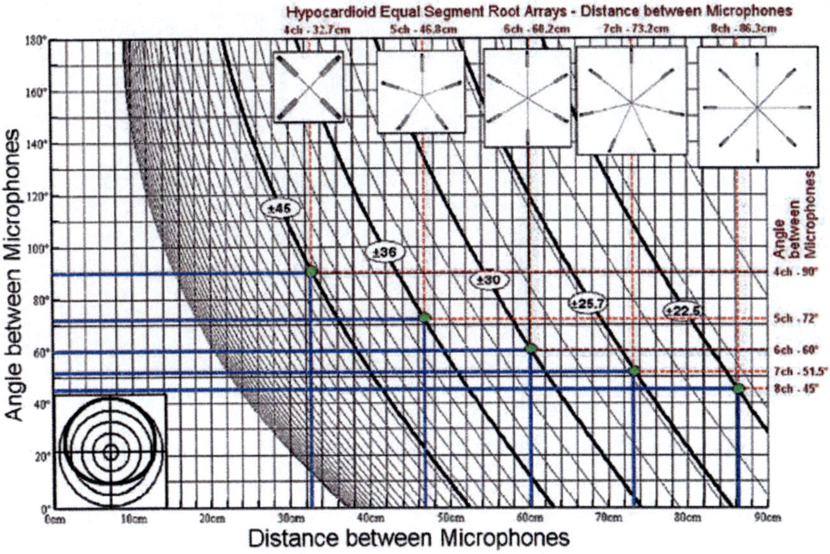

Fig. 6.3 The SRA diagram for hypo-cardioid microphones for equal segment microphone arrays (from Williams 2012b)

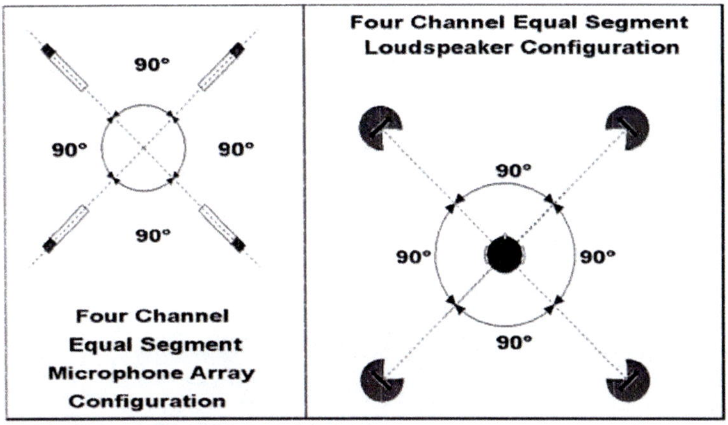

Fig. 6.4 The Quad Square (from Williams 2012b)

> "…Using just the Quad Square it is also possible to create a mixdown to stereo that will fold the surround sound segments into the front stereo sound image—this is called a 'Twisted Quad' mixdown…" and is described in (Williams 2005).

"…The following mixdown algorithm should be used—LS is mixed with the Right channel, and RS is mixed with Left channel. The percentage of RS and LS

Fig. 6.5 Close-up of central (hypo-cardioid) quad square, which is used in the core position of the full array (photo by Mickael Kennessi; from Williams 2012b)

mixed with the front Left and Right channels respectively will depend on the ratio of direct to reverberant sound that is desired.

The next stage is to introduce the satellite microphones on the long arms placed on the bisector or each of the angles within the Quad Square. These microphones are placed on arms of about 1.20 m long…" as shown in Fig. 6.6. The theory of this configuration is explained in detail in (Williams 2007) and (Williams 2008).

> "… In order to simplify the explanation of the configuration it can be considered that each satellite microphone will produce two new coverage segments to the clockwise and anti-clockwise sides of each satellite microphone. The sum of two adjacent segments for example between the Left and Centre (the satellite) microphones, and the Centre (the satellite) and Right microphone will be equal to the total segment coverage between the Left and Right microphone, as shown in Fig. 6.7…"

> "…In other words the Left Front Segment Coverage must be a total of 45° coverage, and the Right Front Segment Coverage must be also 45°. The Total Coverage of the Front Segment therefore adds up to 90° (i.e. ± 45° when talking about a stereo segment). The only way to obtain such a small segment coverage (45° or ±22.5°) for the LFSC and RFSC is to increase

Fig. 6.6 The M.A.G.I.C array (Microphone Array Generating Interformat Compatibility) using the 'WilliamStar' microphone array support system (from Williams 2012b)

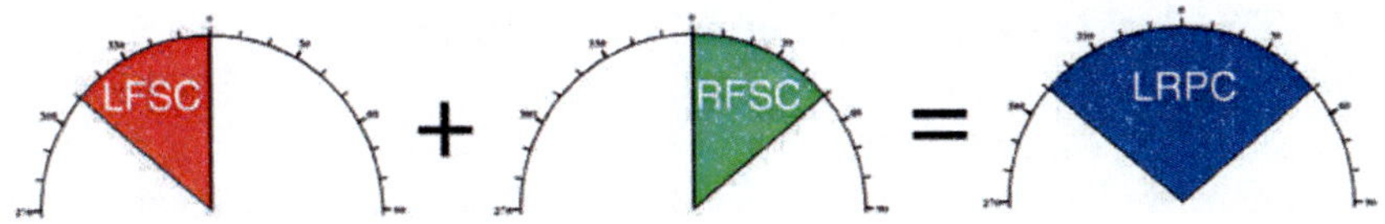

Fig. 6.7 Segment coverage of the front segment (from Williams 2012b)

> the distance between the Left & Centre, and Right & Centre microphones. The segment coverage in both cases will unfortunately be offset to the left and to the right. The only way to bring the segment coverage so that it corresponds to the actual angle between the microphones is to apply a 3 ms delay to the Centre microphone…"

This is again explained in detail in (Williams 2007, 2008).

> "…When this construction is applied to each of the four segments, it will create an eight channel master recording array,…" as shown in Fig. 6.6.

"…The main advantage of this MAGIC configuration is that it will be compatible with many different reproduction formats:

- Stereo—by using just the Left and Right microphones
- Or Twisted Quad stereo by mixing the LS and RS into the Right and Left microphone channels respectively
- Quadraphony using only the Quad Square of microphones
- 5 Channel reproduction using the Quad Square plus the Centre satellite microphone
- 7 Channel reproduction using the Quad Square plus the Centre, Left Median and Right Median satellite microphones.
- 8 channel reproduction using the Quad Square plus all the satellite microphones…"

"…Remember, this type of microphone array is a univalent microphone/loudspeaker system each loudspeaker receives a signal from one and only one microphone.

The system for the 1st Layer of sound recording described above uses hypo-cardioid microphones. This is by no means the only possibility, cardioid or super-cardioid microphones can also be used as long as the distance parameter is adjusted to take this into account. It may also be interesting to use hypo-cardioid microphones for the Quad Square and only cardioids for the satellite microphones. This configuration using cardioids (for the satellite microphones) tends to produce a better balance in bass frequency response when using the 5 or 7 channel reproduction configurations, otherwise the bass response may become a little over-powering, especially in 7 channel reproduction…".

6.2 The 2nd Layer—Addition of Height Information

As described in (Williams 2012a):

"...no height information can be generated by loudspeakers placed only in the horizontal plane—the complex pattern of localization cues that can be used in Binaural Technology to produce this height information is completely destroyed by the high level of acoustic crosstalk in loudspeaker reproduction.

We therefore need to resort to a 2nd layer of loudspeakers to initiate a natural perception of height. However localization information generated by this 2nd Layer must not be in conflict with the main horizontal plane localization information. Two approaches are possible:

- eliminate as far as possible the interaction between the two planes of sound catchment using either Figure of Eight (or Super-cardioid microphones), with the maximum attenuation angle of the directivity pattern directed towards the direct sound source
- if interaction exists, then it is necessary to construct the 2nd layer of sound catchment in such a way as to generate localization information that does not conflict with the main horizontal plane.

The first approach was the one adopted for the GOArt project in Göteborg, and also for a previous pilot recording of contemporary music at the Watford Colosseum in London.

However in critical listening tests conducted after this series of recordings, it was found that it was not possible to completely eliminate any interaction between the two layers of catchment.

The second layer was made up of four figure of eight microphones pointing at 90° to the horizontal plane and spaced at 52 cm between each capsule. With the Schoeps CCM 8 Figure of Eight microphones the directivity patterns are actually pointing upwards at 90° to the body of the microphone..." as shown in Fig. 6.8.

"...They are positioned at 0°, 90°, 180° and 270° with respect to the compass rose and the 2nd layer is about 1 m above the main 7 channel MAGIC array, as shown in Fig. 6.9...".

6.2.1 *The Reason for Choosing 52 cm Between the 2nd Layer Capsules*

"...The Figure of Eight directivity patterns are all orientated vertically, so we can consider that this creates a 'Time-Difference-only' sound recording system. From any of the SRA diagrams we can see (by extrapolation) that 52 cm between the capsules (with 0° between the axes of directivity) means that we have a Stereophonic Recording Angle (SRA) of ±45° for each coverage segment (a total of 90° coverage for each segment). For this to be absolutely true the microphone array must be

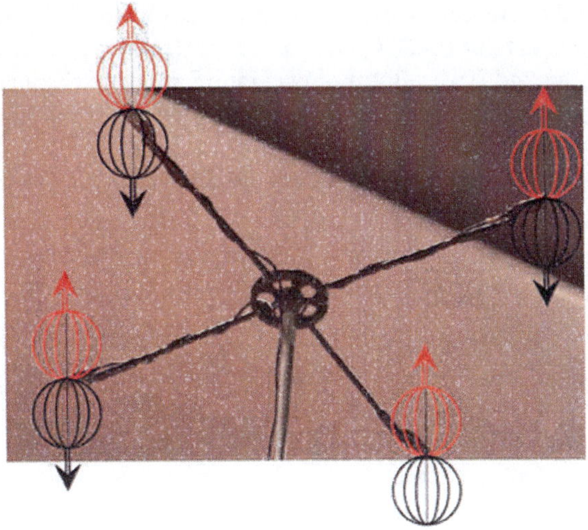

Fig. 6.8 The 2nd layer—the vertical fig-8 cross (photo by Michael Kenessi, from Williams 2012b)

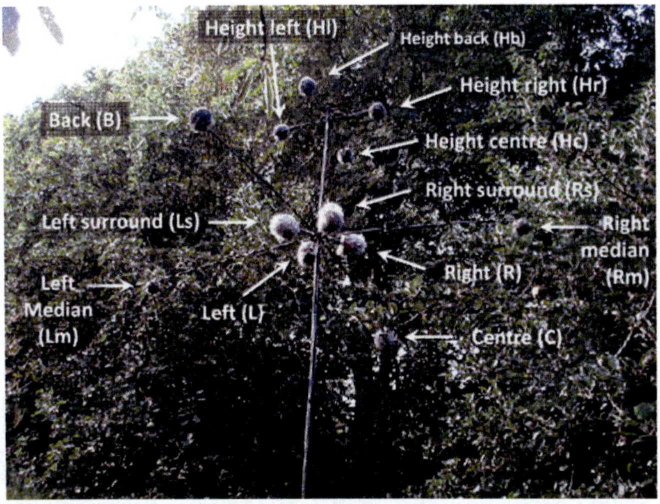

Fig. 6.9 The combined 1st and 2nd layer—the full 12-channel array (from Williams 2012b)

placed in plane wave propagation conditions—in other words the sound source must be relatively distant (at least five times the distance between the microphones). This indeed was the situation in all the recordings where this system was tested—the 2nd layer capitation was essentially reverberation information. We can consider that, in the full 12 channel microphone array configuration, there could be redundancy between the four satellite MAGIC array microphones and the height microphones.

In fact a perfectly satisfactory 3D sound image is obtained by using only eight microphones—the central Quad Square part of the MAGIC array and the Figure of Eight Cross of the 2nd Layer. But if only eight microphone channels are recorded then this would mean that the master recording would not contain information that was compatible with the 5 channel multichannel format or the 7 channel Blu-Ray format. So in practice all 12 channels must be recorded for full compatibility…".

6.3 The Primary Isosceles Triangle Structure

"…The relative positions of the main MAGIC layer and the 2nd or top layer is such that each 2nd layer microphone forms an Isosceles triangle with two of the microphones in the Quad Square. We will call the 2nd layer microphones HC (Height centre at 0°), HL (Height left at 270°), HR (Height right at 90°) and HB (Height back at 180°).

As can been seen in Fig. 6.10, the primary set of isosceles triangles are formed by:

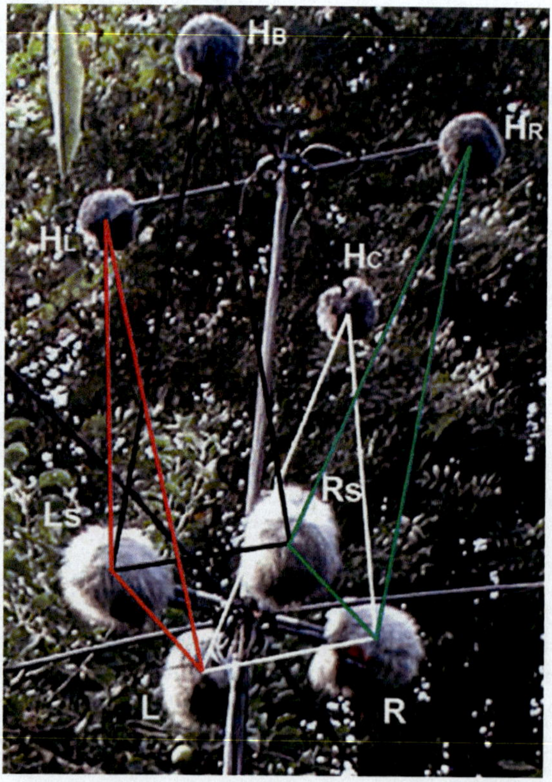

Fig. 6.10 The primary isosceles triangle in the 12 channel microphone array axes of the vertical layer switch (from Williams 2013)

- L, R and HC in the front (in white)
- LS, L and HL on the left hand side (in red)
- R, RS and HR on the right hand side (in green)
- Rs, LS and HB at the back (in black) …" (from Williams 2012b)

In (Williams 2012a), Williams expressed the impression that "…localization in the vertical plane using a standard horizontal array of loudspeakers was extremely difficult to achieve. This was based on examination of the HRTF characteristics in the vertical plane. In Binaural Technology, using headphones or earphones very carefully matched to the individual listener's HRTF, vertical localization can be quite satisfactory. Acoustic crosstalk in loudspeaker listening makes this process almost impossible. However very precise crosstalk cancellation, and listener HRTF matching can produce a similar effect but it is highly listener position dependent (within a few millimeters). …"

6.4 Psychoacoustic Experiences with the M.A.G.I.C. System

"…The introduction of a 2nd layer of loudspeakers already introduces a natural vertical localization sound source. This means that we could expect virtual localization between loudspeakers to fulfill the role of creating vertical dimension localization and therefore with the complete configuration of loudspeakers, 3 dimensional space reproduction.

In the design of the 3D Multiformat Array used in this series of recordings in Göteborg, some difficulty in vertical localization was expected when loudspeakers were mounted one above the other, but it was considered probable that localization from diagonal pairs of loudspeakers would be projected only onto the horizontal plane and not the vertical plane.

Careful psychoacoustic testing of vertical and diagonal localization was to produce some very interesting results. But first of all a test recording was made in the anechoic chamber in the Acoustics Department of the University of Göteborg. A series of Level Difference and Time Difference signals were generated using a standard horizontal microphone pair (25 cm/90° cardioids)—the sound source moving around the microphone pair in the horizontal plane.

Afterwards, in the listening tests, the two signals were routed to either just the vertical pair of loudspeakers or to just a diagonal pair of loudspeakers.

No precise localization was experienced in the vertical plane, however very precise localization was observed along the diagonal plane. In the first case this confirmed what was expected…" (and described in Williams 2012a),

"…whereas the second case was a complete surprise. In the second case localization was expected to be projected onto the horizontal plane with no vertical component, but in reality the localization followed the line between the diagonal loudspeakers and was a clear and realistic reproduction of the sound source.

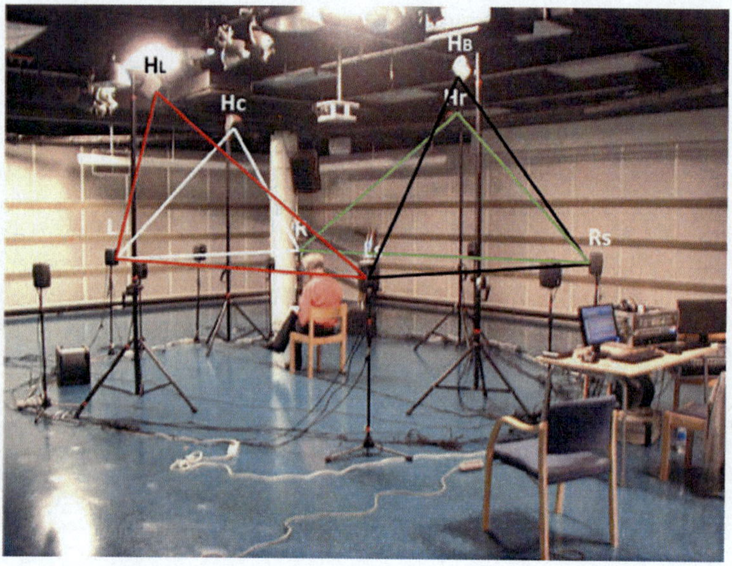

Fig. 6.11 The primary isosceles triangle structure in replay in a temporary listening room installed at the Applied Acoustics Department of Göteborg University (from Williams 2012b)

These observations, concerning the localization characteristics of the loudspeaker configuration, completely justify the primary isosceles triangle structure of the experimental 3D Multiformat Microphone Array.

This means that we can expect no reliable sound source localization on loudspeakers that are situated vertically one above the other, but that loudspeakers placed in an isosceles triangle structure around the listener …" (as shown in Figs. 6.11 and 6.12), "…will produce reliable virtual localization of sound images. Of course the microphone array structure must be the mirror image of the loudspeaker structure. But this does not mean that the microphones have to be in exactly the same orientation as the loudspeakers, but the general univalent triangular structure must be the same. …" (from Williams 2012b)

6.5 The 7 Channel Listening Experience

In a previous VDTonmeistertagung (see Williams and Le Du 2000) Williams made a demonstration of the compatibility of the 7 channel multiformat microphone array (M.A.G.I.C) with 4 or 5 channel systems. In that demonstration he showed how it was possible to change from 4 to 5 to 7 channels without any matrixing, by only muting the channels that were not required. In this demonstration there was little appreciable change in the total surround sound image when switching from 4 to 5 channels. However the one remark that was made, was that there was an increase in

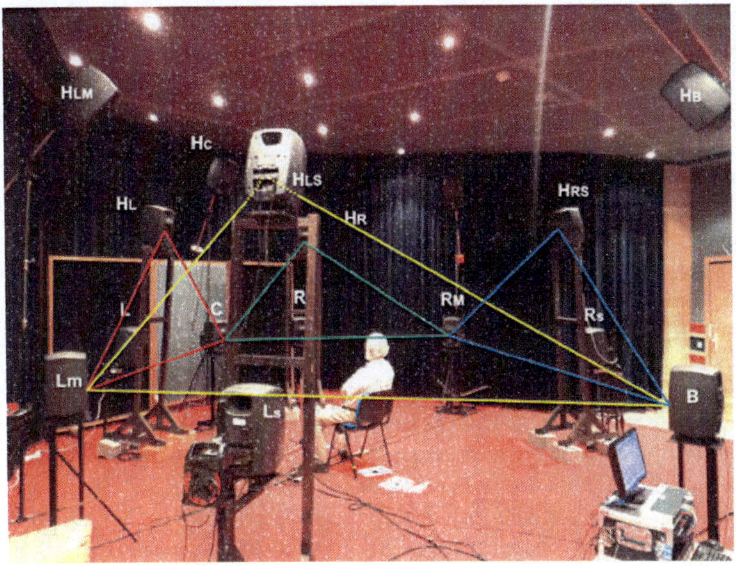

Fig. 6.12 The isosceles triangle structure, temporary setup at Galaxy Studios, compatible with 12 channel AURO 3D (from Williams 2013)

bass reproduction with the 7 channel system. This remark has been frequent in the many demonstrations that he has done of this system (Fig. 6.13).

> "...The reason seems to be that below a certain aliasing frequency the reproduction passes from individual loudspeaker spherical wave front propagation to a combined loudspeaker cylindrical propagation response. The actual energy reaching the listener is therefore greater in the cylindrical propagation situation compared with the spherical propagation situation..." (Fig. 6.14)

"...The aliasing frequency occurs when the half wavelength is equal to the distance between the loudspeakers. If the distance between the loudspeakers is about 1.70 m then the aliasing frequency is about 100 Hz.

In the 3D Multiformat Microphone Array the passage from a 3D eight channel system to a 3D twelve channel system (that is when the C, Lm, Rm and back loudspeakers are added to the eight channel system) the increase in bass response could explain the slight improvement, or at least preference, of listeners for the full 3D twelve channel system.

A simple solution to this problem would be to change the directivity pattern of the C, Lm, Rm and B microphones from hypo-cardioid to cardioid thereby decreasing this bass frequency imbalance. The reason being that the cardioids have less bass response than hypo-cardioids, therefore in the extreme bass response there would be less contribution from the cardioids, compared to the hypo-cardioids, in cylindrical wave front propagation.

In these listening tests it was found that time alignment of the sources was extremely critical—this time alignment can either be obtained by changing the physical position of the loudspeaker (the distance from the loudspeakers to the listener must be exactly the same) or by electronic delay so that the arrival time at the listener, is exactly the same, for an identical signal from each loudspeaker.

The complete 12 channel array does have some redundancy in that some of the channel components are not necessarily needed. The HC channel is reproduced with loudspeakers that are positioned immediately above the C channel loudspeaker. This also applies to the HL channel with respect to the Lm channel, the HR channel with respect to the Rm channel, and the HB channel with respect to the B channel. It was found that the system performed well as a 3D reproduction format when only eight channels were reproduced—L, R, LS, RS, HC, HL, HR, and HB. That is the extremities of each Isosceles triangle in the structure. Only a slight improvement or preference was observed when the C, Lm, Rm and B channels were reintroduced, but this is probably due to this artificial increase in bass response when all hypo-cardioid microphones are used in the 1st layer array.

This means that perfectly satisfactory reproduction of the 3D sound-field is possible with only 8 channels (or even 7 channels if we use the 7.0 Blu-ray format—but of course no back channel is available in this case).

If the listening tests are carried out in a 22.2 loudspeaker configuration then only certain loudspeakers should be used. The correspondence between the 12 channel reproduction system and specific loudspeakers in the 22.2 configuration is shown in Table 6.1. …" (from Williams 2012b). (Rem.: only the loudspeakers with channel numbers 1-6, 9-12, 15 and 19-21 are considered compatible with the twelve channels of the 3D Multiformat Microphone Array) (See also Fig. 6.15)

6.6 The Listening Tests

> "…For the listening tests, MAGIC array with height channel recordings were used from four different churches (GOart Project). No audience had been present during the recording sessions, therefore the listeners had no previous knowledge of the techniques used to create the recordings…"

"… They were however asked to attend live listening session in the churches afterwards, listening to the same extracts as had been recorded. (No recording equipment was present for these later sessions.) They were asked to fill out a questionnaire concerning their appreciation of the sound of the organ and the church environment.

The listening panel was then invited to attend, individually, listening tests at the Applied Acoustics Dept. of Göteborg University where they were able to hear the original recordings. They were again asked to fill out a questionnaire concerning their perception of the sound of the organ and its acoustic environment.

In each church, three specific positions of the microphone array were recorded, with small variations in position according to the organ structure and acoustics of each

Table 6.1 Table of equivalents between the 22.2 loudspeaker configuration and the 12.0 multi-format configuration (from Williams 2012b)

Channel number	Channel name	Label	Azimuth	Elevation	Distance (m)
1	Front Left	FL	45	0	1.55
2	Front Right	FR	–45	0	1.55
3	Front Centre	FC	0	0	1.55
4	Low Frequency Effect 1	LFE1	30	0	1.55
5	Back Left	BL	135	0	1.55
6	Back Right	BR	–135	0	1.55
7	Front Left Centre	FLc	30	0	1.55
8	Front Right Centre	FRc	–30	0	1.55
9	Back Centre	BC	180	0	1.55
10	Low Frequency Effect 2	LFE2	–30	0	1.55
11	Side Left	SiL	90	0	1.55
12	Side Right	SiR	–90	0	1.55
13	Top Front Left	TpFL	45	45	1.55
14	Top Front Right	TpFR	–45	45	1.55
15	Top Front Centre	TpFC	0	45	1.55
16	Top Centre	TpC	0	90	1.55
17	Top Back Left	TpBL	135	45	1.55
18	Top Back Right	TpBR	–135	45	1.55
19	Top Side Left	TpSiL	90	45	1.55
20	Top Side Right	TpSiR	–90	45	1.55
21	Top Back Centre	TpBC	180	45	1.55
22	Bottom Front Centre	BtFC	0	–39	1.55
23	Bottom Front Left	BtFL	45	–39	1.55
24	Bottom Front Right	BtFR	–45	–39	1.55

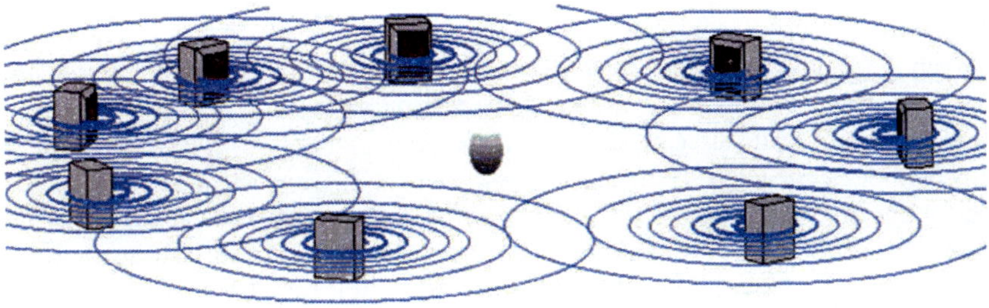

Fig. 6.13 Above the aliasing frequency: individual loudspeaker spherical wave front propagation (from Williams 2014)

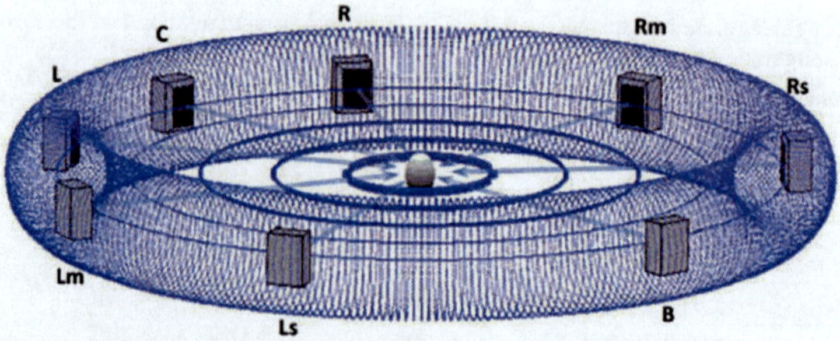

Fig. 6.14 Below the aliasing frequency: coupled loudspeaker toroidal ring wave front propagation (from Williams 2014)

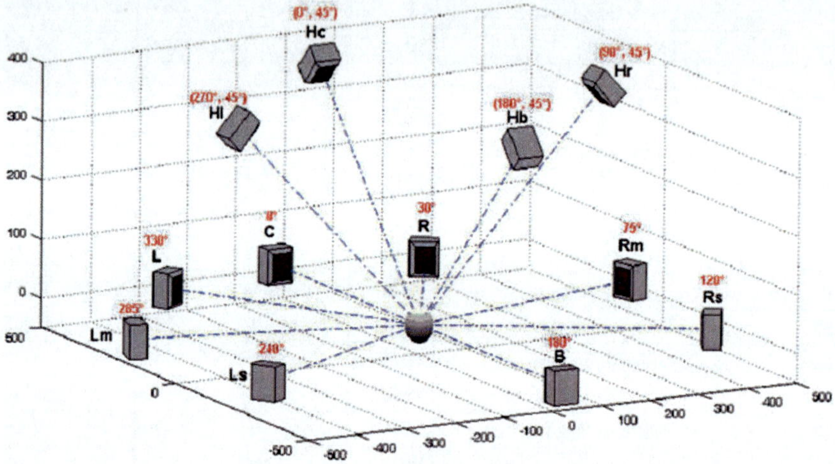

Fig. 6.15 Schematic of 12 channel loudspeaker replay set-up, compatible with 12 channel AURO 3D (from Williams 2014)

church. There was a close correspondence between the live listening impressions and the 3D reproduction listening tests. One person in particular on the listening panel was able to locate the three positions of the recording array to within 50 cm, without any previous knowledge of the actual positions used. This was indeed a remarkable performance, and confirmed that this person had an exceptional knowledge of the timbre of the organ (of the Ôrgate Nya Kirka), coupled with a considerable experience of the acoustics of the church—he was in fact responsible for the manufacture, tuning and voicing of the pipes in the organ! However this remarkable identification of the recording positions would not have been possible if the microphone array recording and reproduction system itself did not produce an exceptionally realistic reproduction of the organ and its surrounding acoustic environment…”.

"...Only one criticism could be made of the listening system in that the loudspeakers had a bass roll-off frequency at about 38 Hz, so the real bass frequency reproduction of the organ was not totally satisfactory. LFE channels were tried but proved unacceptable.

However the mechanical assembling of the recording system was considered too long (from 2 to 3 h)—further work has to be done in simplifying the assembly procedure. The other limiting factor was the considerable expense of a complete twelve channel microphone array and support system. Although this was considered a major difficulty for a budget that was to be allocated for this project, it was recognized that this was certainly necessary if a very high quality recording was to be achieved, both with respect to the timbre restitution and also the 3D spatial reproduction. It was also considered that the specification of the listening room would have to be upgraded if a permanent listening room was to be installed. ..." (from Williams 2012b).

6.7 Experiments in Vertical Localization and Mic-Array Proposals

In Williams (2016a) results on the very interesting challenge of vertical virtual localization with microphone arrays and loudspeaker arrays related to his Isosceles Triangle approach are presented:

"...The basic principles of design for height reproduction are applicable to most of the surround sound microphone array structures that make use of both Time Difference and Level Difference, to generate virtual surround sound localization in the horizontal plane.

The research into elevation reproduction can be divided into two segments:

1. the first 45° of elevation,
2. the last 45° of elevation, from +45° to +90°, 90° being immediately above the listener. There is no reason why this could not be extended to 3 segments or more, but economy of channels is a major consideration, and localization with 2 segment coverage is perfectly satisfactory. In the reproduction of virtual sound, it is not true to say that the more channels we have, the better is the 3D sound reproduction. This may be the case in object oriented mixing but not in microphone array recording and reproduction.

The research for localization within the first 45° of elevation can itself be divided into two phases:

1. the research necessary to establish the relative contribution of Time Difference and Level Difference to localization in elevation
2. the optimum position of loudspeakers in the reproduction configuration, necessary to reproduce the most accurate localization.

It was also considered necessary to maintain the same type of univalent structure in relation to the microphone array recording system, and the loudspeaker reproduction set up.

The addition of a Zenith microphone produces optimum results in the upper segment from 45° to 90°, but is not absolutely necessary if we are only concerned by the reverberation field around and above us, and not interested in the precise and accurate localization of direct sound in the complete hemisphere above the listener. But with the introduction of a Zenith channel and with close similarity between the recording and reproduction configurations, we will of course satisfy both 3D (or hemispherical) virtual localization in the and reproduction of direct sound sources in the elevation (for example for birds or helicopters, or even large musical instruments with a strong height component such a church organ, etc. etc.)…".

6.7.1 *The Psychoacoustic Parameters for Vertical Virtual Localization in the First 45° Segment of Elevation*

A previous paper (see Williams 2013) Williams has described the choice of parameters that are necessary for good localization in the first 45° of elevation.

"…Two basic characteristics must be considered:

1. Height information is mainly captured by a Time Difference system
2. An Isosceles array structure with respect to the horizontal microphones and the first elevation level of microphones will produce optimum results…" as shown for a recording array in Fig. 6.10, and the loudspeaker configuration for reproduction in Fig. 6.11.

> "…Vertical Level Difference information was found to be almost without interest at all, and produced either no height localization at all, or a very mediocre result compared with Time Difference information present…"

6.7.2 *The Psychoacoustic Parameters for Vertical Virtual Localization in the 45° Segment of Elevation from + 45° to +90°*

"…Two basic options are available to experiment the quality of localization in this last segment:

1. A microphone array structure using only Time Difference information—nicknamed 'The Witches Hat Approach or Witches Hat Localization'
2. A microphone structure that uses both Level Difference and Time difference information—nicknamed 'The Top Hat Approach or Top Hat Localization'.

It is hoped that the image that is projected by these two approaches is one of

(A) Time Difference only, in both the lower elevation and upper elevation segments for the Witches Hat.
(B) a combination of Time Difference in the first elevation segment, coupled with Time and Level Difference for the second upper segment of elevation…".

6.7.3 *The "Witches Hat" Localization System*

"…'Witches Hat' Localization uses only Time Difference information for localization in both upper segments, from 0° to +45°, and from +45° to 90°…" as shown in Fig. 6.16.

> "…This is of course valid only for a microphone array structure using a horizontal array, a first elevation array and a top Zenith microphone. The height of the Witches Hat is meant to represent the distance that must separate the horizontal array and the first elevation array to produce localization in the first elevation segment (Time Difference parameter is predominant), and the distance that must separate the first elevation array from the top Zenith microphone (again Time Difference parameter is predominant). Figure 6.17 shows a practical 8+4+1 microphone array under test…" (Fig. 6.17)

"…Both laboratory tests and on-location recording were carried out to study the characteristics of this type of microphone array structure. Recordings were made:

1. in a studio environment with a loudspeaker as a sound source and a Witches hat microphone array structure.
2. on-location, with a the occasional helicopter passing overhead, and general environmental sound. These recordings were made on a small island, so there is also the sound of barges passing by, in the horizontal plane.
3. On location during the Lossiemouth Airshow in Scotland with aeroplanes of all types passing around and above the microphone recording array.

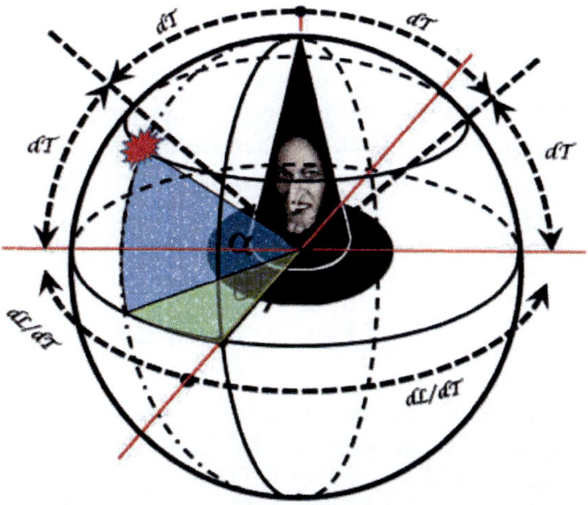

Fig. 6.16 "Witch's Hat" localisation in the upper hemisphere (from Williams 2016a)

Fig. 6.17 8+4+1 Array in "Witch's Hat" configuration (from Williams 2016a)

It was found after multiple listening tests, that this design structure gives very good localization in the first elevation segment, but very poor results in the top elevation segment. If the Zenith microphone is muted, then first elevation segment localization is excellent but the top localization follows reproduction in the square generated but the first elevation array structure. A helicopter flying above the array will give the impression of approaching the array system from the desired direction, but immediately it passes overhead it will follow one side of the square and then the other, and then will give a correct direction restitution as the helicopter flies away in the opposing direction. This effect goes completely unnoticed when recording music, as we are not really looking for perfect localization of the reverberant field above…".

6.7.4 The "Top Hat" Localization System

"…Top Hat Localization uses Time Difference only information for localization in the first elevation segment from 0° to 45°. Whereas the second elevation segment

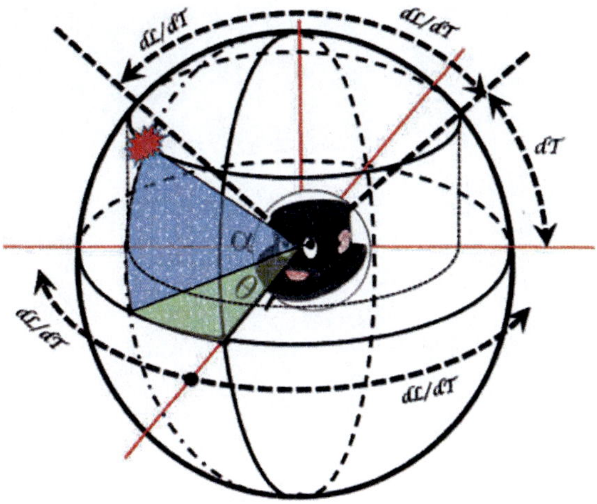

Fig. 6.18 "Top Hat" localisation in the upper hemisphere (from Williams 2016a)

uses a hybrid system, where both Time Difference and Level Difference contribute to localization as shown in Fig. 6.18…"

> "…In other words the horizontal surround sound and the 2nd segment use the same parameters to obtain Localization from 45° to 90°, i.e. both Time Difference and Level Difference, as the horizontal surround sound layer. The two surfaces—the horizontal surround sound and the top elevation segment use the same parameters for localization, whereas the segment from 0 to 45° uses only Time Difference to obtain good localization. Figure 6.19 shows this type of array structure…"

> "…This structure would seem logical if we consider that top elevation plane is also horizontal or -more correctly- parallel to the horizontally plane—the brim of the hat and the top of the hat! (Rem.: One should not be deceived by the direction of the 1st elevation layer of the microphone in Fig. 6.19, as the directivity axis is actually at 90° to the body of the microphones.)..."

Figure 6.20 shows the test recording setup with a Blue Tooth loudspeaker as a sound source being moved around the microphone array by Michael Williams.

Fig. 6.19 A "Top Hat" 8+4+1 3D microphone array (from Williams 2016a)

Fig. 6.20 360° Rail, blue tooth loudspeaker sound source, and 13 channel Top Hat Array under test (from Williams 2016a)

"…According to the listening tests it turned out that the 'Top Hat' design of array structure gives excellent localization results in both segments of elevation.

Complete coverage and quality localization were achieved in the whole of the hemisphere above the microphone array—hence the name of "Integral 3D" has been given to this type of coverage. The laboratory tests indicate that the precision of localization in the complete hemisphere around the array and at various angles of elevation is also completely satisfactory. The exact quantitative verification of this localization function during recording and reproduction presents a number of difficulties. This is due to the necessity to measure exactly the physical sound source position, but even more so, to measure the perceived localization position. An experiment to determine this correspondence function between the two situations is under consideration…".

6.7.5 "Integral 3D" Compatibility Tests

"…During tests carried out at the Chalmers University in Gothenberg in 2013, it was established that surround and elevation localization was completely acceptable if the horizontal part of the array structure was reduced to only 4 channels (a "4+4" array). On the other hand if compatibility with all other surround and 3D systems was required then the full 12 channel array structure (an "8+4" array) was necessary.

However this 4 channel horizontal array structure—leading to a "4+4+1" 3D array calls for more analysis.

It is widely accepted that a quadraphonic surround sound array produces a more or less acceptable reproduction for surround sound, but it can be criticized because the linearity of reproduction in each segment is far from satisfactory because each segment is quite wide at 90°. This is improved considerably by the use of a 5 channel surround sound array with 72° segments, as well as, of course, the 6 channel surround array. Also, it would be hard to deny the superiority of the 7 channel Blu-ray configuration. When adding in a back channel to produce an eight channel surround array then the result is almost perfect, that is for surround sound.

In experiments during GOART Gothenberg project we were able to study the compatibility characteristics of a "4+4" microphone 3D array, with the 8 channel surround reproduction configuration.

It was found that excellent surround sound could be obtained if the upper square of 4 microphone were positioned at 45° to the lower 4 channel array, and the upper 4 channel array was reproduced by routing each microphone to the lower 8 channel loudspeaker array. In other words, if we consider that the quad array is made up of Left, Right and LS and RS, then:

1. Height Centre (at 0° azimut and 45° elevation) is routed to the Centre of the 8 channel loudspeaker surround configuration elevation) is routed to the back of the 8 Channel loudspeaker surround configuration
2. The Height Right Median channel (at 90° azimuth and 45° elevation) is routed to the Right Median loudspeaker of the 8 channel loudspeaker configuration
3. The Height Left Median (at 270° azimuth and 45° elevation) is routed to the Left Median loudspeaker of the 8 channel loudspeaker configuration.

(Rem.: The compass angles rotate in a clockwise direction) The arrangement described above fits with the loudspeaker layout..." in Fig. 6.13.

"...For the "4+4" 3D configuration the Centre, Left, Right, Lm, Rm, LS, RS and Back loudspeakers are all active. Signals from HC, HL, HR and HB are folded into the Centre, Rm, Lm and Back loudspeakers of the surround sound configuration.

The extraordinary thing is that the 3D reproduction is quite satisfactory, and the surround sound configuration is almost perfect. The surround sound configuration is almost indistinguishable from the natural 8 channel surround sound recording and reproduction configuration.

This is the case whether we work with a 4+4 array, a 5+5 array, a 6+6 array, etc., as long as the microphones in elevation have an isosceles triangle structure in relation to the horizontal array structure.

Most people nowadays have opted for a 5 channel structure for the horizontal array. It would seem logical to adopt a 5+5 3D array structure to help compatibility from one context to another.

The addition of a Zenith microphone to the previous microphone array structure does not change the compatibility characteristics, except of course that the Zenith microphone is eliminated (muted) from the projection that is made onto the horizontal array structure.

The full dual set of quintuple arrays plus the Zenith channel becomes a 5+5+1 integral 3D. This type of array will therefore be compatible with any of the lower order 3D and surround sound formats…".

6.7.6 From "Integral 3D" to "Comfort 3D"

"…This development comes from one remark made during the listening tests at Galaxy Studios in 2015. One of the listening panel said, quite spontaneously, that reproduction of the sound source at 15° and 30° degrees of elevation seemed more comfortable than reproduction in the horizontal plane. This remark deserves very much deeper analysis:

Mono sound recording restricts the sound reproduction to one loudspeaker. And we do not look for any space around that loudspeaker. However in Stereo Sound reproduction we are so taken by the stereophonic spread of sound that we do not easily realize that sound that comes from outside the Stereophonic Recording Angle (SRA) is still being reproduced as mono sound on either the left or right loudspeakers—in most cases reverberation from all around the stereo pair.

Sound from above and below the stereo array is either spread out between the two loudspeakers, or again reproduced as mono sound on the left and right loudspeakers. When we move to surround sound, then again the horizontal spread of sound is very satisfactory, but does not remove the fact that sound from above and below the array is condensed onto the horizontal spread of sound.

When we move on to the so called 3D reproduction, we again are so taken with the 3D reproduction spread or more correctly the hemispherical reproduction of sound, that we do not perceive easily the fact that the lower sound field is again being projected onto the horizontal surround sound spread.

The remark that sound is more comfortable at 15° or 30° degrees elevation then takes on a new meaning. It is simply that the indirect sound architecture is being distributed both above and below the direct sound source. The sound source is generally distributed over 10° or15° above and below the horizontal plane of recording. Therefore the recording of the direct sound which is usually in the horizontal plane of the array system, does not have a component below the horizontal plane. It is obvious that one cannot record and then reproduce the direct sound at an angle of around 10° or 15° of elevation as we are too used to perceiving this sound along the horizontal plane…"

"…However why not introduce an array that will record and reproduce sound both above and below the horizontal plane. We do not need to consider the sound from directly under the array as there will be no significant sound from that direction, so the 'Voice of the Devil' channel is not required.

And so we have the extension to 'Integral 3D', that now becomes 'Comfort 3D', the lower segment under the horizontal plane can now record and reproduce the full architecture of the sound source, to the left, to the right, to the top, and to the bottom. We then obtain 'comfortable' sound reproduction—the sound is reproduced with its

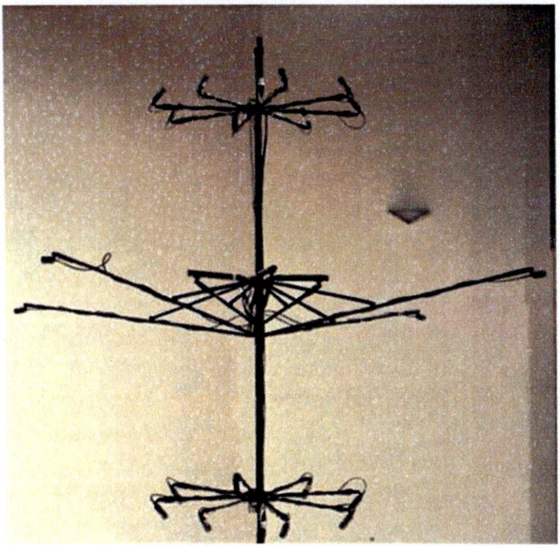

Fig. 6.21 The "Comfort 3D" microphone array (from Williams 2016b)

full architecture, or the full extent of the surface of acoustic radiation. The lower elevation array and the upper elevation array are, of course, both oriented at 45° to the horizontal plane array.

The story is not completely finished, as we still have to rely on the brain to reconstruct depth from various acoustic clues, like direct-to-reverberant sound ratio, or the expected timbre variations with distance, etc. But at least we have at least 75% of the 3D sound field covered…" (from Williams 2016a) (Fig. 6.21).

6.7.7 MMAD 3D Audio

In (Williams 2022a) the author writes:

> "…I am specifically researching into the accurate reproduction of sound source localization (or as near as is possible), its associated acoustic architecture, as well as the natural acoustic environment in the complete hemisphere (or sphere) around us. This, to me, is the summum of good quality spatial sound recording. Many pseudo height microphone array recording systems have been created recently, specifically for use in the context of the cinema or home-cinema industry. These pseudo height systems are simply a combination of two surround sound systems placed one above the other—essentially two layers of surround sound reproduction. It must be understood that these systems have absolutely nothing to do with creating any real impression of height. Although the result in reproduction is far from satisfactory for the discerning listener of 3D, it seems to have been accepted, in the context of cinema or home-cinema reproduction, in giving an impression to the public of something happening above their heads—the listener gets the impression of what is now being called 'an immersive' sound-field…"

"...The MMAD 3D Audio system, on the other hand, is designed to generate a realistic virtual 3D Audio localization of sound in the upper hemisphere, and in both upper hemisphere and in the lower hemisphere for the most recently developed systems. The full array can be divided into 3 sections, the surround sound array in the horizontal plane around the listener, the 1st height layer at around 35° to the horizontal plane, and the Zenith microphone at the top of the complete array.

Although the division of the upper hemisphere into two segments of 45° would seem logical at first, hindsight has shown that this will cause considerable difficulty when we come to linking with the upper Zenith Zone. However, a division of 35° for the Height segment (the segment between the Surround Sound Array and the Height Array), and 55° for the Zenith Zone (the segment between the Height Array and the upper Zenith microphone) will be nearer the mark. We will need to analyze each of these segments separately at first...".

Williams then describes combinations of different microphone directivities for the 'Surround Sound Array' (main layer array) and the 'Height Array', ranging from hypo-cardioid/hypo-cardioid, via cardioid/cardioid to hypo-cardioid/super-cardioid and cardioid/super-cardioid combinations. In these configurations, the Zenith-microphone always assumes the same microphone characteristic that is used for the other microphones of the Height Array.

From the selection of systems listed above, the combination of Cardioid-cardioid shall be examined in more detail below:

All Cardioid Array—The Height Segment

Williams writes:

"...As most recording engineers are well equipped with Cardioid microphones, the first worked example concerns an all-Cardioid array. First of all, looking at the initial Height segment of 35°, i.e. 0° to 35° of elevation. The starting point for this configuration is two parallel Cardioid microphones one above the other at a distance of 126 cm., as show in Fig. 6.22—the value of 126 cm is highlighted in red in the Segment Coverage Angel (SCA) Diagram...".

"...It is important to emphasize that this configuration uses only Time Difference to generate height information. In the multiple listening tests that have been carried out to study the validity of this method of generating height information, Time Difference generation has proved to be absolutely fundamental in the process of being able to record and reproduce viable height restitution. It is unfortunate that this important parameter of Time Difference has been completely neglected in most other experimentation in height localization systems. It must also be noted that the microphone array is of course only part of the complete process of a 3D audio system, the loudspeaker configuration must also follow some basic rules in relation to this type of recording system, for instance: univalent microphone/loudspeaker channels and especially isosceles triangle loudspeaker structure.

The other important configuration parameter is the distance and angle between the microphones, this again can be determined from..." Fig. 6.22,

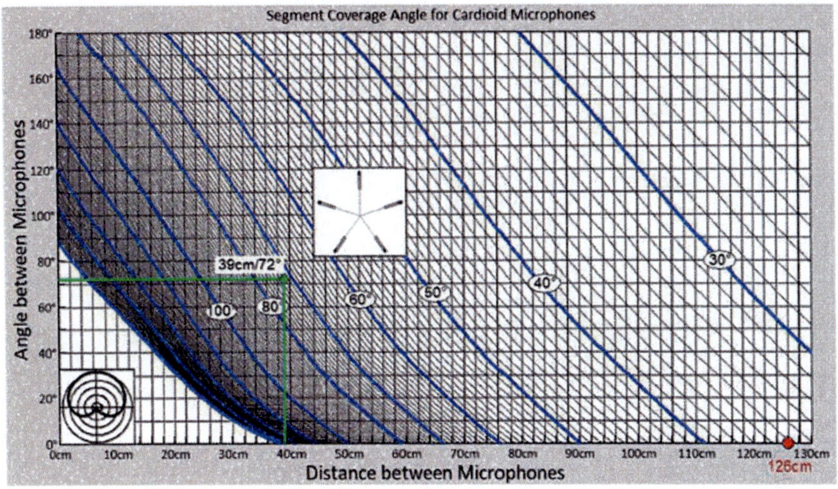

Fig. 6.22 Segment coverage angle diagram for Cardioid microphones (from Williams 2022b)

> "…and is highlighted in green on the left side—both the 5 channel surround array and the 5 channel height array are 39 cm/72°, i.e. 39 cm between the microphone capsules and 72° between the axes of microphone directivity. A plan and elevation view of this array is shown in Fig. 6.23…" (from Williams 2022a).

"…The surround sound array is therefore formed by Cardioids in a 5 Channel configuration (drawn in black) with a wingspan of 72.8 cm. The Height array is also formed by Cardioids in a 5-channel configuration (drawn in blue), again with a wingspan of 72.8 cm. The first elevation array is rotated through 36° degrees so that

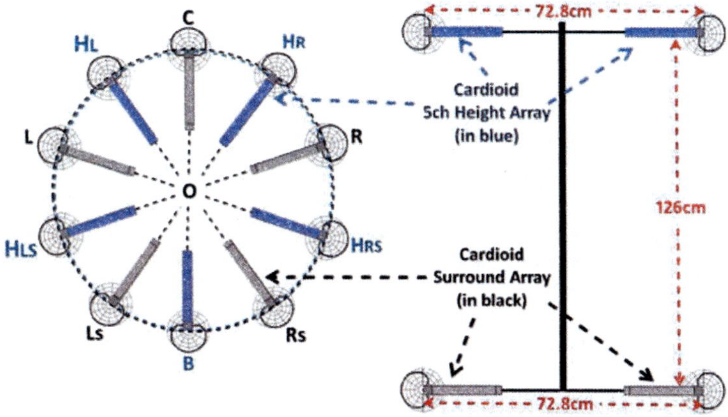

Fig. 6.23 Cardioid surround sound array (5-channel) plus cardioid height array (5-channel) with rotation of 36° (from Williams 2022b)

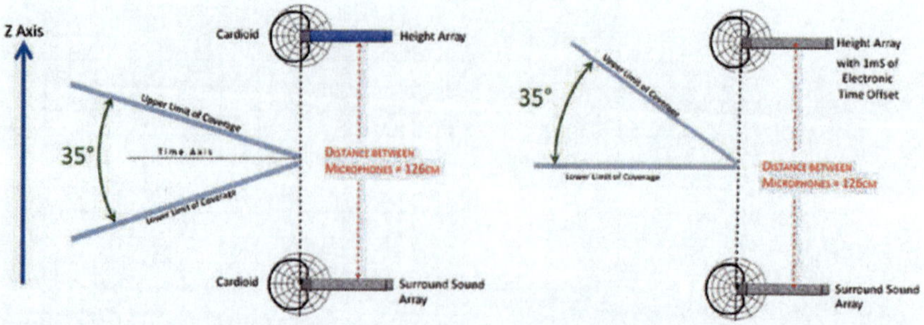

Fig. 6.24 (Left): microphones of cardioid surround sound array and cardioid height array with no 'segment steering'; (right): microphones of cardioid surround sound array and cardioid height array with 1 ms Electronic Time Offset (ETO) (from Williams 2022b)

individual microphones in the elevation array form isosceles triangles with respect to the lower surround sound array microphone pairs.

As the wingspan of the Height Array is the same as the Surround Sound Array, there is no initial Microphone Position Offset,…" as shown in Fig. 6.24 (left part).

"…In order to align the lower coverage limit with the horizontal reference plane, which will rotate the upper coverage limit to 305°, as shown in Fig. 6.24 (right part), we need to apply clockwise steering of 17.5°. This is achieved by applying 1 ms Electronic Time Offset (ETO) to the Height Array i.e. the first elevation array microphones have to be delayed by 1 ms with respect to the lower surround sound array microphones.

If we apply Electronic Time Offset of 17.5° or 1ms to the Zenith Zone Segment Coverage, we can steer the Zenith Zone coverage segments so that they are aligned with the vertical Zenith microphone axis. The final stage is just to join the height array to the zenith zone array to assemble to a complete cardioid 3D audio array, as shown in Fig. 6.25. The array solutions presented are, of course, just half of the total system for complete 3D Audio recording and reproduction. Without equal attention being paid to the loudspeaker configuration [rem.: conforming to the Isosceles layout (see Figs. 6.11, 6.12 and 6.15)] we cannot expect satisfactory results…" (from Williams 2022a)

6.8 Summary

> "…The M.A.G.I.C. array (with height channels) 12 channel 3D microphone setup, as defined in…"(Williams 2013)

"…is based on the isosceles triangle structure.

This type of array has proved capable of producing a realistic and robust 3D sound field. If we adopt a minimalistic approach to the number of channels needed in reproduction within lower order configurations, it is obvious that there is a certain

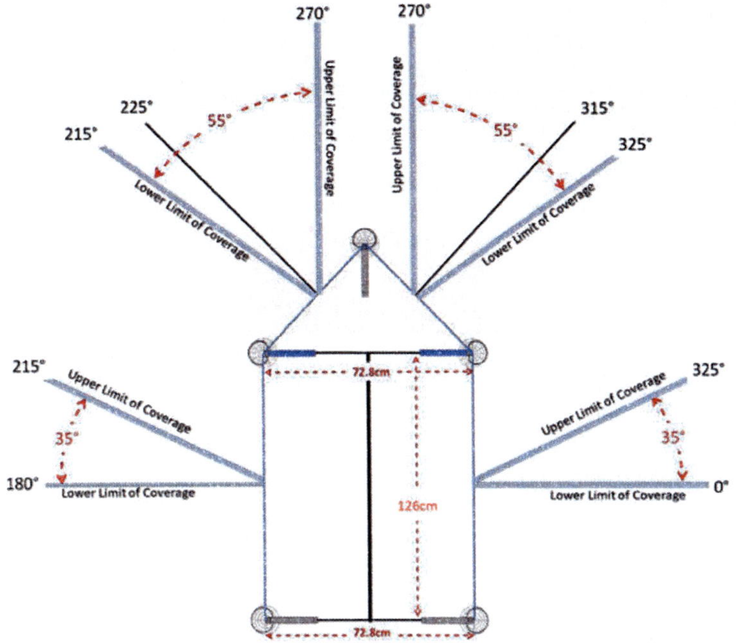

Fig. 6.25 Total coverage of a complete 5-channel cardioid/cardioid 3D Array with ETO (Electronic Time Offset) being applied to each segment (from Williams 2022b)

amount of redundancy in the overall 12 channel array. However this is consistent with the aim of maintaining complete compatibility of the overall master recording array with most of the present-day lower- order/channel reproduction systems.

Standard reproduction systems using 2 channels (Stereo), 4 channels (Quadraphony), 5 channels (so called multichannel), 7 channels (Blu-ray), 8 channels (Octophony) or the 3D reproduction formats, are directly compatible with the 12 channel 3D array, without any mixing or matrixing—only the selection of the specific channels is required. …" (from Williams 2014).

> "…With the introduction of a 3rd (i.e. lower) layer of 8 microphones, the MAGIC array has evolved to become a 25 microphone "Comfort 3D" array which also covers sound-pickup (and consequently also reproduction) from the lowest part of the bottom hemisphere…" (see Williams 2016a).

As part of the MMAD 3D (Multichannel Microphone Array Design for 3D) system Williams introduces the principle of Electronic Time Offset (ETO) between the Height Array and the Surround Sound Array (i.e. main layer array) in order to steer the segment coverage angle in a desired direction, thereby obtaining perfect critical linking between the various parts of the system.

> "…For the Height Array and Surround Sound Array microphones with directionalities ranging from hypo-cardioid to super-cardioid can be used, always including the use of a

Zenith microphone for optimized 3D audio sound recording and reproduction in conjunction with his specific Isosceles loudspeaker layout…" (see Williams 2022a)

More details on Michael Williams' latest developments can be found in (Williams 2022b).

References

Williams M (2004) Microphone arrays for stereo and multichannel sound recording (vol I) Il Rostro, Milan (see www.williamsmmad.com/publications)

Williams M (2005) The whys and wherefores of microphone array crosstalk in multichannel microphone array design. Paper 6373 presented at the 118th audio engineering society convention, Barcelona, May 2005

Williams M (2007) Magic arrays, multichannel microphone array design applied to multi- format compatibility. Paper 7057 presented at the 122nd audio engineering society convention, Vienna, 2007

Williams M (2008) Migration of 5.0 multichannel microphone array design to higher order MMAD (6.0, 7.0 & 8.0) with or without the inter-format compatibility criteria. Paper 7480 presented at the 124th audio engineering society convention, Amsterdam, May 2008

Williams M (2012a) Microphone array design for localization with elevation cues. Paper 8601 presented at the 132nd audio engineering society convention, Budapest, April 2012a

Williams M (2012b) 3D and multiformat microphone array design for the GOArt project. In: Proceedings to the 27. Tonmeistertagung des VDT, Cologne, Nov 2012b, p 739

Williams M (2013) The psychoacoustic testing of the 3D multiformat microphone array design, and the basic isosceles triangle structure of the array and the loudspeaker reproduction configuration. Paper 8839 presented at the 134th audio engineering society convention, Rome, May 2013

Williams M (2014) Downward compatibility configurations when using a univalent 12 channel 3D microphone array design as a master recording array. Paper 9186 presented at the 137th audio engineering society convention, Los Angeles, Oct 2014

Williams M (2016a) Microphone array design applied to complete hemispherical sound reproduction—from Integral 3D to comfort 3D. Paper presented at the 140th audio engineering society convention, Paris, June 2016a

Williams M (2016b) The basic philosophy of the 3D microphone array for recording and reproduction. Paper presented at the 29th Convention of the Verein Deutscher Tonmeister, Nov 2016b

Williams M (2022a) MMAD 3D audio—designing for height—practical configurations. Paper presented at the 152nd convention of the audio engineering society, May 2022

Williams M (2022b) Williams MMAD, stereo, surround & 3D audio. (see www.williamsmmad.com/publications)

Williams M and Le Du G (2000) Loudspeaker configuration and channel crosstalk in multichannel microphone array design. In: Proceedings of the 21. Tonmeistertagung des VDT, Hannover, Nov 2000, pp 347–383

Chapter 7
DTS:X®

Abstract In 2023, DTS is celebrating its 30-year history of sound evolution focused on surround and immersive sound for film and music. DTS was at the forefront with bringing digital 5.1 surround sound to the cinema, then translating that experience to the home, starting with DTS Digital Surround™ which is a proprietary codec first used on LaserDisc—and has since extended into advanced stream types such as 'DTS:X® Master Audio' (for Blu-ray Disc/UHD) with full immersive object-based lossless encoding, and DTS:X immersive for streaming media. In this chapter DTS speaker configuration options both for Cinema Theatre, as well as Home Cinema are presented. We present DTS:X content production workflows for cinema, optical media and streaming, and touch on MDA (Multi-Dimensional Audio) and IAB (Immersive Audio Bitstream, see SMPTE ST 2098–2). The basic possibilities of the DTS® Creator Suite software are explained, and we take a look at its components, including the DTS Renderer, and the DTS Headphone:X and DTS Monitor Plug-Ins. The DTS Neural:X Upmixer is presented, as well as the powerful DTS:X® CODEC. The chapter concludes with a short look at the DTS® Cinema Tool for creation of professional DCPs (Digital Cinema Package), DTS:X® Mediaplayer for QC of DTS home media bitstreams and the DTS:X® Bitstream Tools.

Keywords 3D · Immersive · Surround · DTS · DTS:X · DCP

7.1 The History of DTS—Digital Theatre Systems

Free citation from (XPERI 2020):

> "…DTS history goes back to the early 1990s, when—after a demonstration of DTS's digital audio technology—Steven Spielberg decided to release his blockbuster movie "Jurassic Park" with the first DTS digital soundtrack. Only three years later, the Academy of Motion Picture Arts and Sciences recognized DTS with a Scientific and Engineering Award for the Design and Development of the DTS Digital Sound System for Motion Picture Exhibition."

The DTS soundtracks for theaters were recorded onto CD's using the AptX ADPCM codec with timecode synchronization to a track on the film. This usually needed 4 or 5 CD discs per movie, so a special stacking disc changer with 6 AptX

E. Pfanzagl-Cardone, *The Art and Science of 3D Audio Recording*,
https://doi.org/10.1007/978-3-031-23046-2_7

decoders and a link to the timecode reader on the projector was required. This format was not released to consumers.

The initial consumer DTS format for movies was Laser disc, (1997) where the DTS 5.1 (Digital Surround aka Coherent Acoustics) audio was embedded in place of the stereo PCM audio track. Following that, DTS started appearing on DVD discs in 1998.

> "As the years progressed, DTS expanded into the home entertainment space, developing surround sound solutions to enable Hollywood studios to deliver a theatrical experience to consumers' homes." (free citation from XPERI 2020)

Starting (also 1997) DTS 5.1 44.1 kHz music CD's were produced under the DTS Entertainment label using the (consumer) DTS Coherent Acoustics codec. Playback was possible on CD players equipped with a build-in or an attached external DTS-decoder. This was 1.234 Mb/s 5.1 audio used to deliver the first digital surround music format to the homes.

Free citation from (XPERI 2020):

"... Part of the ongoing success of the DTS codec throughout the years was that it had been made an optional audio stream for DVDs (Digital Versatile Disc) as well as mandatory on the BD (Blu-ray Disk) standard, accelerating the availability of content even further. (see Fig. 7.1) The first Blu-ray disc with DTS was released in 2005.

As mentioned above, DTS had started developing tools and solutions to enable the creation of content in the DTS format. The availability of those tools, the ease of their use and the attractive price point made it a go-to solution for content production and authoring facilities (see Fig. 7.2).

DTS:X was launched in theatres with the release of the movie "American Ultra". The first Blu-ray release featuring DTS's new immersive audio format was "Ex-Machina" in 2014.

7.2 DTS:X® Immersive Audio Cinema Format

As an immersive, object-based audio format, DTS:X is changing the way audio is created in the studio and delivered to consumers at home. DTS:X can use channels as well as audio-objects with metadata to place sounds in a three-dimensional space. In this fashion, the correct 3D movement of audio-objects can be carried all the way to the consumer's living room and played back in virtually any speaker layout. Flexibility and scalability are the new highlights of this technology."

As of August 2022 the DTS:X content 'eco-system' consists of over 450 DTS:X theatrical releases, more than 200 home-theatre releases, created in 130 DTS:X production facilities, providing 'Hollywood' as well as regional content to over eleven hundred DTS:X equipped movie theatres worldwide.

From (DTS Inc 2021a):

Fig. 7.1 DTS-encoded audio on DVD and BD (graphic from XPERI 2020)

"… Over the last few decades, film sound has gravitated to the 5.1 'surround sound' channel format that DTS helped popularize. In spite of the widespread adoption of 5.1, the movie-going experience still saw continuing differentiation in the form of additional 7.1 channel formats, along with increased quality standards, that have helped to push the audio experience even further.

The move to immersive sound formats, which are adding height and ceiling speakers to cinemas, confirms that differentiation marches on. The concept of the 'PLF' (Premium Large Format) cinema has firmly taken hold, where a multiplex of

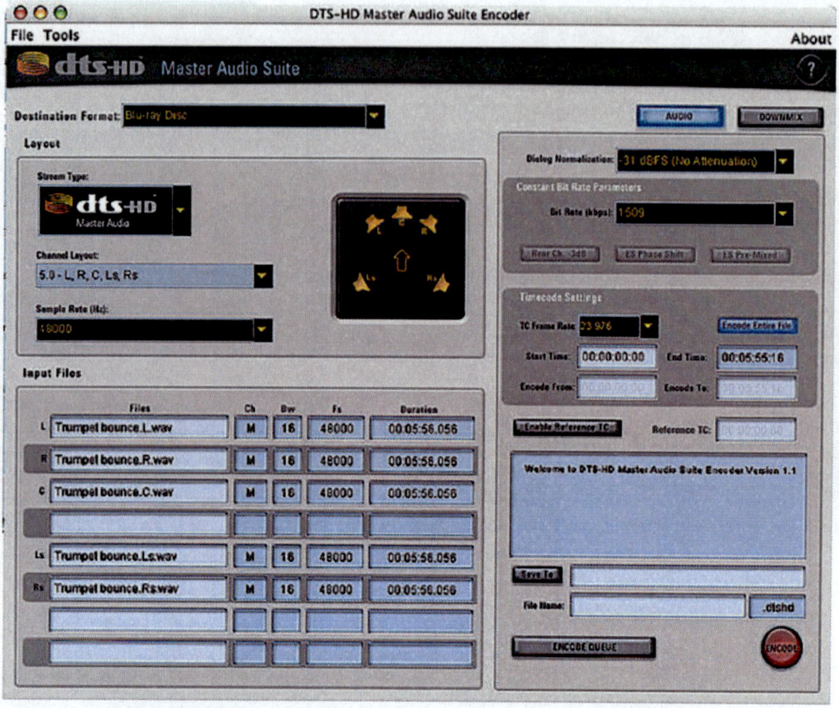

Fig. 7.2 DTS-HD master audio suite encoder software (graphic from XPERI 2020)

a dozen or more screens will have one or two PLF rooms and the rest of the rooms being of more conventional design.

Mixing stages are often physically large enough to help film makers experience their work in a cinema-like environment, but small enough—and fortified with higher performance sound systems and acoustics—to allow a more critical listening with regard to soundtrack balance and quality. Mixers must be able to hear the best possible expression of their art, so they know how the mix will play in the finest cinemas in the world.

Stages designed for immersive mixing often embody the state of the art in immersive sound reproduction in terms of the number, placement, and the quality of speakers, as well as the quality of electronics and room acoustics.

Additionally, the mixers never ignore the fact that they must also ensure that there is a version of the soundtrack that still plays perfectly in the vast majority of 5.1 cinemas.

7.3 DTS:X Theatre Speaker Configuration Options

Each DTS:X Cinema is designed with optimal acoustical coverage and performance for a given room as the guiding principle. Depending on the size and the dimensions of the room, height and/or ceiling speakers are used to achieve the best immersive effect possible.

DTS:X Cinema speaker configurations can be divided into two components: the base layer and the height layer. The base layer covers all the speakers in a typical 7.1 cinema, plus any speakers added in front to fill the gap between the surround speakers and the screen (also referred to as 'wide screen' or proscenium speakers). The height layer covers all the speakers added to support height effects, anywhere above the base layer.

Once all of the base and height speaker locations have been decided, a Speaker Configuration File is created that allows the DTS:X Renderer to generate the appropriate speaker feeds for each speaker, speaker groups and arrays. Every speaker configuration is uniquely designed for a given room.

7.3.1 Base Layer

In current 5.1 or 7.1 cinemas, as illustrated in Fig. 7.3, the surround speaker 'arrays' spanning the walls are each driven by a single channel of the soundtrack. These arrays provide wide coverage but have limited ability to convey small-scale sounds or precise sound movement. DTS:X Cinemas require additional surround speaker coverage to improve resolution and localization of the object-based soundtrack. The right side of Fig. 7.3 illustrates one example of a DTS:X Cinema which uses additional speakers along the side of the cinema, subdivided into two smaller arrays, Lssa and Lssb, within a '9.1' base channel configuration. This offers a modest but useful degree of improved directionality over 7.1 cinemas, especially in the important "proscenium" area near the screen edges.

Ideally, every speaker in a DTS:X Cinema will be provided its own signal, as shown in Fig. 7.4; this is in addition to being able to address groups of speakers as arrays for presenting channel-based content. This approach allows audio objects to move freely among the speakers, conveying as much positional detail as possible, while maintaining the broad coverage of array-based presentation.

7.3.2 Height Layer

All DTS:X Cinemas require the use additional height speakers on the walls, the ceiling or both, to improve the sense of immersion and to support specific overhead effects.

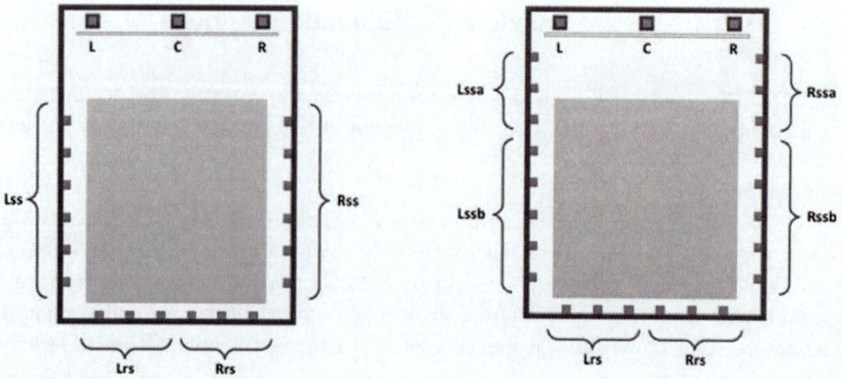

Fig. 7.3 Left side—a standard 7.1-channel cinema speaker layout supporting 4 surround signals, right side—base layer of a DTS:X Cinema with a "9.1" array supporting 6 surround signals (from DTS Inc. 2021a)

Fig. 7.4 Base layer of a DTS:X Cinema with addressable speakers and arrays supporting 26+ surround signals (from DTS Inc. 2021a)

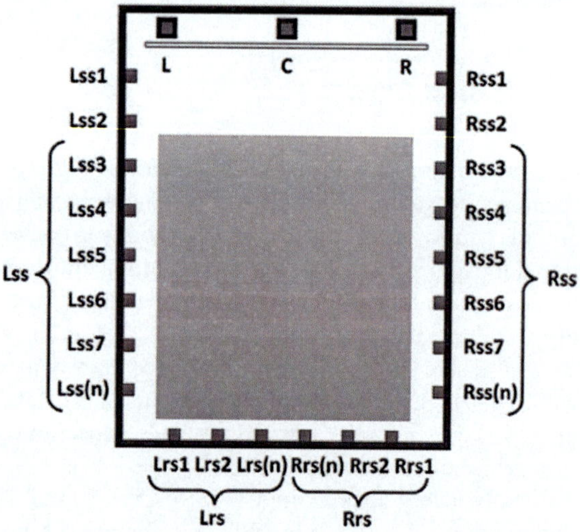

The examples for speaker placement options illustrated in Figs. 7.5, 7.6, 7.7 and 7.8 attempt to show actual speaker counts and locations for typical size cinema installations. Speaker count may vary accordingly for larger or smaller cinemas. Other options may exist for non-standard cinema applications.

In Figs. 7.6 and 7.7 the height speaker density approximates or equals that of the base layer, which is good practice.

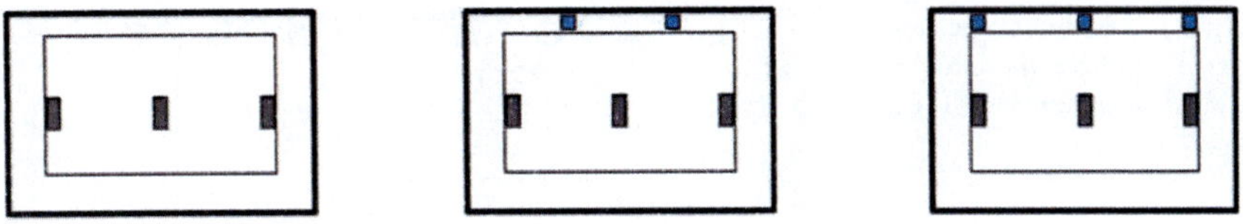

Fig. 7.5 Front view showing screen wall speaker options (from DTS, Inc 2021a, b, c, d)

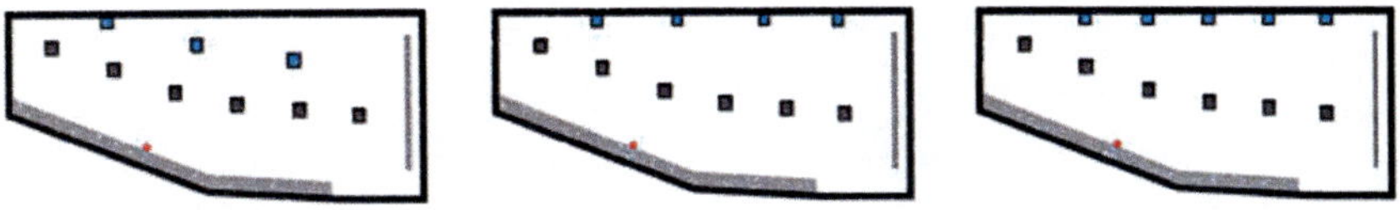

Fig. 7.6 Side views showing side wall speaker options for cinemas that cannot use ceiling speakers (from DTS, Inc 2021a, b, c, d)

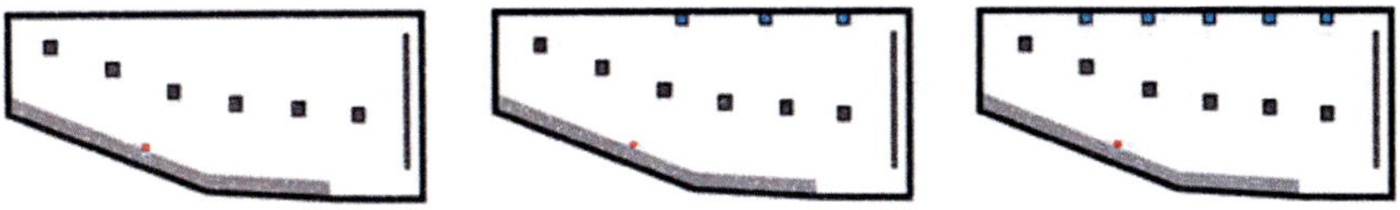

Fig. 7.7 Side views showing side wall speaker options for cinemas that cannot use ceiling speakers (from DTS, Inc 2021a, b, c, d)

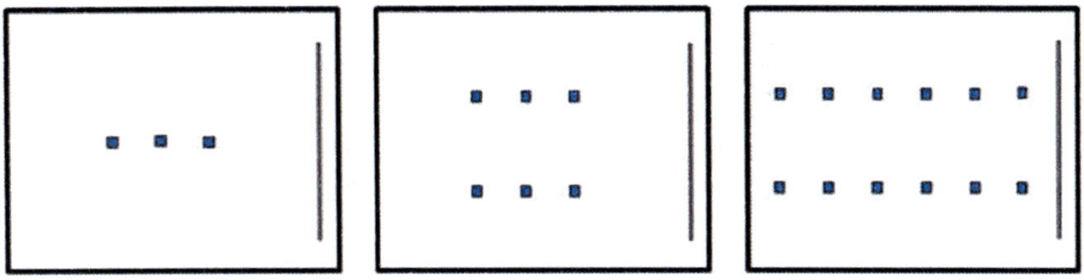

Fig. 7.8 Ceiling speaker options (from DTS Inc. 2021a)

7.3.3 Base Layer Speaker Spacing Requirements

To ensure even coverage for the audience, Fig. 7.9 shows the maximum spacing of base layer surround speakers:

At a distance of 1/4th of the room's width, the angle between adjacent surround speakers on the side walls must not exceed 30°, and the angle between adjacent speakers on the rear wall must not exceed 40°. It is assumed that the speakers on the walls are uniformly spaced.

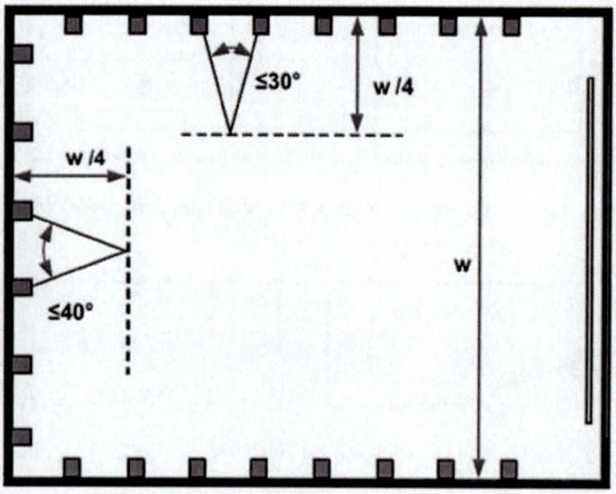

Fig. 7.9 Maximum speaker spacing relative to room width (top view) (from DTS, Inc 2021a)

In order to optimize the coverage of each surround speaker across the audience, it is recommended that the speakers be angled horizontally toward the audience if possible.

7.3.4 Height Speaker Position Requirements

In order to provide effective height effects, a vertical displacement angle of at least 20° needs to be achieved relative to the surround speakers located forward of the Preferred Listening Position (PLP).

In order to achieve effective height effects, a maximum lateral displacement angle of 90° needs to be achieved relative to the PLP (Fig. 7.10).

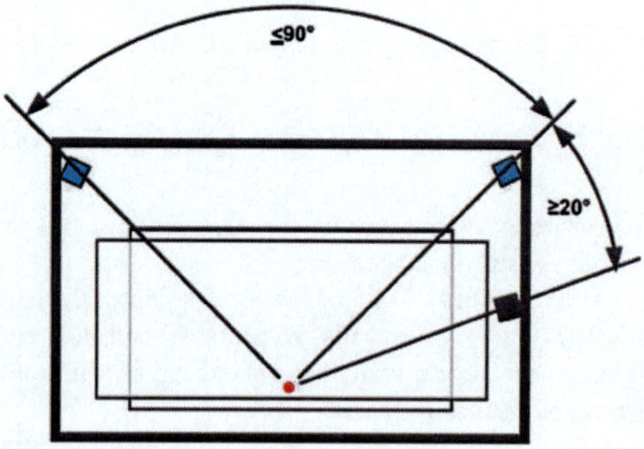

Fig. 7.10 Sidewall speaker minimum elevation requirement; ceiling speaker maximum lateral spread requirement (from DTS Inc. 2021a)

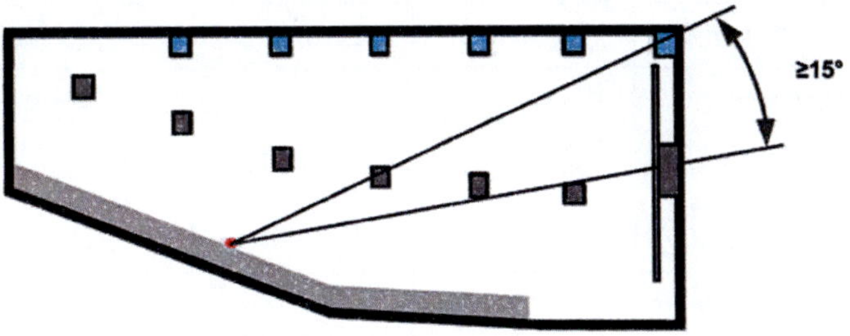

Fig. 7.11 Front wall speaker minimum elevation requirement (from DTS, Inc 2021a, b, c, d)

In order to provide effective height effects from speakers placed on the front wall, a vertical displacement of at least 15° needs to be achieved relative to the main L/C/R speakers (Fig. 7.11). In order to enhance the coverage of each height speaker across the audience, it is recommended that the speakers be generally aimed toward the center of the audience if physically possible.

7.3.5 Speaker Cluster Options

When the cinema system has several more speakers than the reproduction system can support with individual audio feeds, groups of speakers may be defined and fed from a single audio output. A typical example occurs when the DTS:X renderer is built into the media block, which normally has a maximum of 16 audio output channels.

However, it may also be desirable to use small speaker clusters with high output renderers for reasons such as power handling and to reduce the maximum loudness for listeners seated near the room walls.

Figure 7.12 shows how a 15.1 sound system can be created by clustering the speakers into subgroups of three or less speakers. In these illustrations, each cluster (outlined in red) represents one audio output channel." (from DTS Inc. 2021a).

7.3.6 Overall System Bass Requirements

The immersive effect is greatly enhanced when sounds maintain a consistent timbre as they move about the cinema. This implies that the bass response of the smaller surround speakers must be considered. A response down to 40 Hz for the surrounds will help ensure they blend well with the screen speakers. If the native response of the surround speakers roll off above that point, supplemental subwoofers in the rear of the cinema, along with appropriate bass management processing, will be necessary.

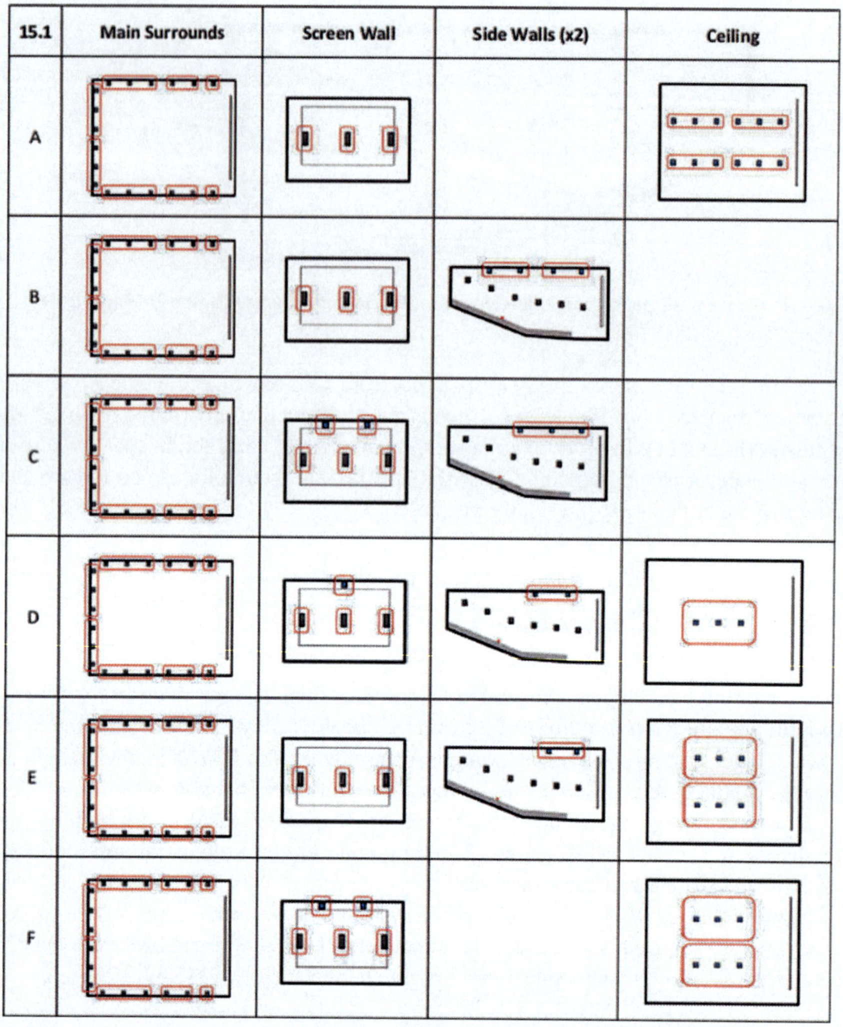

Fig. 7.12 Examples of channel clusters for 15.1 DTS:X systems (from DTS Inc. 2021a)

7.3.7 *B-Chain Considerations*

The B-chain is the portion of the playback system that determines how the sound is presented in the physical room. Figure 7.13 illustrates a system using a cinema processor capable of rendering a signal for each speaker in a premium cinema.

The B-chain is also where speaker groupings are defined. Sounds may be delivered to a single speaker, groups of speakers or they may be intended to address an entire

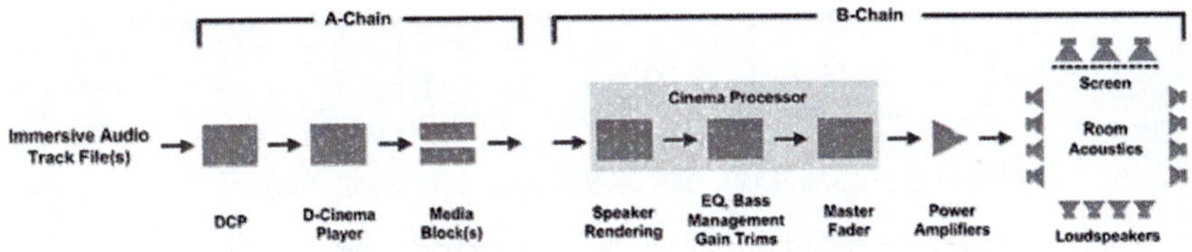

Fig. 7.13 Overall DTS:X system for external rendering cinemas (from DTS Inc. 2021a)

array. The DTS:X renderer outputs separate signals for individual speakers, speaker groups and arrays.

Figure 7.14 shows how the various outputs of the DTS:X renderer are mapped by the cinema processor to individual speakers and speaker arrays (such as illustrated in Fig. 7.4). For example, notice that the direct speaker outputs from the DTS:X renderer, Lss3 through Lss(n), pass through their respective EQ, delay, and gain controls, and then drive their respective amplifier/speaker. In addition, the Lss 'array' signal (part of the bed channel configuration) is distributed across all of those same speakers by means of the summing stages feeding the power amplifiers. This signal has its own EQ, delay, and gain controls to help ensure proper acoustic blending.

In systems that do not have sufficient decoder outputs to address each speaker individually, the same grouping process will be used to define appropriate speaker clusters for the best spatial effect.

EQ and channel loudness calibrations must comply with industry standards. This helps ensure not only good sound quality, but that the sound accurately conveys the intentions of the creators. SMPTE states that channel-based and object-based audio signals recorded at −20 dB FS nominally reproduce at 85 dB SPL at the reference listening position. To achieve consistency in frequency response, auditoriums adhere to accepted standards for electroacoustic response, e.g. SMPTE ST 202. (see SMPTE ST202 1998) (Fig. 7.15)

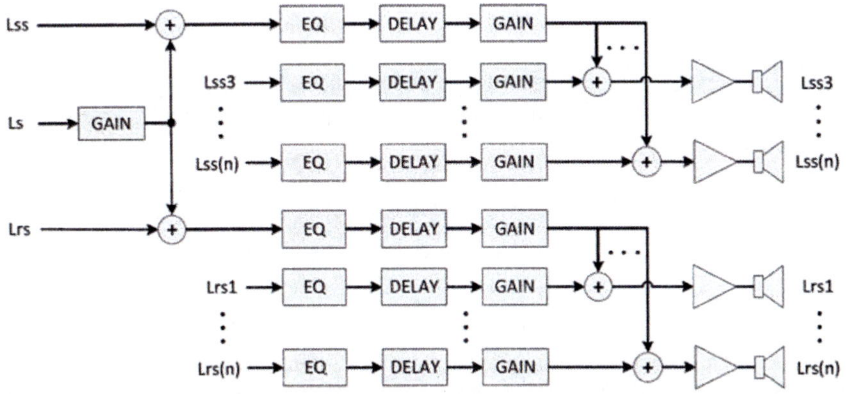

Fig. 7.14 Surround speaker array B-chain structure (partial) (from DTS Inc. 2021a)

Fig. 7.15 DTS:X cinematic speaker layout including ceiling and front screen height speakers (graphic from XPERI 2020)

Another factor gaining importance with object audio soundtracks is the need for extended bass response in the surround speakers. Ideally, the bass response would be the same as that of the screen speakers so as to allow sounds to maintain the same timbre and weight as they move about the cinema. This is even more challenging for surround speakers in object audio systems as they may be used alone or in small numbers for certain sound effects, which can stress their output capabilities.

The recommended solution is to use some form of bass management in the surround channels. In typical bass management, the speaker signals are split around 100 Hz or less, with the high frequencies feeding the speakers and the low frequencies diverted to at least one subwoofer. Often there are two subwoofers, one in each upper rear corner of the room, representing the bass from that side of the room. Other methodologies may be applied as long as the frequency response and loudness criteria are met.

DTS offers B-chain configuration templates that can be provided for DTS:X cinema installations by contacting DTS' parent company Xperi directly." from (DTS Inc. 2021a).

7.4 DTS:X Home Cinema Speaker Configuration Options

For home cinema installations DTS:X offers encoding up to 7.1 in terms of surround sound, and 7.1.4 (plus additional discrete objects that can also act as additional channels, if static) for 3D audio (see Sect. 7.10). DTS:X Encoder further down for more details). DTS:X Decoders can then dynamically render to any speaker layout with as many output channels as supported by the AVR system (see Figs. 7.16, 7.17 and 7.18).

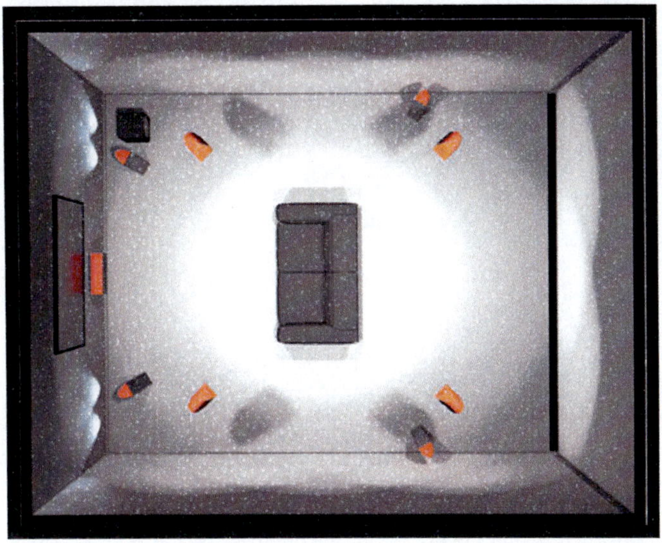

Fig. 7.16 DTS:X 5.1.4 Home Cinema Speaker Setup (top view); (graphic from XPERI 2020)

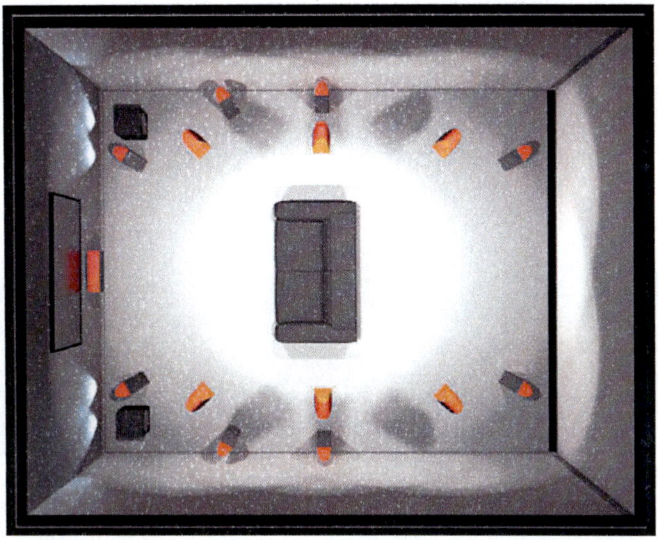

Fig. 7.17 9.2.6 Home Cinema Speaker Setup (top view); (graphic from XPERI 2020)

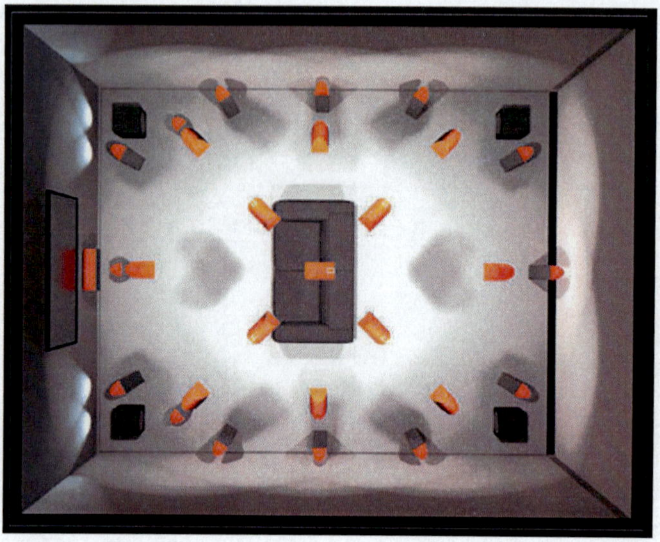

Fig. 7.18 12.4.13 Home Cinema Speaker Setup (top view); (graphic from XPERI 2020)

The devices included in the DTS:X home devices eco-system range from Blu-ray and streaming content via the 'source devices' (game-console, smart-TV/mobile) to the 'sink devices' (AV-receiver/sound-bar)—see Fig. 7.19.

Free citation after (XPERI 2020): "…In terms of streaming: a DTS:X streaming encoder is available (SDK [Software Development Kit] and CLI [Command Line Interface] solution for an efficient delivery of immersive audio and IMAX Enhanced content), as well as open source solutions (verified solutions for FFMPEG and Bento 4) (see Fig. 7.20).

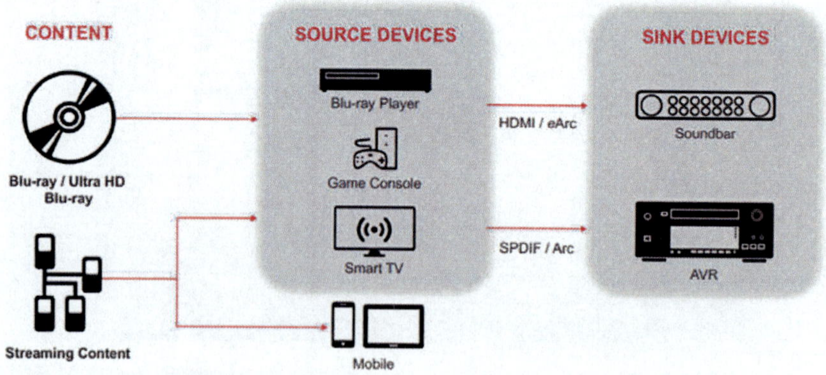

Fig. 7.19 The DTS:X home devices eco-system (graphic from XPERI 2020)

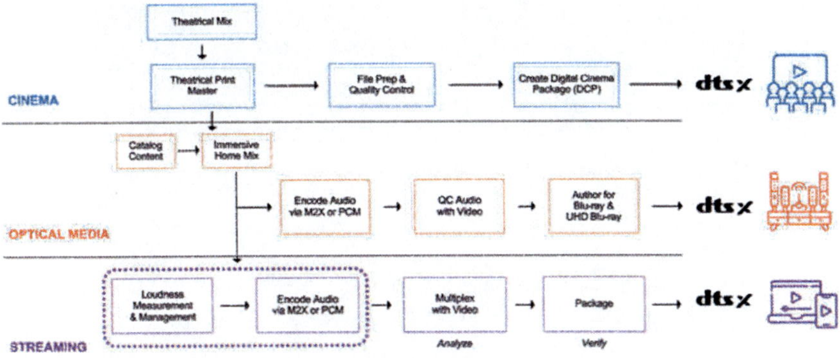

Fig. 7.20 DTS:X content production workflow

7.5 The DTS® Content Creator Software

For content production, the *DTS Content Creator* is available (see DTS Inc 2021b):

"…DTS® Content Creator is a set of software plug-ins and stand-alone applications for digital audio workstations that allow the user to position audio objects in a 3D space and render to an immersive speaker layout. Each audio object can be placed in a virtual 3D space. Audio objects can be positioned within the boundaries of a virtual room and panned around in all directions to create an immersive, object-based program. The program can then be exported as a bitstream (.mda or.mxf/IAB) for playback in MDA (Multi-Dimensional Audio), IAB (Immersive Audio Bitstream, see SMPTE ST 2098-2) and DTS:X supported devices or packaged for further distribution.

- **Bed Mix**—A collection of audio objects associated with a standard speaker configuration (5.1, 7.1, 11.1, etc.), where each audio object behaves as a traditional audio channel and is hardwired to a particular speaker. A combination of bed and positional objects make up the DTS:X program.
- **Audio Object**—A single audio waveform for either one channel of a bed mix or the sound of one object plus associated metadata describing the position of the object, the type of content, and any other relevant information.

For 3D audio rendering, DTS:X uses VBAP ("Vector Base Amplitude Panning", see Pulkki 1997), a method for creating a three-dimensional audio environment using any number of arbitrarily placed loudspeakers. Traditional two-speaker amplitude panning methods have been extended to use three speakers to create a triangle within which an object can be placed and moved around. Amplitude panning has been reformulated to use vectors so as to reduce computational complexity.

DTS Creator is used to position audio objects in three-dimensional space (see Fig. 7.21) and is available as mono, stereo and multichannel configurations.

The '*object*' icon in the panning user interface can be freely moved depending on the Pan Mode to a XYZ coordinate position. The '*Object Link*' icon is used to

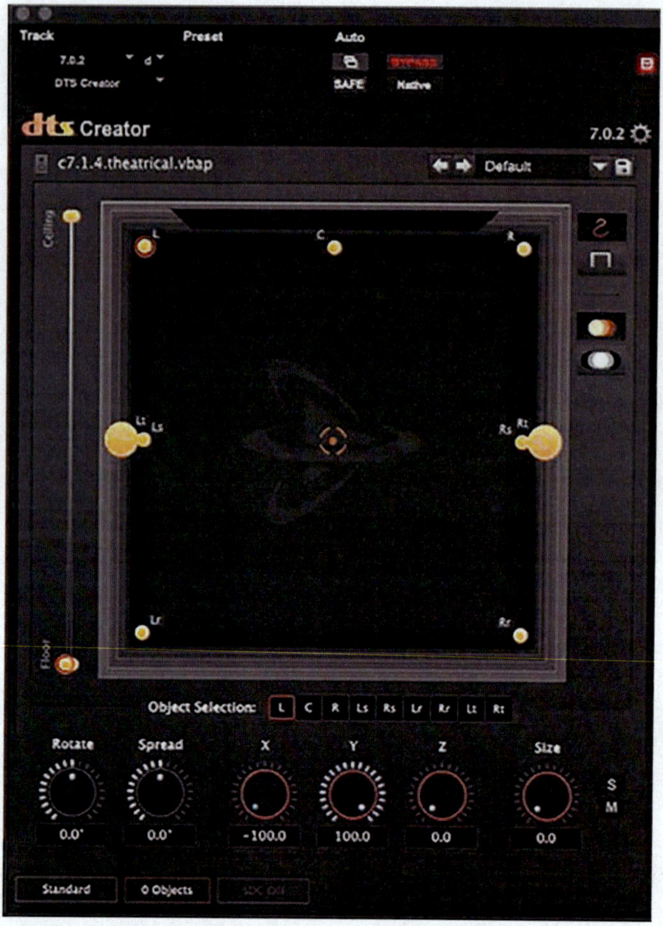

Fig. 7.21 DTS Creator for 7.0.2 (with 7 base layer speakers, no subwoofers, 2 height speakers) (graphic from DTS Inc. 2021b, p. 10)

control object position and its relative position to the *object pivot*. Rotate and Spread controls can be used to alter the position of the object relative to the object link.

7.6 DTS Renderer Input

The DTS Renderer Input is used to input un-rendered audio and associated metadata to the DTS Renderer (see Fig. 7.22). Each DTS Creator instance needs to be assigned to a DTS Renderer Input on the same local system or on a remote system. Audio assignment is realized using standard host (Pro Tools) audio routing while metadata routing happens from within the plug-in. (see Fig. 7.23).

The Audio connection between *DTS Creator* and *DTS Renderer Input* is realized using standard Pro Tools audio routing. For example, a mono DTS Creator instance on a mono track uses a mono bus or mono sub-path to create a connection between a track's output and the track's input on which DTS Renderer Input is instantiated. The Metadata connection is established from within DTS Renderer Input (see Fig. 7.23).

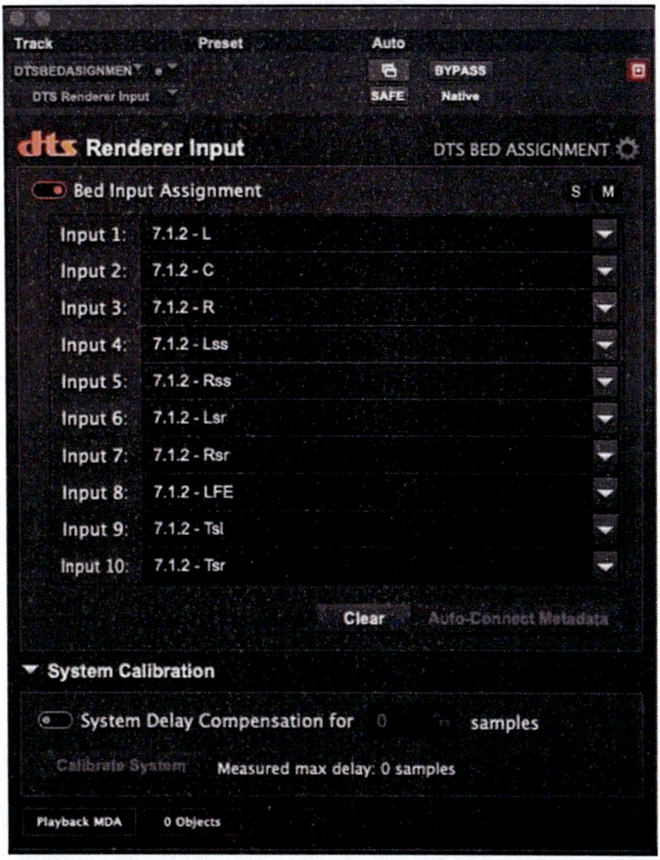

Fig. 7.22 DTS Renderer input plug-in (graphic similar to graphic in DTS Inc. 2021b, p. 14)

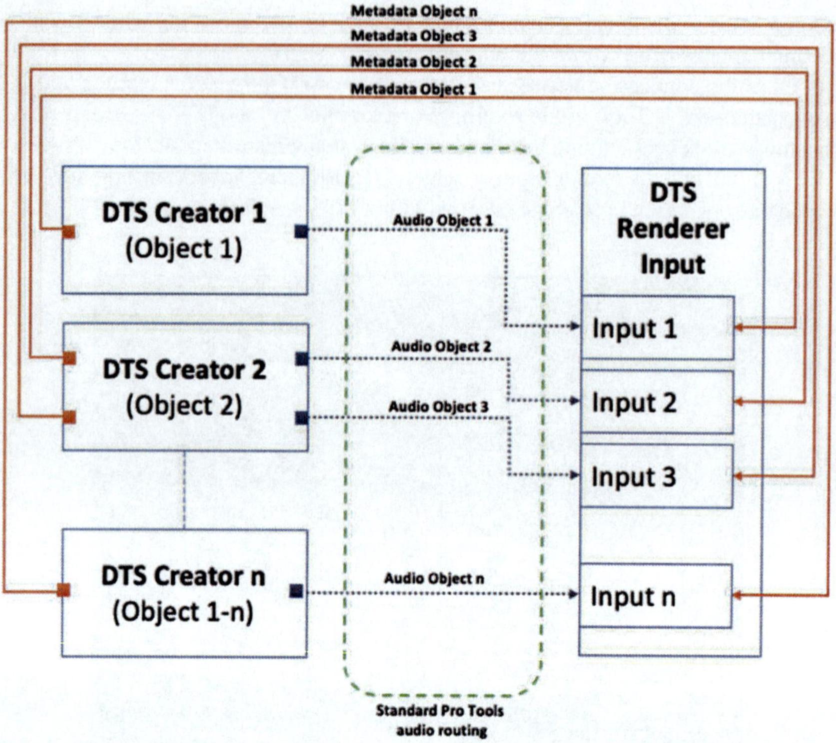

Fig. 7.23 Metadata and audio connections (graphic from DTS Inc. 2021b, p. 15)

7.7 DTS Monitor

DTS Monitor is used to output the rendered audio based on the chosen Speaker or Monitor Configuration (see Fig. 7.24). Additionally, it provides monitoring functionality and controls processing modes of the DTS Renderer. DTS Monitor is also where *MDA (Multichannel Digital Audio) and IAB (Immersive Audio Bitstream)* data can be exported.

DTS Content Creators supports three different bitstream types: MDA, IAB for Cinema and IAB for Home which can be selected from the Export menu.

While MDA is exported as a raw bitstream, IAB for Cinema (IAB Application Profile 1) and IAB for Home (SMPTE ST 2067-201) are wrapped into their respective MXF containers.

The DTS renderer can render audio objects to virtually any speaker layout. The desired speaker layout is selected in the *Speaker Configuration* pull-down menu located in the center of the Monitoring Section in DTS Monitor.

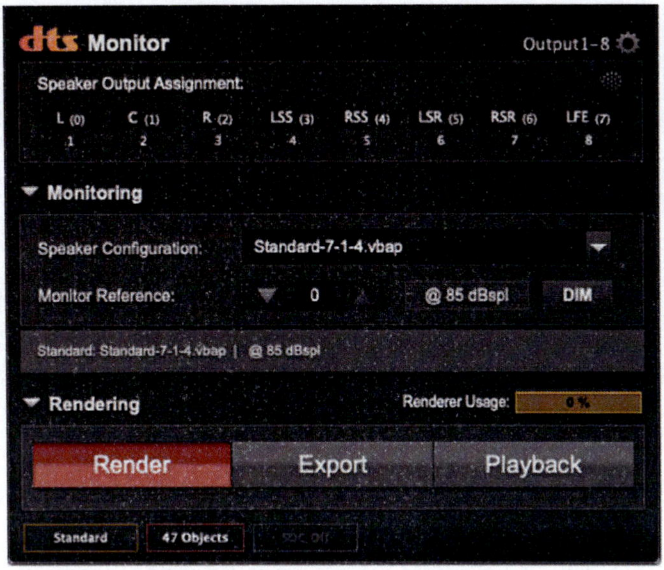

Fig. 7.24 DTS monitor plug-in (graphic from DTS Inc. 2021b, p. 20)

7.8 DTS Headphone:X—Headphone Monitor

The DTS Headphone:X plug-in is fully integrated into the latest version of the DTS Content Creator tool set. Each audio channel and object can be assigned to individual sub-mixes to create a binaural version of the full immersive, object-based mix (Fig. 7.25).

7.9 DTS Neural:X Upmixer

The DTS Neural:X Upmixer is a plug-in to up-mix any Stereo, 5.0, 5.1, 7.0 or 7.1 content to one of the supported immersive DTS:X layouts (5.1.4, 7.1.4, 7.1.5, 9.1.4).

DTS Neural:X Upmixer can be instantiated on the Pro Tools track formats: Stereo, 5.0, 5.1, 7.0 and 7.1. Adding DTS Neural:X to one of these track types will automatically change the track output to 3rd Order Ambisonics to support the up to 16 channel output configuration (of which a maximum of 14 channels are used at the same time; see Table 7.1).

The resulting immersive mix has excellent channel separation and downmix compatibility which makes this a versatile and easy to use tool for immersive content production and mixing (see Fig. 7.26).

Fig. 7.25 DTS headphone:X binaural renderer (graphic courtesy of DTS Inc)

Table 7.1 Fixed output channel layout order in the DTS Neural:X Upmixer PlugIn (table reproduced after table on p. 8 in DTS Inc. 2021c)

Outputs	1	2	3	4	5	6	7	8	9	10	11	12	13	14	15	16
5.1.4	L	C	R	–	–	Ls	Rs	LFE	Lh	Rh	Lhr	Rhr	–	–	–	–
7.1.4	L	C	R	LSS	Rss	Ls	Rs	LFE	Lh	Rh	Lhr	Rhr	–	–	–	–
7.1.5	L	C	R	LSS	Rss	Ls	Rs	LFE	Lh	Rh	Lhr	Rhr	–	–	Ch	–
9.1.4	L	C	R	LSS	Rss	Ls	Rs	LFE	Lh	Rh	Lhr	Rhr	Lw	Rw	–	–

7.10 DTS:X® Encoder

"…The DTS:X® Encoder (Fig. 7.27) supports the following destination formats:

- **Blu-ray Disc/UHD BD** that can be used to create DTS encodes suitable for Blu-ray and Ultra HD Blu-ray use.
- **DVD** that can be used to create DTS encodes suitable for DVD use.
- **Digital Delivery** that can be used to create a variety of DTS encodes suitable for digital delivery formats.
- **Type 1 Certified Content** that can be used to create IMAX Enhanced for Blu-ray and Ultra HD Blu-ray discs.

On the input side, the DTS:X® Encoder can handle WAV (.wav), Broadcast WAV (.bwav), AIF/AIFF (.aif), IAB for home in MXF …" (cited from DTS Inc. 2020a).

" … The DTS:X Encoder main screen is made up of several sections, which include the 'Input table' where the Encode Layout (i.e. speaker layout, e.g. 7.1.4) is selected. The 'Encode Layout' section is a graphical representation of the speaker

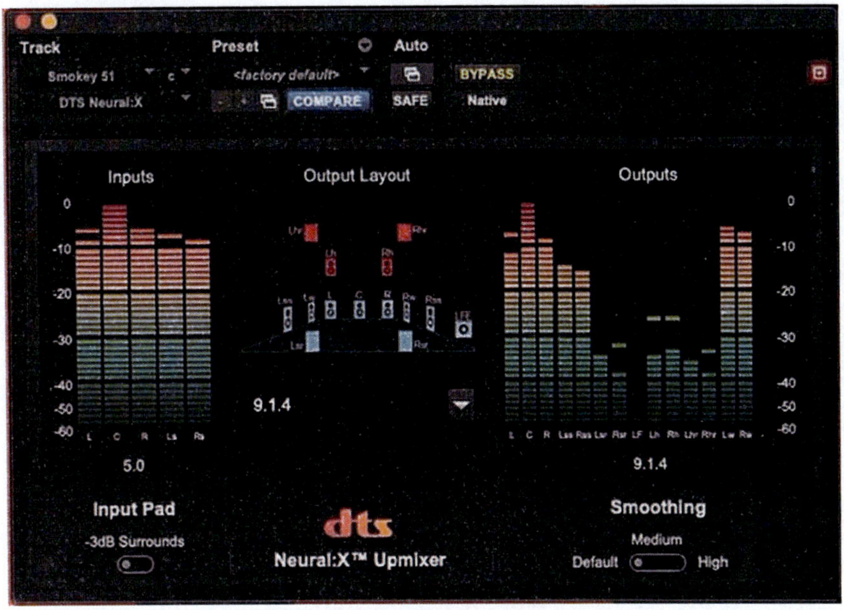

Fig. 7.26 DTS Neural:X Upmixer (graphic courtesy of DTS Inc.)

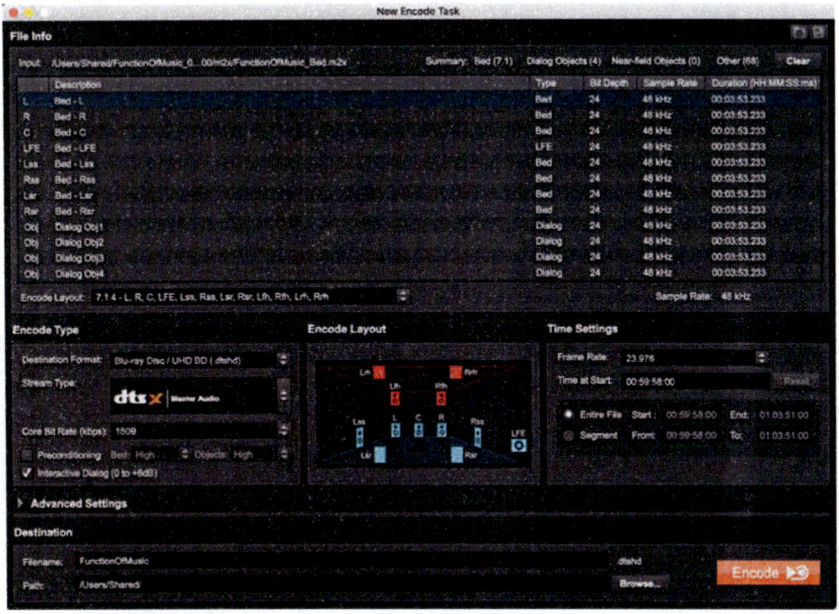

Fig. 7.27 DTS:X Encoder main screen (courtesy DTS Inc.)

layout chosen. In the Encode Type Section the destination format is chosen (e.g. Blu-ray Disc/UHD BD),

Stream Type lists the available DTS stream types available based on the selected destination format, which are:

DTS:X® Master Audio, DTS-HD® Master Audio, DTS-HD® High Resolution Audio, DTS Digital Surround ES™, DTS Digital Surround™ | 96/24, DTS Digital Surround™, DTS Express™ (Digital Delivery only).

In the *Time Settings* Section standard frame rate can be selected, ranging from 23.976 to 60 frames per second.

In the Advanced Settings Section Dialogue Normalization parameters can be set as well as DRC – "Dynamic Range Compression" (which can be set to: None, Film Light, Film Standard, Music Light, Music and Speech), along with other options.

The DTS:X® Encoder provides the following channel layouts from 12.1 down to mono 1.0 (Table 7.2):

- 12.1—L, R, C, LFE, Lss, Rss, Lsr, Rsr, Lfh, Rfh, Cfh, Lrh, Rrh
- 7.1.4—L, R, C, LFE, Lss, Rss, Lsr, Rsr, Lfh, Rfh, Lrh, Rrh
- 7.1.3—L, R, C, LFE, Lss, Rss, Lsr, Rsr, Lfh, Rfh, Crh
- 7.1.2—L, R, C, LFE, Lss, Rss, Lsr, Rsr, Lfh, Rfh
- 7.1.2—L, R, C, LFE, Lss, Rss, Lsr, Rsr, Lhs, Rhs
- 5.1.5—L, R, C, LFE, Ls, Rs, Lfh, Rfh, Cfh, Lrh, Rrh
- 5.1.4—L, R, C, LFE, Ls, Rs, Lfh, Rfh, Lrh, Rrh
- 5.1.3—L, R, C, LFE, Ls, Rs, Lfh, Rfh, Crh
- 7.1—L, R, C, LFE, Lss, Rss, Lsr, Rsr
- 7.1—L, R, C, LFE, Ls, Rs, Lsr, Rsr
- 6.1—L, R, C, LFE, Ls, Rs, Cfh
- 6.1 ES Matrix (pre-mixed)—L, R, C, LFE, Ls, Rs, Cs
- 5.1—L, R, C, LFE, Ls, Rs
- 4.0—L, R, C, Cs
- 2.0—L, R
- 2.0—Lt, Rt
- 1.0—C (Mono)

Playback compatibility from channel-based DTS:X® encodes to DTS-HD® and legacy decoders is made possible by either using the default downmix coefficients made available in the codec, or by creating custom downmix coefficients to 7.1, 5.1 or 2.0 (Stereo). …" (free citation from DTS Inc. 2020a).

"…For the creation of DCP bitstreams for cinema theatres there is the DTS® Cinema Tool Software, which is a stand-alone application for preparing MDA and IAB (SMPTE ST 2098-2 (2019))

Immersive Audio Bitstream) bitstreams for DCP (Digital Cinema Package) authoring and D-Cinema distribution." (free citation from DTS Inc. 2021d); (see Fig. 7.28).

There are two more tools within the DTS:X palette:

Table 7.2 DTS Speaker Channel ID and associated definitions (from DTS Inc. 2020a, p. 41ff)

Channel ID	Definitions
L	Left Speaker
R	Right Speaker
C	Center Speaker
LFE, LF or SW	Low Frequency Effects/Subwoofer
Ls	Left Surround Speaker
Rs	Right Surround Speaker
Cs	Center Surround Rear Speaker
Lsr	Left Surround Rear Speaker
Rsr	Right Surround Rear Speaker
Lss	Left Surround Side Speaker
Rss	Right Surround Side Speaker
S	Surround back
Lc	Left Center Speaker
Rc	Right Center Speaker
Lfh	Left Front Height Speaker
Cfh	Center Front Height Speaker
Rfh	Right Front Height Speaker
LFE 2	For second low frequency effects/Subwoofer
Lw	Left Wide Speaker
Rw	Right Wide Speaker
Oh	Overhead Speaker
Lsh	Left Surround Height Speaker
Rsh	Right Surround Height Speaker
Crh	Center Rear Height Speaker
Lrh	Left Rear Height Speaker
Rrh	Right Rear Height Speaker
Clf	Center Low Front Speaker
Llf	Left Low Front Speaker
Rlf	Right Low Front Speaker
Lst	Left Side Top
Rst	Right Side Top
Lsd	Left Side Down
Rsd	Channel Right Side Down
Lt	Left Total
Rt	Right Total

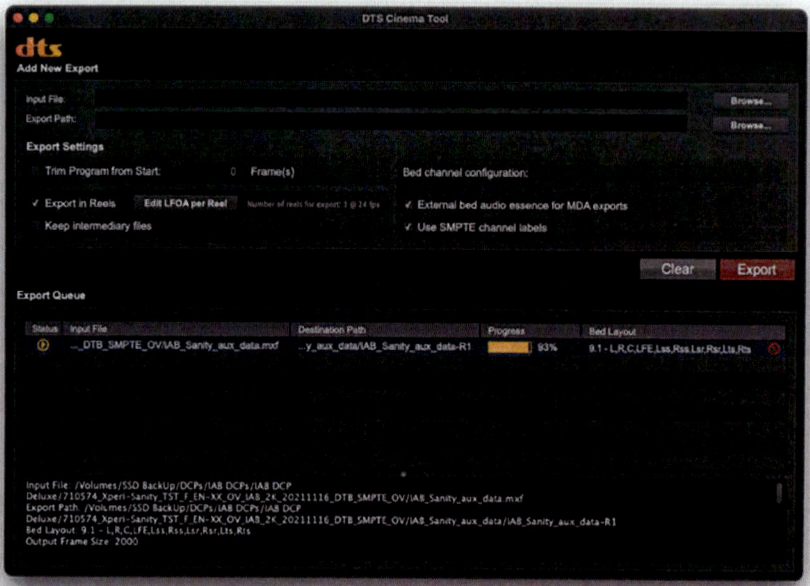

Fig. 7.28 DTS® Cinema Tool (from DTS Inc. 2021d)

7.11 DTS:X® Mediaplayer

"…—enables you to play DTS encoded audio against video prior to final multiplexing and authoring. It supports real-time playback of :X and DTS-HD stream types synchronized with AVC and HEVC video at up to 4 K resolutions. Also, it includes native support of CFF, MOV and MP4 file formats with frame accurate sync, random access seeking and the ability to monitor legacy downmix modes." (free citation from (DTS Inc. (2022)); (see also DTS Inc. 2020b)

7.12 DTS:X Bitstream Tools

"…—helps you to save time editing without the need of an additional encode pass, and validate DTS bitstreams. The DTS:X Bitstream Tools support MDA files as well as previously encoded DTS:X and DTS-HD bitstreams." (free citation from (DTS Inc. (2022)); (see also DTS Inc. 2020c).

References

DTS Inc. (2020a) DTS:X® Encoder User Guide. whitepaper V.2.2, Nov 2020
DTS Inc. (2020b) DTS:X® Mediaplayer User Guide. whitepaper V.1.7, Nov 2020
DTS Inc. (2020c) DTS:X® Bitstream Tools User Guide. whitepaper V.1.7, Nov 2020
DTS Inc. (2021a) DTS:X® Cinema Guidelines. whitepaper V.2.22, Feb 2021
DTS Inc. (2021b) DTS® Content Creator User Guide. whitepaper V.1.1, Jun 2021
DTS Inc. (2021c) DTS® Neural:X™ Upmixer User Guide. whitepaper V.1.0, Jul 2021
DTS Inc. (2021d) DTS® Cinema Tool User Guide. whitepaper V.1.0, May 2021
DTS Inc. (2022) DTS:X® Encoder Suite 2.2. PDF Overview-sheet
Pulkki V (1997) Virtual sound source positioning using vector base amplitude panning. J Audio Eng Soc 45(6):456–466
SMPTE ST 202M (1998) Standard for motion pictures—dubbing theaters, review rooms, and indoor theaters—B-chain electroacoustic response.
SMPTE ST 2098-2:2019 (2019) Immersive audio bitstream specification.
XPERI (2020) DTS:X Content Production. Tools and Solutions to nourish the DTS:X content ecosystem. PDF for public presentation

Chapter 8
SONY "360 Reality Audio"

Abstract "360 Reality Audio" is Sony's 3D audio format. For the purpose of 3D audio mixing Sony has come up with the 360 WalkMix Creator ™ Software (by Audio Futures, Inc.), which is available as a plug-in for most common DAWs (Apple Macintosh, as well as Windows). In contrast to the 'in a shoe-box' approach for sound-source positioning, which is common with competitor plug-ins, 360 WalkMix Creator™ is based on positioning all sound sources on a (virtual) sphere. With the 360 WalkMix Creator™ it is SONY's aim to enable musician and creators to produce music easily and creatively in an immersive spherical sound field using the 360 Reality Audio Music Format. In this chapter we take a look at the 'Compact view' and 'Focus view' mixing desk style representations of the sound objects in 360 WalkMix, with associated parameters like azimuth, elevation and width and at the practical implication of how to let sounds objects move along the sphere. Associated loudspeaker layouts range from 2.0 stereo up to 13.0 (which is actually a 5.0.5+3B layout, made up of 5 main layer speakers, 5 height speakers and 3 bottom layer speakers).

Keywords 3D audio · Sphere · 360ra · Immersive · Sony · WalkMix

The reasons in SONY's 360 Reality Audio system to go for a spherical approach in terms of sound source positioning, relative to the listener, may have to do with the possibility of a more continuous and seamless reproduction of sound sources which are moving along the surface to this sphere, as opposed to a layout in which speakers are arranged along the walls (and ceiling, etc.) of a traditional rectangular type of room, as is usually the case.

E. Pfanzagl-Cardone, *The Art and Science of 3D Audio Recording*,
https://doi.org/10.1007/978-3-031-23046-2_8

Fig. 8.1 360 WalkMix Creator's 'Compact view' mode: Conga stereo-stem positioned horizontally right hand side of the listener (graphically adapted video still from 360 RA 2022)

However, it should also be mentioned though that this does not constrain the user to physically also arrange his multichannel replay speakers along a (virtual) sphere, as it is possible to define different distances for each speaker to the 'sweet spot'/center of the virtual sphere (more details on this below).

Also, it seems that Sony may have a historic inclination towards 'spherical sound capture' considering the spherical microphone arrangement by Akitaka Ito, which he has patented for Sony (see Ito 2001). This microphone arrangement consists of 8 microphones, arranged along the surface of a sphere, with six of them positioned on a horizontal circle, one microphone at the top of the sphere, and the other one at the bottom. With the help of an additional signal processor, horizontal as well as vertical panning of a sound source can be achieved (more details can be found in Chap. 4 of Pfanzagl-Cardone 2020).

On the input side, 360 WalkMix Creator can handle up to 128 sound objects in the form of mono or stereo files in 48 kHz format; i.e. either 128 mono files or 64 stereo-files. The position of these audio files along the sphere is defined by the user via the parameters Azimuth and Elevation: 0 degree is defined as in front of the listener and the left hemisphere is associated with positive azimuth values up to +180°, while the right hemisphere is associated with negative azimuth values down to −180°.

A similar logic applies with elevation: upwards from the horizontal plane (which is defined 0 to be 0 degree elevation) elevation will be positive, up to a max of +90°, while below the horizontal plane will be negative, down to −90°.

Figure 8.1 shows an example of a stereo recording (stem-file) of Congas, which are currently positioned along the horizontal plane, right hand side of the listener.

8.1 Object- or Compact-View

In 360 WalkMix Creator there are two views in which sound objects and related parameters can be viewed, positioned and edited. The first one is the OBJECT view ('Compact' view), as pictured in Fig. 8.1, which displays the sphere with all sound objects in two view-ports on the left and on the right side. Possibilities for display are 3D view (which can be rotated), as well as fixed Top, Front, Left, Right and Back views. On the very right hand side in the OBJECTS section the available sound objects are listed by name and with associated level. The positioning information of a selected object can be entered below in the PROPERTIES section, where also parameters like object color, gain, azimuth, elevation and width can be edited.

To spread the panning of the two channels of a stereo stem file further apart, the chain symbol can be unlocked to perform this operation. When locked again, for azimuth as well as elevation, the mean-value between the two L/R file positions will be applied.

In the lower left hand corner the number of sound objects already in use is displayed. On the upper right hand side the plugin can be switched on and off, the master volume is displayed as well as the CPU performance. When pressing the expand button in the upper right hand corner, the plugin will switch to the second screen which is called FOCUS view and displays all sound objects along with their parameters in a more 'mixing desk' style arrangement (see Fig. 8.2).

8.2 Focus View

Also in 'Focus view' the sound objects are displayed in two different view-ports on the sphere but at the same time all of their positioning and volume parameters are displayed below in a 'mixing desk style', with the master fader on the very right hand side.

8.3 Available Loudspeaker Layouts

The loudspeaker layouts available in 360 WalkMix Creator range from stereo 2.0, via Quad and several 5.x and 7.x surround (with height) layouts up to 13.0 (which is actually a 5.0.5+3B layout). The loudspeaker layout description used by Sony is as follows: the first number stands for the number of loudspeakers in the horizontal plane (i.e. main layer, or front layer), the second number is used for the number of LFE speakers [rem: Sony does not foresee any use of LFE-speakers as part of their standard setups, so this value is always 0], the third number stands for the amount of speakers in the height layer (or 'upper layer'). After that, loudspeakers in the bottom layer (or 'lower layer') are displayed with a '+' in front of them, e.g. the most

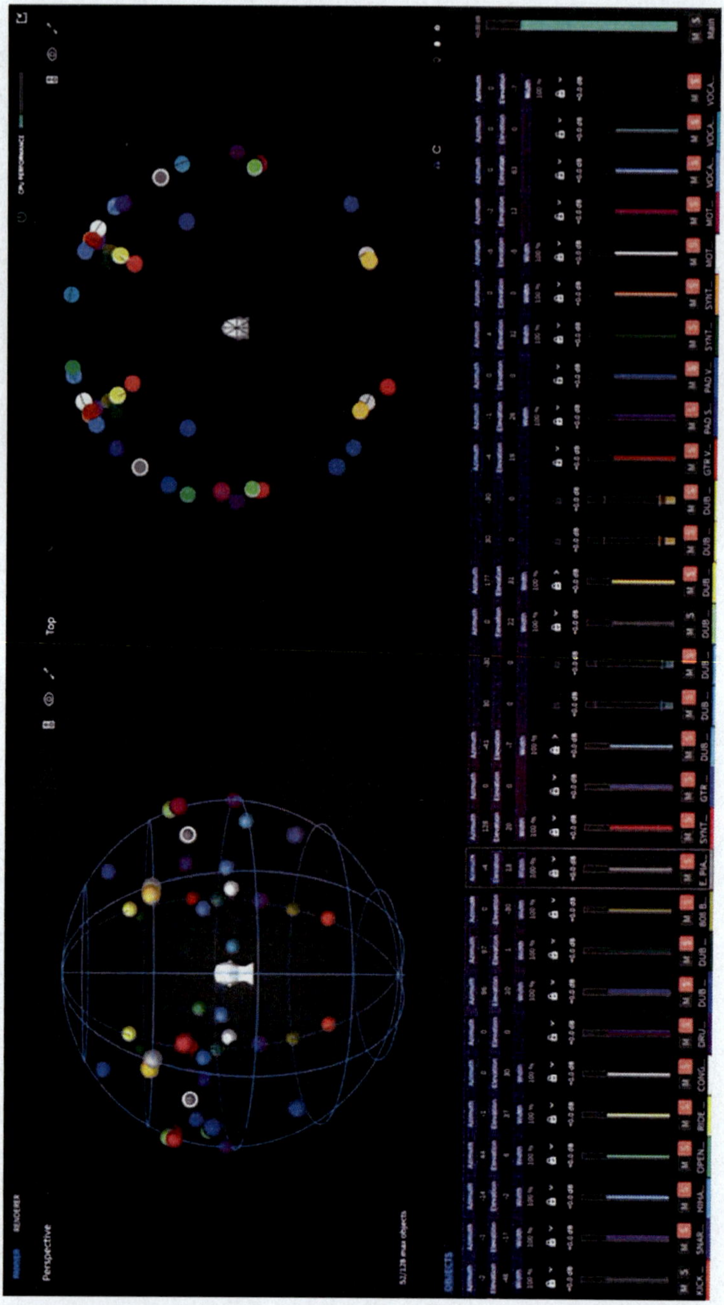

Fig. 8.2 360 WalkMix Creator's 'Focus view' mode displays the sound objects on the sphere along with their positioning and volume parameters in a 'mixing desk style' (graphically adapted video still from SONY 2022)

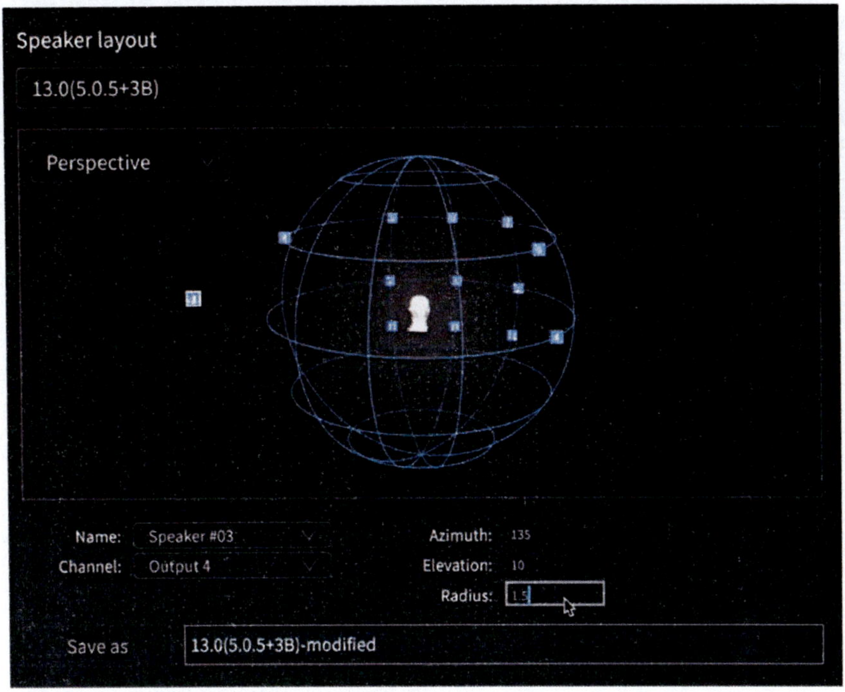

Fig. 8.3 The most elaborate loudspeaker layout is '13.0', as currently available in 360 WalkMix Creator™ (video still from SONY 2022)

elaborate loudspeaker setup, as proposed for regular use by Sony, would be the 13.0 type (which is more precisely described as 5.0.5+3B, hence 5 speakers in the main layer, 5 speakers in the height layer and 3 speakers in the bottom (lower) layer) and can be seen in Fig. 8.3.

The list of available layouts is as follows: 2.0, 4.0, 5.0, 5.0.2, 5.0.2+2B, 5.0.4, 5.0.4+2B, 7.0, 7.0.2., 7.0.4, 7.0.4+2B, 9.0.4+2B, 13.0 (i.e. 5.0.5+3B).

On the lower right hand side in Fig. 8.3 the actual physical position of the loudspeakers in use can be defined in relation to the virtual sphere by entering their azimuth and elevation parameters, as well as a distance parameter 'Radius', as in a real-world setup these loudspeakers may have to be arranged along the flat walls of a rectangular room. In Fig. 8.3 we see that e.g. the left surround speaker ('Speaker #03') is positioned at a distance of 1.5 m from the listener head/sweet spot. Associated delay compensation for all loudspeakers involved in the system can be adjusted within the plugin or externally, depending on the need of the specific setup.

As for the loudspeaker numbering it is worth to take note of Sony's approach to assign the lowest number to the center speaker in each layer. This means that the center loudspeaker of the 'main'- (or 'front'-) layer is assigned #0 and then the numbering ascends from L to R and front to rear in the following order for a 5.0 loudspeaker setup: C = 0, L = 1, R = 2, LS = 3, RS = 4.

In Fig. 8.4 side view and rear view for a 7.0.4 loudspeaker layout (without any bottom layer loudspeakers) are pictured. Note that the loudspeaker numbering follows the above described logic.

Figure 8.5 holds all the information in respect to azimuth and elevation of the loudspeakers of the 'height'- (or 'upper'-), 'main'- (or 'front'-) and 'bottom'- (or 'lower'-) layer for a 5.0.5+3B setup.

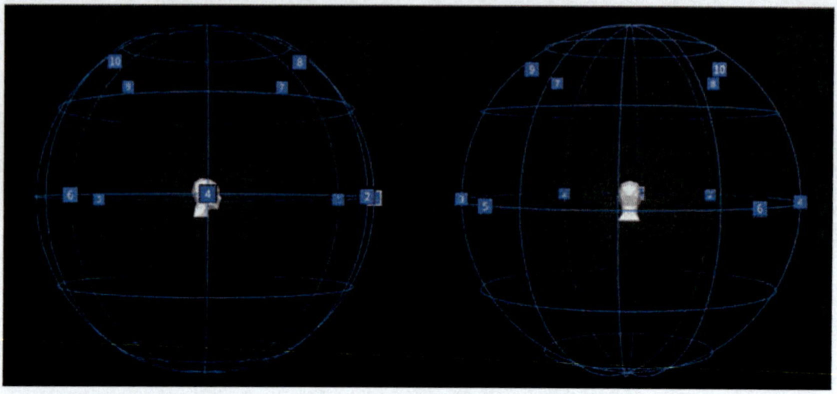

Fig. 8.4 '360 WalkMix Creator ™' loudspeaker layouts for 7.0.4 side view (left) and rear view (right) (adapted from SONY 2022)

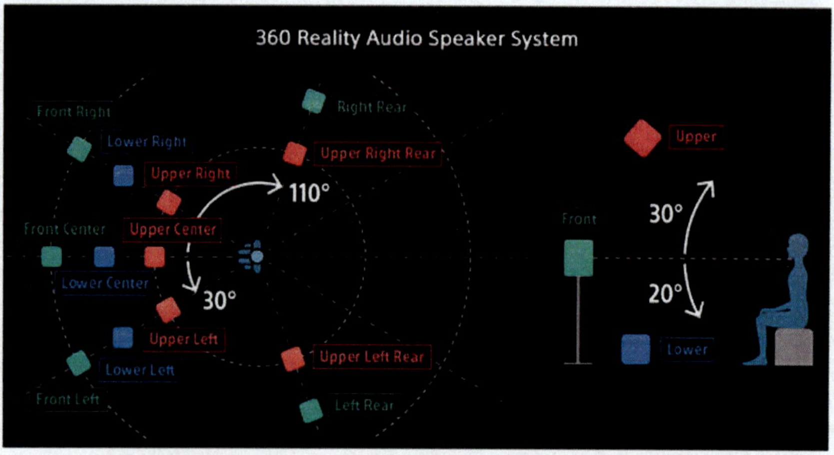

Fig. 8.5 Loudspeaker position azimuth and elevation data for SONY's suggested 13.0 (i.e. 5.0.5+3B) loudspeaker layout (graphic from www.360ra.com/faq)

8.4 Practical Aspects of the 360 WalkMix Creator™

To start working with the 360 WalkMix Creator in a DAW, the plugin needs to be inserted into every audio channel which should be part of the 3D mix. Preferably the 360 WalkMix should be inserted into the last insert slot on each channel, so that all plugins and processing will be placed before that in the signal chain, in order to also be effective for the final 3D mix (see Fig. 8.6).

Also, in Pro Tools e.g., a stereo master fader (and associated channel) will need to be assigned, where the 360 WalkMix Creator plugin needs to be inserted as well and defined 'Master'. From this master channel the final result of the 3D audio render will be taken and routed to the assigned replay loudspeakers (or headphones, in case of binaural simulation only) (Fig. 8.7).

In order to achieve sound source movement in the horizontal and vertical domain, automation of azimuth and elevation data is possible via the respective DAW's parameter automation features. Figure 8.8 shows an example of the stereo conga stem with azimuthal automation data to let the sound rotate in the horizontal plane.

As can be seen in Fig. 8.9, with a lot of sound objects used in the 3D audio mix the visual display of all elements can get quite dense in 360 WalkMix Creator. However, the possibility of display via two different perspectives, as well as the color coding and also the associated names let arise no doubts and the possibility to also blend in the positions of the loudspeakers (see 3D view on the left side) is also helpful in terms of orientation.

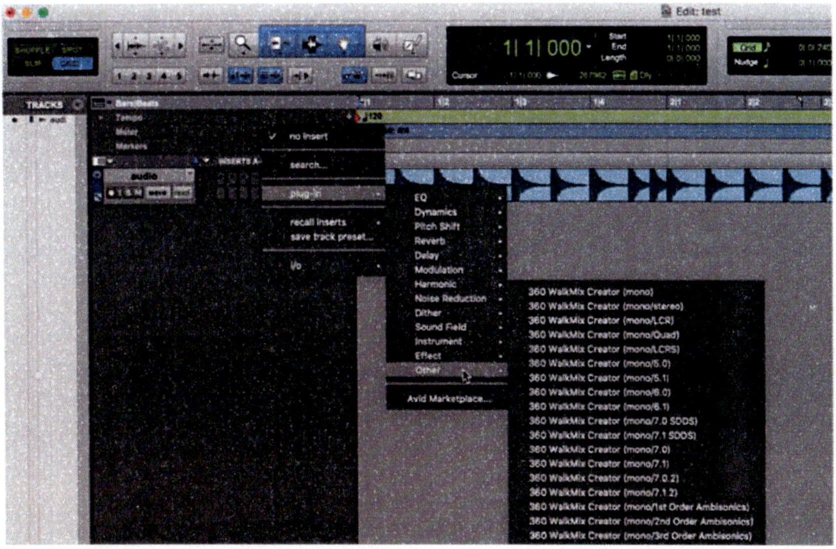

Fig. 8.6 The '360 WalkMix Creator™' plugin needs to be inserts into all audio-channels which should be used for the 3D audio mix (example for Pro Tools DAW), (detail of video still from SONY 2022)

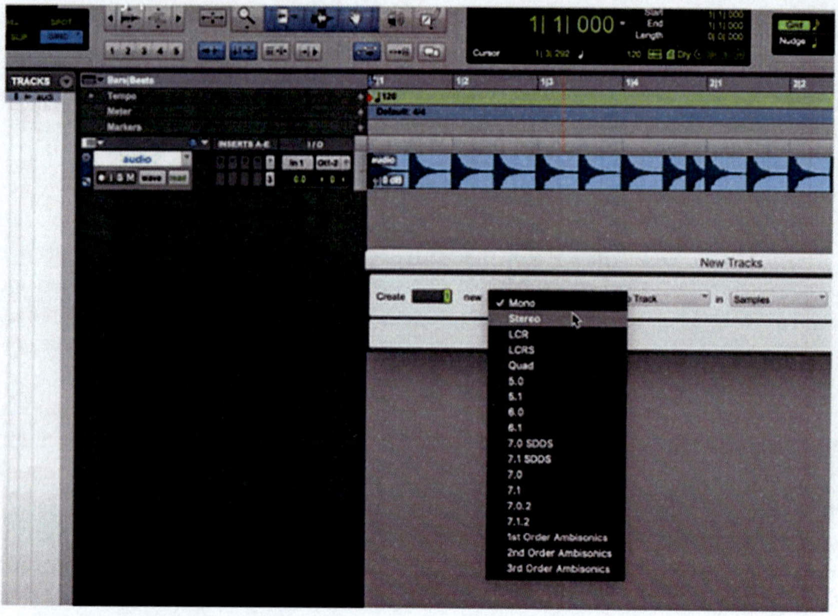

Fig. 8.7 The '360 WalkMix Creator™' plugin also needs to be inserts into a stereo MASTER channel from which the output of the 3D audio render is obtained and routed to the assigned loudspeakers (example for Pro Tools DAW); (detail of video still from SONY 2022)

Once the mix is completed to the satisfaction of the sound engineer, the rendering of the final outcome can be achieved by switching to the RENDERER (see switch close to the top left corner). Depending on the rendering setting, the 360 WalkMix Creator™ generates MPEG-H/.mp4 files and/or multiple sets of.WAV files with instructional meta data that defines audio positioning and automation.

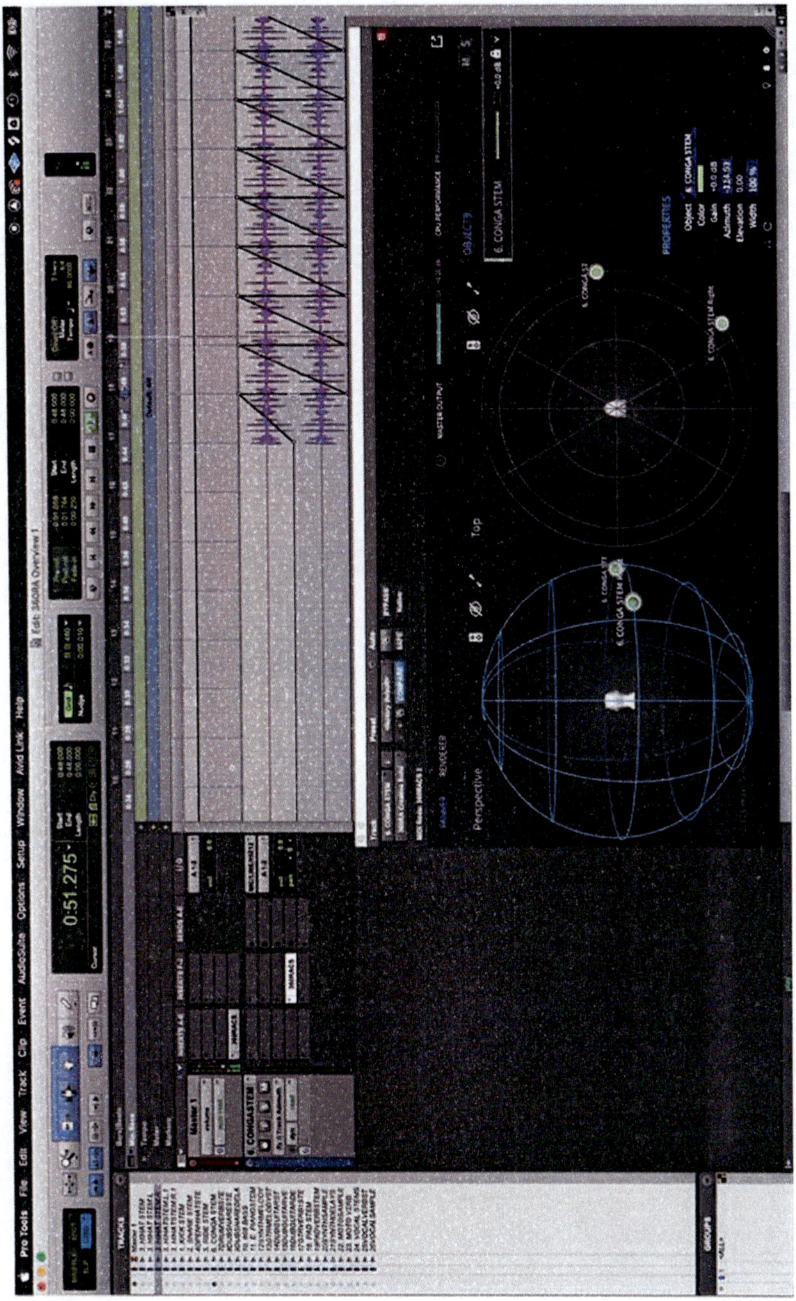

Fig. 8.8 The conga stereo-stem from Fig. 8.1, with azimuth automation data in Pro Tools to let the sound rotate in the horizontal plane (detail of video-still from 360 RA 2022)

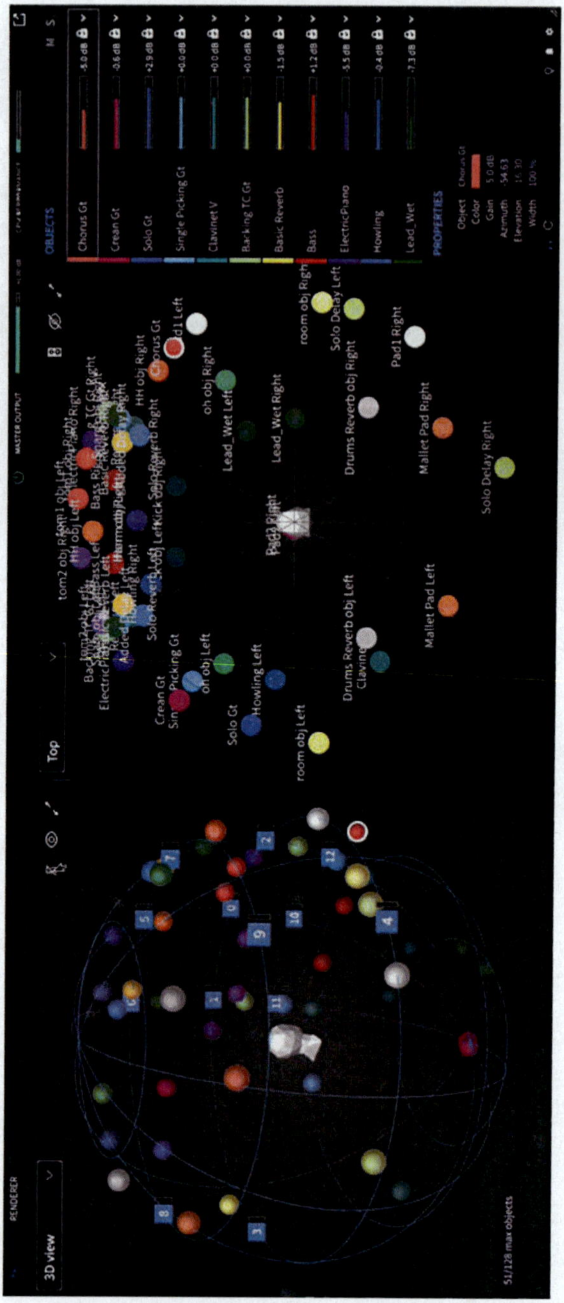

Fig. 8.9 A mixing session example in '360 WalkMix Creator™' with 51 sound objects and loudspeaker positions indicated on the left hand side '3D view' representation of the virtual sound sphere (detail of video still from 360 RA 2022)

References

Ito A (2001) Surround sound field reproduction system and surround sound field reproduction method. EU Patent EP 1,259,097 A2

Pfanzagl-Cardone E (2020) The art and science of surround and stereo recording. Springer-Verlag, GmbH Austria. https://doi.org/10.1007/978-3-7091-4891-4

A/V-Source References

360RA (2022) 360 WalkMix creator getting started tutorial. Video-tutorial created 23.06.2022 https://youtu.be/n9B9BaLcByM. Accessed 15 Aug 2022

SONY (2022) How to use 360 WalkMix Creator™. Video-tutorial created 30.06.2022. https://youtu.be/sTBVYTHnlJE. Accessed 16 Aug 2022

Chapter 9
Recording Microphone Techniques for 3D-Audio

Abstract First, a few case studies are presented of 3D audio recordings with large and small orchestra, string quartet, church organ and solo instruments. Consecutively the following 3D audio microphone techniques are presented in detail: AB-BPT-3D decorrelated array (Blumlein-Pfanzagl), Bowles-Array, Ellis-Geiger Triangle, Geluso MZ and 3DCC technique, 2L technique (Lindberg), OCT-3D (Wittek and Theile), ORTF-3D (Wittek and Theile). During the course of the presentation of more than 20 different 3D microphone systems various important aspects are analyzed such as sufficient channel separation (signal correlation and de-correlation) versus cohesive overall 3D image; creative decisions for finding the right spot for optimized diffuse-sound pickup by the microphones; generating effective localization and height envelopment in 3D audio systems; sufficient sonic separation between the 3D audio main and height layer; mapping the respective microphone arrays to various loudspeaker layouts, ranging from 8.1 surround to Dolby Atmos. The chapter concludes with the case study of a 6DOF (6 Degrees Of Freedom) recording for VR-purposes at Abbey Road Studios and a small selection of binaurally based 3D recording techniques.

Keywords 3D audio · Immersive audio · Mic arrays · Height information · Microphone technique · Signal correlation · Spatial hearing

As there is no clear proposal for a 'standard' 3D-Audio microphone array, which would deliver 'ideal' sonic results for the Auro-3D or Dolby Atmos systems it makes most sense to take a look at a number of case studies, with sound-sources of variable size; among those are also some carried out by staff from "Galaxy Studios":

Some parts of the following section are cited from (Van Daele and Van Baelen 2012):"…An Aurophonic recording typically has 8–12 microphones. This is only 4–6 extra microphones compared to standard 5.1 Surround Sound recording setups. The choice of microphones depends on the localization info that the engineer wants to record. Depending on the recording venue, several rig sizes can be used:

- Large (with uncorrelated sound in the lower frequencies)
- Medium
- Small (where lower frequencies become mono)

E. Pfanzagl-Cardone, *The Art and Science of 3D Audio Recording*,
https://doi.org/10.1007/978-3-031-23046-2_9

Fig. 9.1 Auro 9.1 medium-sized mobile rig for (outside) location recording (from Van Daele and Van Baelen 2012)

Several microphone set-ups have been developed and used to make Aurophonic recordings under varying circumstances. The basic idea, however, is always the same: a regular (5-channel) surround sound microphone setup is used as the basis, completed with only four (or preferably five) additional (height) microphones placed above the main microphones and one microphone to record the 'Voice of God' channel (see Figs. 9.1 and 9.2. Only using these 3 layers (lower, height and top) the reproduction of a natural space can be achieved with a minimum of 10 speakers, while also providing compatibility with the 5.1 surround sound standard" .

9.1 Music Recordings with Large Orchestra

"For a large orchestra, in a good, large hall, a standard surround sound setup can be used based on a Decca Tree (wide-spaced AB-pair with Center microphone), combined with a wide-spaced AB-pair for the surrounds (LS, RS).

Above the main microphones (L, R, LS, RS), the height microphones (HL, HR) are added at about half the distance between the main L/R-microphones. In a good hall these microphones can be omnidirectional, but when less room acoustics are wanted, cardioid microphones can be used as well, directed downwards towards the

Fig. 9.2 Auro-3D small-sized mobile mic-array as boom for on-set recording (from Van Daele and Van Baelen 2012)

orchestra. The rear height microphones (HLS, HRS) should be placed at the same height as the front height microphones, high above the surround microphones (LS, RS)" (from Van Daele and Van Baelen 2012). [Rem.: It is interesting to note that in the photo in Fig. 9.3 the rear microphones of the main layer, as well as of the height layer are pointing towards the sound source and not towards the back-wall of the hall. According to Wilfried Van Baelen the decision of whether these microphones should be directed towards the sound source or rather in the opposite direction, in order to only pick up diffuse sound (including rear wall reflections), will depend entirely on the acoustics of the recording room and the sonic impression intended by the sound engineer or Tonmeister.]

9.2 Music Recording with Small Orchestra, Grand Piano Solo

"For these occasions, smaller setups are likely to be more appropriate. Good results have been obtained using an ORTF LCR-setup, with a medium-spaced AB-pair for the surround-channels. These are then augmented with height microphones placed above the ORTF-setup and above the AB-pair. For the choice of microphones and

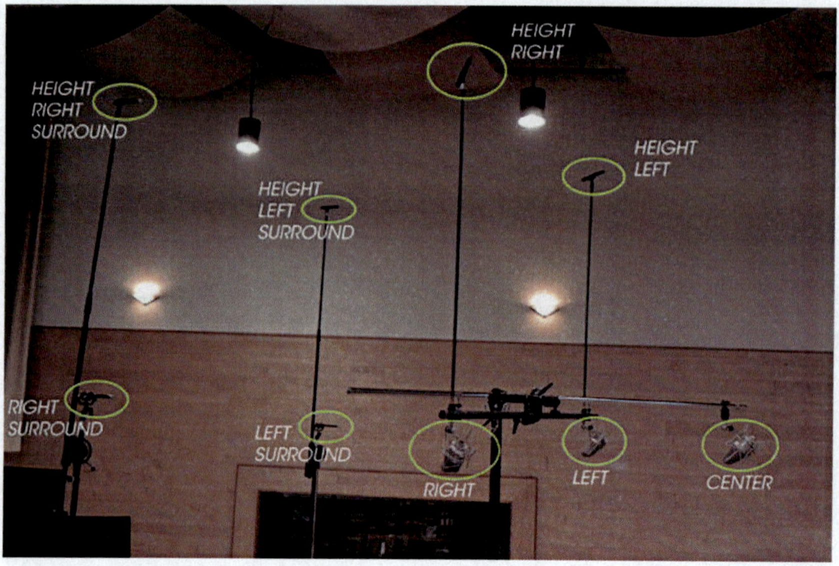

Fig. 9.3 Auro-3D mic setup for large-orchestra recording (courtesy Van Baelen, Galaxy Studios)

their specific placement the same rules apply as for the large setup…" (from Van Daele and Van Baelen 2012).

Another example of microphone setup for a large orchestra recording shall be given below (free citation from Nipkow 2012): "…Some Tonmeisters have reported an increase in perceived naturalness of 'timbre' (sound-color) of musical instruments in Auro-3D recordings in comparison with 2D-surround and stereo recordings (see Nipkow 2010 and Zielinsky 2011). These recordings have used large-AB style omni microphone setups, with proportions equivalent to the Auro-3D loudspeaker setups." According to (Nipkow 2012) this enhanced sonic property can be attributed to the fact that the signals of the 'height' microphones (HL, HR) pick up diffuse- but also direct-sound from the sound source at a distance of several meters, which—on replay—is being interpreted as early reflections and/or direct sound by the human listener. Interestingly the same positive effect—concerning timbre - cannot be achieved by replaying only captured diffuse sound from height speakers. Also, replaying coherent signal-information through both the front loudspeakers L, R as well as through the height speakers HL, HR in an attempt to create 'phantom image' localization between the speakers L/HL or R/HR does only result in sound-coloration, according to (Pesch 2010) and (Theile and Wittek 2011).

In Fig. 9.4 an Auro-3D microphone setup is pictured which uses large-AB style spacing of omni microphones arranged according to the proportions of a typical Auro-3D loudspeaker replay setup.

According to (Nipkow 2012) one of the main goals in Auro-3D is to capture the 'room-sound' (i.e. diffuse sound) in an almost de-correlated manner, which means

Fig. 9.4 Auro-3D mic setup for large-orchestra recording at HMT, Hannover (from Nipkow 2012)

a signal-coherence of (or around) zero. In order to achieve this, he proposes to use either largely spaced omnidirectional microphones, or microphones with strong directionality (microphones with super- or hyper-cardioid pattern and appropriate angling) (see also acoustic 'fluctuation', Griesinger 1998)

For the HLS and HRS channels the capturing of direct sound should be avoided, according to (Nipkow 2012). Therefore he suggests the use of omnidirectional microphones far away from the sound source, or microphones with a strong directional pattern which are aimed away from the sound source. Figure 9.5 shows omnidirectional HLS and HRS (along with HL and HR) microphones, used for an organ recording in a church in Switzerland.

Figure 9.6 shows the very 'creative' placement (presumably derived from Tonmeister-experience, as well as constraints in terms of 'practicability') of microphones for the Auro-3D recording of a church organ in Hofkirche in Luzern, Switzerland, as documented in (Nipkow 2012). Note that there is a clear distinction between microphones used for a more direct pickup of the instrument itself (i.e. 'spot-microphone' in green) and the overall 'room sound' with the Auro-3D main microphone system (L, R, LS, RS, HL, HR, HLS, HRS). Due to the large microphone spacings and the somewhat arbitrary layout/placement of the microphones during post-production, (an 'artistic'/psychoacoustic) time-alignment of the microphone-signals in the hard-disk recording software may be required to avoid echo-effects or simply make the overall sound 'more compact' to fit the much smaller spacings of the Auro-3D replay loudspeaker setup.

Fig. 9.5 Large-AB microphone setups for HL, HR and HLS, HRS as used at Hofkirche in Luzern (from Nipkow 2012)

The assignment of microphone signals to loudspeakers is according to the following 'rules': signals with the highest amount of direct sound should be routed to speakers L, R and HL, HR. Microphone signals with the least distracting direct sound components can be routed to HLS, HRS, in order to avoid irritation in the listener. The above mentioned need for time-alignment of the signals of the 'room microphones' (as opposed to the frontal 'main microphones' L, C, R) depends on the musical source material: with percussive music a correction is often necessary to avoid distracting or unnaturally sounding echo effects. If the music is not percussive in nature, the time-delays between microphone pairs (e.g. L, R to LS, RS but also L, R to HL, HR or HL, HR to LS, RS) can lead to a pleasant and lively spatial impression (as well as to the above mentioned 'fluctuations'; see (Griesinger 1998)).

9.3 Music Recording with String Quartet

An interesting experiment was undertaken by (Dietel and Hildenbrand 2012) in which the authors have recorded a string quartet in Auro-3D, using a 5.0 surround array of omnidirectional microphones in a L, C, R arrangement, combined with two largely spaced omnis for the surround channels LS, RS. This horizontal layer was combined with three different 'height' layers:

1. four largely spaced omnis, positioned roughly 1 m (3feet) above the omni capsules L, R, LS, RS of the base layer,
2. slightly behind this, a layer of 4 super-cardioids, pointing towards the ceiling in order to only pick up diffuse sound, and

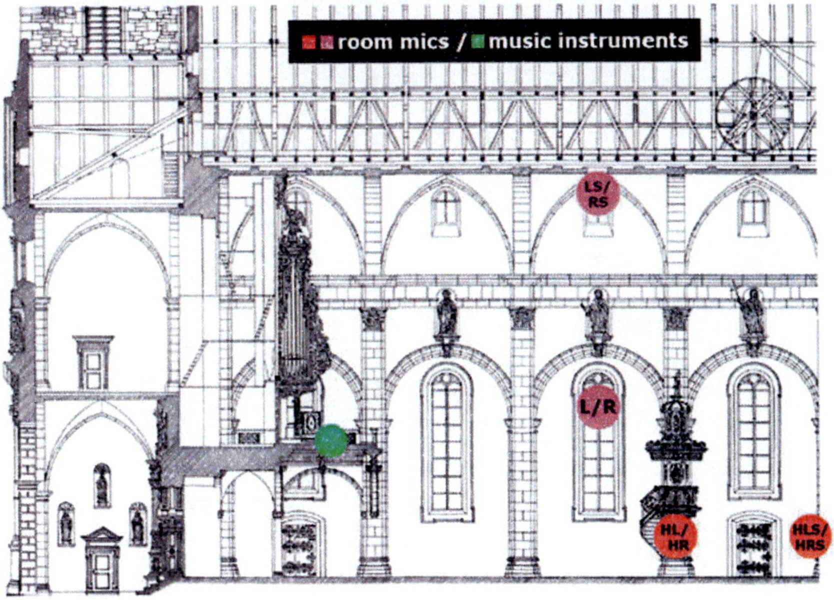

Fig. 9.6 Schematic of microphone-positions used for the Auro-3D recording of a church organ (from Nipkow 2012)

3. even further behind: four cardioid microphones, the front pair of which was arranged in an 'equivalence'-style recording, while the rear pair was pointed mainly towards the back walls (see Fig. 9.7. and Fig. 9.8).

9.4 Music Recording with Church-Organ

The same microphone arrangement was also used to record an organ in a church (unfortunately, also with this recording no information was given concerning the critical distance of the recording venue).

In a subjective listening test with 42 participants (mainly students and practicing sound-engineers/Tonmeisters) the base layer was combined with all of the three different height layers and ratings were to be made concerning

(a) the perceived distance of the sound source,
(b) overall preference of a recording,
(c) perceived elevation of the sound source
(d) strength of musical/acoustic envelopment of the listener.

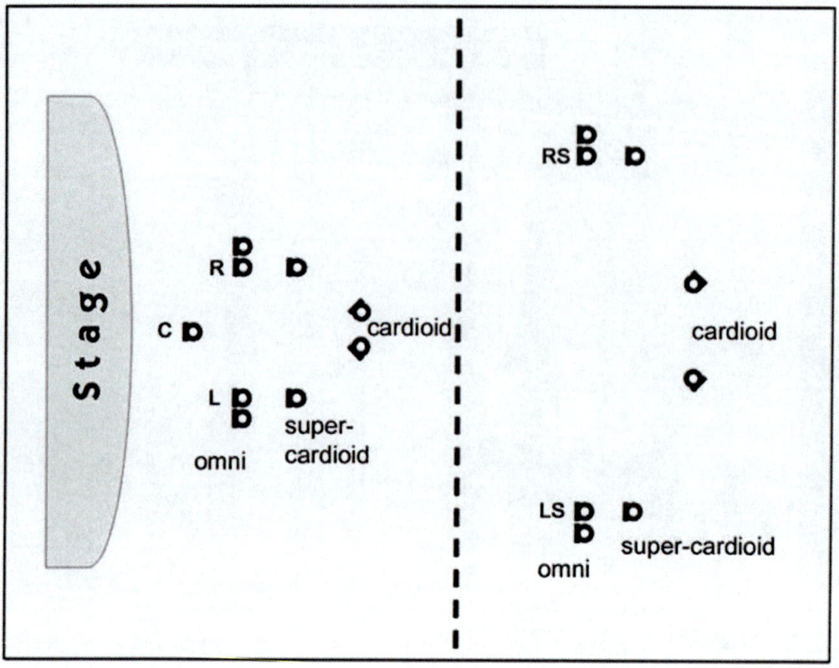

Fig. 9.7 Auro-3D—Schematic of 5-channel omni-mic base-layer combined with three different height-layer setups (from Dietel and Hildenbrand 2012)

Fig. 9.8 Front arrangement (left) and back arrangement (right) of Auro-3D 5-channel omni-mic base-layer combined with three different height-layer setups of Fig. 9.7 (from Dietel and Hildenbrand 2012)

Not too surprisingly, the playback which used the signals of the super-cardioids for the height-layer (i.e. only diffuse sound) were perceived to sound 'most distant'. This was true both for the string quartet, as well as the organ recording.

In terms of preference there was no uniform answer: preference seemed to depend very much on individual taste, with a tendency in favor of the super-cardioid layer concerning the string quartet recording, probably due to the more 'realistic' spatial reproduction it was able to provide (according to written feedback of the subjects). In contrast, for the organ recording both height layers, which contained more direct signal (i.e. omnis, as well as cardioids) were preferred; this was probably due to an otherwise overall abundance of diffuse sound in all loudspeaker signals.

The responses regarding perceived elevation were very unevenly distributed and partly opposed to each other. The most obvious differences in perception where found with a comparison between the cardioid and the super-cardioid height layer. The results found through this part of the listening comparison do not fulfill the expectation that signals with a higher percentage of direct signal in the height layer will lead to higher elevation of the perceived sound source. In case of the string quartet recording it was actually the contrary, as the signals with the highest amount of diffuse sound resulted in the highest elevation.

According to (Dickreiter et al. 2008) vertical localization accuracy in humans is between 9° and 17° in the frontal hemisphere, augmented by a factor of up to 3 with increased elevation of the sound source.

The result to the fourth question in the listening comparison by Dietel and Hildenbrand provided clear results: while for the organ recording the overall envelopment was strongest for the omni height-layer [most likely due to large signal de-correlation down to the lowest frequencies due to large capsule spacing], with the string quartet recording the super-cardioid height layer proved best. The super-cardioid microphones do not provide a solid low-end in terms of frequency response, which—apparently—does not matter so much in this case, as the string quartet does not provide much level in the very low register. Instead, the super-cardioid polar pattern managed to provide a high amount of early reflections and diffuse sound, which resulted in evoking a convincing spatial impression in the listeners. For both recordings, the cardioid height-layer was not able to evoke a high amount of envelopment in the listeners. Reasons for this may lie in the much higher signal correlation for low frequencies in the front-pair of cardioids [i.e. poor spatial impression], as well as the fact that the direct-/diffuse sound ration in the cardioid front-pair and rear pair differs too much from each other in order to provide a seamless acoustic connection between front and back for the listener. (Dietel and Hildenbrand 2012)

9.5 Music Recording with Soloist

In order to show the vast amount of possibilities, how to arrive at a technically proper and apparently at the same time sonically valid Auro-3D recording, I would like to present another case study from the paper by (Nipkow 2012)—see Fig. 9.9.

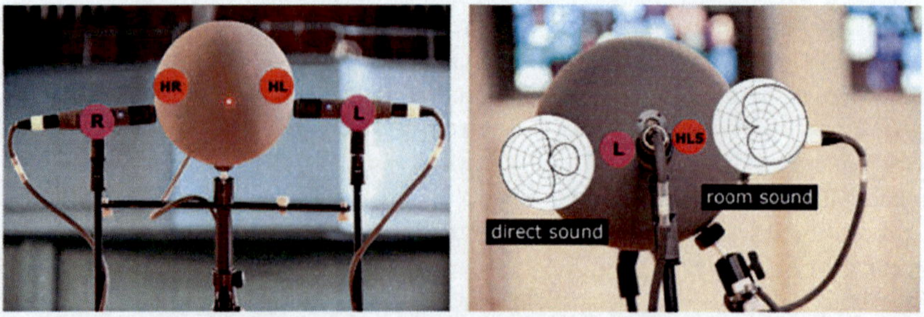

Fig. 9.9 Detailed view on recording setup with Schoeps KFM-6 and Sennheiser MKH-800 Twin (from Nipkow 2012)

In this configuration, the KFM-6 is being used as a baffle that approximates a human head, which—according to Nipkow—holds the added advantage that its physical dimension (due to shadowing effects) results in spectral differences at the built in microphone membranes, which largely resemble those involved in natural hearing. This also helps localization precision. In close vicinity to the membranes of the KFM-6 a L and R Sennheiser MKH-800 TWIN was placed. With appropriate processing of the signals from their front- and rear-membranes, super-cardioid characteristic is achieved for the L and R front channel, while the unprocessed signals from the rear-facing cardioid capsules are used for the HLS and HRS channels (with their null pointing at the sound source), which mainly picks up diffuse sound, coming back from the rear- and side-walls of the church. In this way optimum direct-/diffuse-sound ratios were obtained for the front channel signals L, R as well as HLS, HRS (which should not contain direct sound at all). For the HL, HR channels only the signals from the mic capsules, built into the KFM-6 were used. These receive mainly early (side-wall) reflections and some direct sound from the front, with the side-wall reflections dominating in level over the direct sound. This leads to the violin being localized clearly at stage level and not in an elevated position between the main and the height layer (see Fig. 9.10).

The relatively large spacing of 4.5 m between the sound source and the microphones was chosen according to the wish (and experience) of the musician, mainly in search for a well balanced tonal characteristic and 'overall sound' of the instrument. The rather large distance has the additional positive side-effect that slight movements of the musician do not lead to a 'jump' in terms of localization, which can easily occur with non-stable sound sources in the vicinity of a small-AB microphone setup (or one which uses highly directional microphone patterns). For the LS and RS channel two omni microphones (Neumann KM-183) were positioned next to the side-walls further back in the church (see Fig. 9.11 with microphone stand and RS microphone), preferably with the membranes facing the back wall.

In his paper Nipkow arrives to the conclusion that the signals in the channels HL, HR should differ relevantly from the signals in the front channels L, C, R in order to avoid (unwanted) elevation effects. As we know from research by (Blauert 1974)

Fig. 9.10 Stage and microphone arrangement for solo-violin recording with Schoeps KFM-6 and Sennheiser MKH-800 Twin in a church (from Nipkow 2012)

Fig. 9.11 Right rear omni room microphone (RS), on stand close to side wall, as used in (Nipkow 2012)

it is mainly sound energy in the frequency band around 8 kHz, which is responsible for the perception of 'height' in humans. Drawing from his experience as practicing sound-engineer, Nipkow recalls that an increase in level in this frequency band for microphone signals which contain early reflections results in an enhanced perceived 'brilliance' of the sound, while the same increase with direct-sound signals usually leads to an unpleasingly 'sharp' tonal quality. Therefore it seems desirable that the HL and HR channels are being used to reproduce early reflections from the recording venue, which can lead to an enhanced, more brilliant overall sound. For this purpose Nipkow suggests the use of largely-spaced omni microphones for HL and HR in order to achieve low correlation between their signals, especially in the low-frequency range. However, as shown by his use of the KFM-6 inbuilt microphone signals, it is also possible to use microphone pairs HL, HR with lesser microphone spacing, since low frequencies play only a minor important role in the HL and HR channel, as the human outer ear boosts higher frequencies and uses this information to 'decode' sound source elevation (according to Nipkow). With largely spaced AB microphone pair-signals a high degree of decorrelation is found also for low frequencies, which is beneficial for spatial impression with loudspeaker reproduction. When using the signals of the KFM-6 this useful mechanism is—at least partly—replaced for mid and high frequencies due to the shadowing effect of the sphere, which leads to strong separation of L/R early reflections. This 'head shadowing' effect is of course not effective for low frequencies, but in churches and large concert halls there is the effect of 'fluctuation' (see Griesinger 1998), which applies also to very small capsule spacings. These fluctuations are clearly audible with the KFM-6 (as well as with artificial human heads).

If omni-mics are being put up high above the orchestra one problem may occur though: as also cited in (Nipkow 2012), classical sound-engineer Gregor Zielinsky reports about the problem that wood (and brass-) instrument can have too much level in such height-microphones, due to their strong directional characteristics. This may lead to unwanted localization distortion and these instruments might therefore be heard too much 'from above' (i.e. from the HL and HR loudspeakers). As a countermeasure Nipkow suggests that—under these circumstances—the microphones for pickup of the HL and HR signals may therefore be placed at the same height as the L, R and LS, RS microphones.

As a general 'sound-concept' for the representation of classical music via an Auro-3D loudspeaker system Nipkow suggests that the front L, C, R speakers should manly carry 'direct' sound (as well as diffuse 'room' sound), the height-speaker HL, HR early reflections and diffuse sound, while the rear speakers should normally replay only 'room sound' (i.e. diffuse sound). Diffuse sound components should come from all speakers and be almost de-correlated (i.e. correlation values around—but not equal to—zero), which is shown by using different colors for these signal components in Fig. 9.12.

As visible in Fig. 9.12 the main microphone is used for replay of direct sound and early reflections through the loudspeakers L, C, R of the main layer, while room microphones (assigned to the height layer), positioned along the side walls and close to the stage, pick up mainly early lateral reflections—thus resulting in a

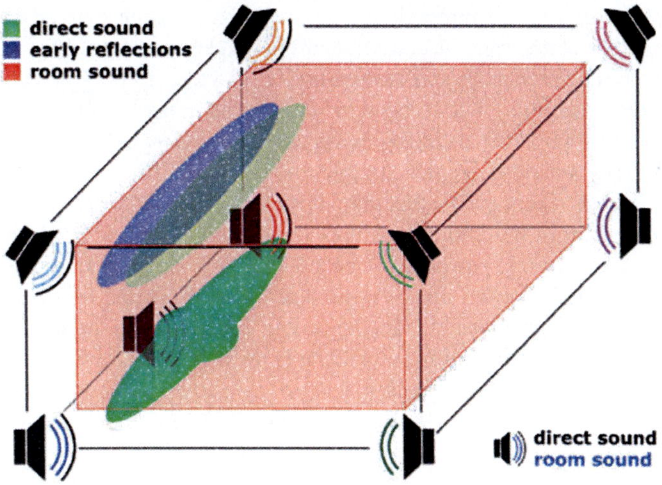

Fig. 9.12 Sound-concept for the loudspeaker-signals of an Auro-3D recording of classical music (from (Nipkow 2012)

'cohesive', sonically 'well connected' main and height loudspeaker layer. The direct sound component leads to a stable localisation of the sound source in the main layer. Due to a small amount of direct sound in the room microphones, used for the height layer, bright sounding instruments seated in the rear section of the orchestra will benefit from perceived 'elevation', which corresponds very well to perception under 'real world'/'concert hall' conditions. The overall sonic result is that of a 'frontal projection' of the sound source through all loudspeakers involved.

According to Nipkow, the strength of this sonic approach lies in the perceived transparency which can be achieved; a clear separation between the direct sound coming from the stage and the sense of musical envelopment, evoked by the diffuse sound, coming from all speakers, and—in addition—an enhanced and very naturally sounding 'timbre' of the instruments, due to early reflections being radiated by the channels HL, HR.

In the following sections we will take a look at various other microphone techniques proposed for the recording of 3D audio, in alphabetical order.

9.6 The AB-BPT-3D System for Decorrelated Signal Recording

The large-AB with centerfill technique AB-BPT (AB-Blumlein-Pfanzagl-Triple), has already been presented in (Pfanzagl-Cardone 2020) in the chapters about stereo and surround microphone techniques (as well as in (Pfanzagl-Cardone and Höldrich 2008)). This arrangement is used as a main layer in a 5.1 surround arrangement, enhanced with appropriate microphones for the height-channels HL, HR and HLS, HRS. The choice of microphone pattern for the height layer microphones will depend very much on room acoustics and ensemble size (and therefore might vary between omni and hyper-cardioid), but as a general starting point the use of super-cardioids, which are angled towards the ceiling, is suggested (see Figs. 9.13 and 9.14).

Whether they should be angled straight up (for more diffuse sound), or slightly forward, to pick up more early reflections from the ceiling right above the orchestra and have their axes of minimum sensitivity (at 126° for super-cardioids, at 110° for hyper-cardioids) pointing towards the sound source in order to minimize pick-up of direct sound, will depend on the desired acoustic effect and on the local room acoustics. If less reflected sound from the ceiling directly above the stage is desired, then pointing the super-cardioids more towards the ceiling corners in the stage area may be necessary. Also, the capsule spacing of the large-AB system, as well as of the surround and height-layer microphones is scalable and the sound-engineer or Tonmeister is challenged to make an informed decision depending on the size of the ensemble and room dimensions. To ensure good low-frequency decorrelation it is recommended to use a capsule spacing lager than the reverberation radius (critical distance) of the venue.

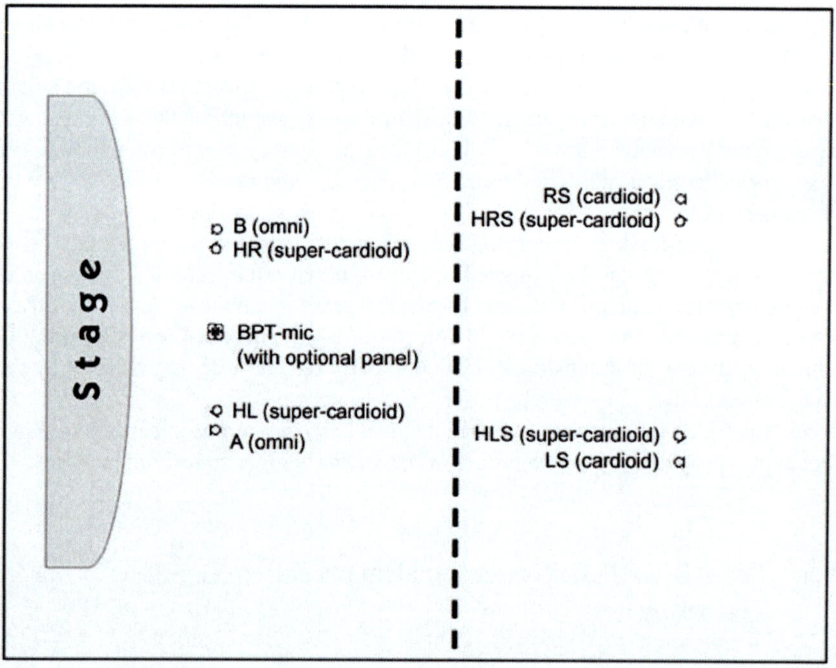

Fig. 9.13 AB-BPT-3D 9.1 microphone system (top view), using graphical elements from (Dietel and Hildenbrand 2012)

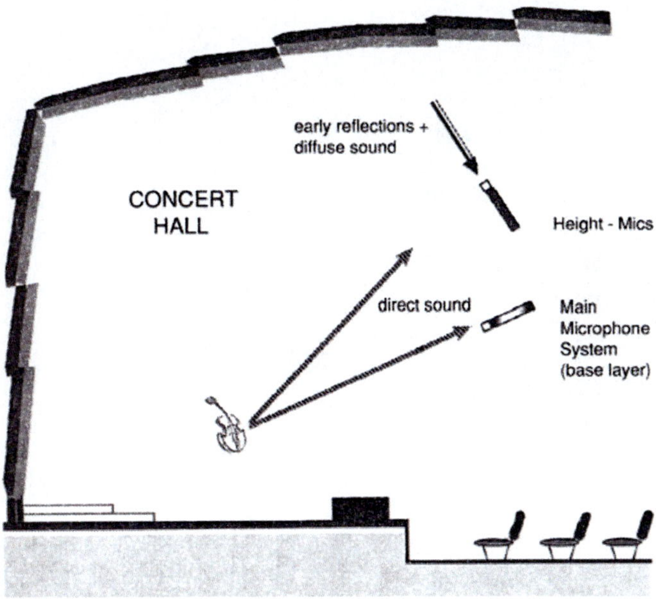

Fig. 9.14 AB-BPT-3D microphone system (side view), using graphical elements from (Fukada et al 1997)

For the rear (main layer) microphones, back-facing cardioids can be used for the base layer, and again super-cardioids for the height layer microphones HLS, HRS.

Usually, if the super-cardioid microphones of the height-layer have a capsule spacing larger that the reverberation radius of the room and are pointing upwards, they should pick up well de-correlated diffuse sound. In case they are paired much closer to each other (for ease of rigging, etc.), it might be necessary to establish a physical opening angle between them which ensures that their signals are maximally de-correlated by using an included angle of 126° for super-cardioids and 110° for hyper-cardioids.

On occasions where a very compact ('one-point' style) microphone array is needed (which may be the case with smaller sound sources like a soloist or small ensemble), which consists of a 7.1 surround base layer, the following arrangement can be used: a back-to-back configuration of two BPT microphones (with an included absorptive panel in between, for better channel separation) provides 6 channels for L, LSS, LRS and R, RSS, RRS. For the Center-Channel (main layer) the use of a super-cardioid or hyper-cardioid is suggested, which should be mounted in proximity to the BPT-microphones, but can be positioned slightly closer to the frontal sound source on stage, for enhanced stability and localization accuracy of the center image (Fig. 9.15).

For the height layer, again the use of 4 super-cardioids is suggested, with the angling suggestions as written above. Since in this case the microphones of the

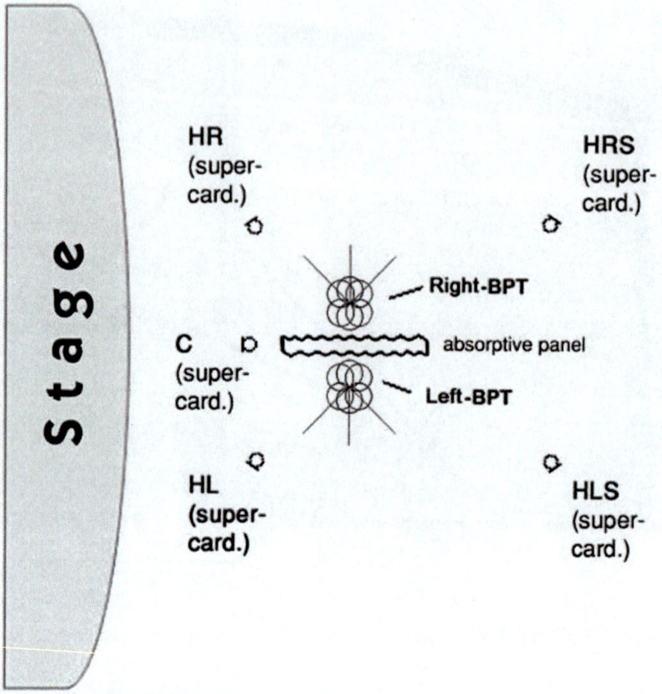

Fig. 9.15 Compact AB-BPT-3D 11.1 microphone system (7.1 + 4.0) (top view) using also graphical elements from (Dietel and Hildenbrand 2012)

height layer will likely have to be arranged relatively close to the BPT back-to-back arrangement, the angling between them might be of even greater importance in order to achieve de-correlated signals. See Fig. 9.16 for a commercially available compact version of the Blumlein-Pfanzagl-Triple, the NEVATON BPT, which contains 3 double-membrane condenser capsules in one housing. Due to its arrangement and underlying sound-capturing principle the Blumlein-Pfanzagl-Triple has also been referred to as "Coincident Decca Tree" by Dr. Piotr Majdak of the OEAW (Austrian Academy of Sciences). This may sound surprising at first, as the 'classical' Decca Tree (or Decca Triangle) is commonly seen to consist of three omnidirectional microphones arranged in a triangular configuration above the head of the conductor, mainly relying on time-of-arrival differences in terms of localization.

However, this common conception is actually wrong, as the early experimentation of the Decca engineers—stretching out over the course of more than 10 years—was actually to find a system which would offer a nice separation of a the 'left', 'center' and 'right' section of the orchestra. For this purpose they tried various versions of 'baffled' triangles (using omnidirectional microphones) as well as triangular configurations with directional microphones (see Chap. 11 in (Pfanzagl-Cardone 2020) for more details). However, the sonic result from the last—now common—version

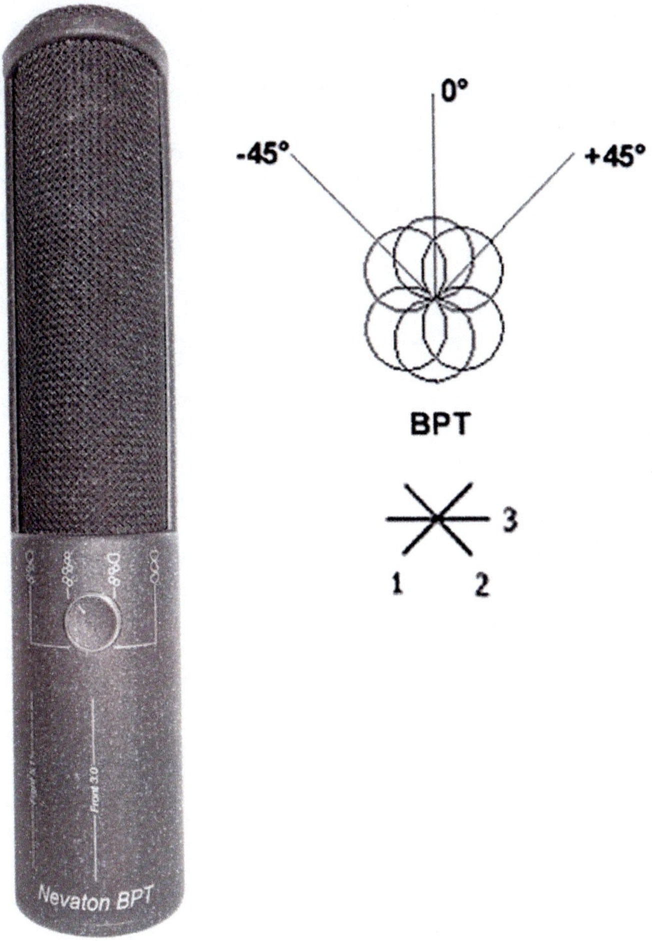

Fig. 9.16 BPT (Blumlein-Pfanzagl-Triple) left: photo of vertically coincident 3-capsule NEVATON BPT-microphone; right: schematic view (graphic is Fig. 4.8 from Pfanzagl-Cardone 2020)

of using e.g. three omnidirectional Neumann M50 microphones, is heavily based on the fact that the M50—due to internal construction principles of the capsule—in fact becomes highly directional towards high frequencies, and also that the Decca engineers have established 'reference points' within the orchestra setup, towards which the M50s have to be pointed, essentially making this whole setup a highly 'directional' one (for more details see (Haigh et al 2021), Chap. 8, p. 128 ff).

9.7 The Bowles Array with Height Layer

In (Bowles 2015) the author describes various microphone techniques with height layers, able to record 3D audio.

After a considerable amount of experimentation (described in (Bowles 215) under real-world sound recording conditions, Bowles has arrived at the following proposal for his 3D microphone system:The main layer of his system for recording 5.1 surround-sound "...consists of omni-directional microphones for left and right front and rear channels. Depending on size of musical group and/or desired imaging, a directional microphone can be used for the center channel. Surround microphones are spread further apart than the left and right pair, pointing backwards at angles approximating speaker spread and angle in the ITU-R BS.775-2 specification. Taking into account head shadowing, this increased distance between the surround microphones helps to establish 'rear imaging'" (from Bowles 2015).

For the height channel layer, Bowles used four hyper-cardioid microphones located directly above their corresponding omnidirectional microphones. "This layer was placed above the main layer; angling the hyper-cardioids at 60° upward so their null axis pointed directly below. I then angled them outward to provide better separation between left and right height channels (with resultant wider imaging in the height layer). These angles helped to define the height layer as distinct from the main layer. Additionally, since hyper-cardioids have a low-frequency roll-off there is less chance of undesirable low-frequency build-up when listening to the full array..." (Bowles 2015).

On his experience in post-production Bowles reports that "....For large ensembles, more 'upward-facing' hyper-cardioid microphones can be used for gathering height information from sound sources. This will help in gathering height information from the center and edges of larger ensembles, as well as from the audience noise in the case of live pop concert recordings.

Further improvements to the current Bowles array:

1. using additional microphones coincidentally-mounted with the hyper-cardioids, then using M/S processing to split the front and rear lobes of each height microphone (finally discarding the rear lobe signal entirely); or,
2. use of DSP-enabled microphones which carry out the same function internally. For the second option one danger is that one model on the market is a shotgun microphone; the resulting polar pattern might be too narrow for proper imaging in the height channel layer. Either of these two options will (1) enable the height channel layer to be isolated even better from the main layer, and (2) enable similarly-isolated recording of a top 'voice of god' layer for Auro-3D encoded Blu-Ray releases..." (from Bowles 2015).

Bowles concludes: "...When considering microphone arrays, it is important to visualize sound sources, and microphone polar patterns (and their frequency response) in three dimensions. This is even more vital when recording in surround-sound and with height channels. Recording the vertical axis also forces us to consider ceilings just as important as floors, side walls and rear walls in acoustic spaces.

Representing this vertical axis accurately is important because of our diminished ability to localize sounds coming from vertical angles greater than 45°—similar to our diminished ability to localize sounds accurately coming from behind.

This 'Bowles array' proves to be a good solution for a surround-sound recording with a height channel layer: it contains information specific to ceiling reflections, with sufficient isolation from floor reflections. This height layer can also be mixed into a traditional surround sound or stereo recording as ambience, without excessive comb-filtering. It is in a true 9.1 playback environment that these recordings really stand out, presenting the listener with a height layer which is clearly defined yet integrates with the surround layer…"(from (Bowles 2015).

9.8 Ellis-Geiger Triangle

With a strong background in recording multichannel surround music for cinema, Dr. Robert Jay Ellis-Geiger has developed a triangle microphone array that is comprised of three double M-S (Mid-Side) sets. According to his publication (Ellis-Geiger 2016): "…Listening tests demonstrated the flexibility of the array for solo, duo through to ensembles. The triangle array is scalable depending on the size of the ensemble and room dimensions." The position and direction of each microphone can be viewed as illustrated in Figs. 9.17, 9.18, 9.19 and 9.20.

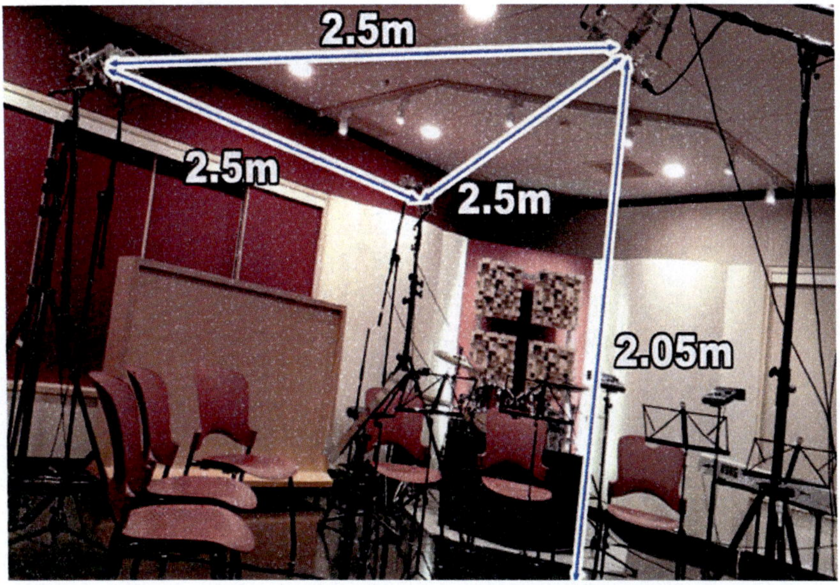

Fig. 9.17 Ellis-Geiger Triangle setup for small ensemble recording (from Ellis-Geiger 2016)

Fig. 9.18 Ellis-Geiger Triangle: Double-MS set (from Ellis-Geiger 2016)

Fig. 9.19 Ellis-Geiger Triangle: front array mic positions and directions (from Ellis-Geiger 2016)

Fig. 9.20 Ellis-Geiger Triangle: rear array mic positions and directions (from Ellis-Geiger 2016)

9.8.1 Front Sets

"…The front double M-S microphone array can be described as two directional microphones facing in opposite directions, combined with a common figure-8 microphone in middle as follows:

- Front Down (FD) (M-S)—this is the combination of the front cardioid microphone facing down towards the musician/s and the figure-8 positioned laterally.
- Front Up (FU) (M-S)—this is the combination of the front cardioid microphone facing up towards the ceiling and the figure-8 positioned laterally …"

9.8.2 Rear Set

"…To capture larger ensembles, the triangle was expanded and the rear double M-S sets were toed into the centre, which resulted in an enhanced immersive experience during playback. The rear double M-S microphone sets can be described as follows:

- Surround Left Down (SLD) (M-S)—this is the combination of the rear left cardioid microphone facing down towards the musician/s and the figure-8 positioned laterally.

- Surround Right Down (SRD) (M-S)—this is the combination of the rear right cardioid microphone facing down towards the musician/s and the figure-8 positioned laterally.
- Surround Left Up (SLU) (M-S)—this is the combination of the rear left cardioid microphone facing up towards the ceiling and the figure-8 positioned laterally.
- Surround Right Up (SRU) (M-S)—this is the combination of the rear right cardioid microphone facing up towards the ceiling and the figure-8 positioned laterally…" (from Ellis-Geiger 2016).

9.8.3 Ellis-Geiger Triangle to Auro-3D 9.1 Mapping

The array is comprised of a 5.1 ITU-R and 4 × height speakers.

ITU-R Speaker Array:

- Front Left, Right: derived from the FD (M-S) set
- Front Centre: FD cardioid
- Surround Left: SLD (Surround Left Down)
- Surround Right: SRD (Surround Right Down)

Height Speakers—The height mapping can be described as:

- HL (Height Left): derived from the FU (M-S) set
- HR (Height Right): derived from the FU (M-S) set
- HSL (Height Surround Left): Back Up Left (BUL) cardioid
- HSL (Height Surround Right): Back Up Right (BUR) cardioid

9.8.4 Ellis-Geiger Triangle Mapping for Dolby Atmos (8.1 + 4 x Height)

Screen Speakers: The mapping of the three screen speakers was the same as for the Auro-3D 9.1 array.

Surround Speakers: The mapping of the Ellis-Geiger Triangle to the surround speakers is as follows: each rear corner of the surrounds formes a stereo pair, which provides amazing location within the surround sound field. The expansions of the abbreviations are as follows:

- SSL: Surround Side Left
- SRL: Surround Rear Left
- SSR: Surround Side Right
- SRR: Surround Rear Right

9.8.5 Ellis-Geiger Triangle and Height Speakers

The mapping of the 4 × height speakers is the same as for the Auro-3D 9.1 array.

About the use of artificial reverb with the height speakers, Ellis-Geiger writes: "…A discrete 4.0 (quad) reverberation processor was used for the height array. It allowed for different pre-delay settings to be set for the Height L, R and Height SL, SR, which greatly enhanced the sense of space in the vertical.

A discrete 5.1 reverberation processor was used for the horizontal Auro-3D array and a discrete 8.1 reverberation processor was used for the Dolby Atmos 8.1 horizontal array…" (from Ellis-Geiger 2016).

9.9 The Geluso 'MZ-Microphone' Technique

The Middle-Z (MZ) Pair

From (Geluso 2012):

"…Based on the principles of middle-side (MS) recording technique, a bidirectional Z microphone is oriented vertically and coincident to a horizontally oriented microphone, which creates a Middle-Z (MZ) pair. In a conventional MS to Stereo decoding matrix we obtain left and right channels as follows (Fig. 9.21):

Left = Middle + Side

Right = Middle − Side

Substituting the Z channel for the side channel we have:

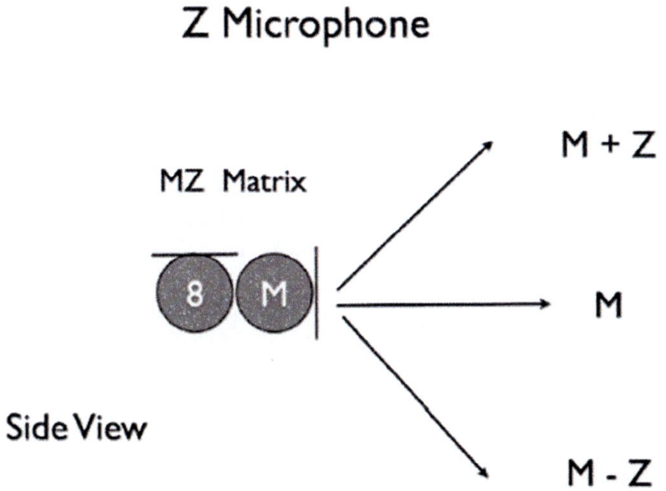

Fig. 9.21 Middle-Z (MZ) microphone pair side view: The Z microphone is a figure of eight receptor oriented vertically with its positive side facing upward. The middle (M) microphone is oriented horizontally 90° to the Z microphone

Height at 45° = Middle + Z.
Height at -45° = Middle − Z.
"…In (Hibbing 1989), the author listed inherent benefits in using a MS system:

- The M microphone may feature any type of directivity.
- The recording angle can be varied to a great extent in post-production.
- The directivity performance of the bidirectional figure of eight S microphone is excellent.
- Compared to XY, The MS technique is less prone to flaws in the microphone construction and thus records at a higher fidelity.

The author [i.e. Geluso] hypothesizes that these MS benefits apply to his proposed MZ microphone technique since a MZ pair is essentially a vertically oriented MS pair.

Bi-directional microphones in figure of eight mode reject sounds arriving 90° off axis better than any other directional or omni-directional microphone (Dickreiter 1989). Hence if the null of the Z microphone is facing the sound stage, excellent separation of the Z channel and its associated horizontal channel can be achieved without having to space the microphones far apart from each other. Since the Z and M microphone capsules are coincident, the MZ microphone technique introduces minimal delay between sounds arriving from the median and horizontal planes. Together this makes the system compact, and thus practical for the recordist. Another advantage is that the vertical pick-up angle can be adjusted in post-production using a MS decoder matrix.

9.9.1 Perception of Height Channels

The way we perceive the location of a sound in the median plane is very different from the way we perceive location in the horizontal plane. Interaural time differences converge as the median elevation angle increases (Blauert 1997). Barbour suggests that a single loudspeaker placed on the median plane is of little benefit for vertical localization, and that a minimum of two height speakers is required to generate effective localization and height envelopment (Barbour 2003). Therefore a useful height capturing microphones system should be able to reproduce at least stereo height channels…" (from Geluso 2012).

9.9.2 Stereo Height Channels

"… To create stereo height channels, we considered two options:

1. A single Z channel is used to generate stereo height channels in conjunction with a coincident stereo microphone system. When summed with left and right

channels independently, the Z channel can virtually steer the left and right signals upward.
Left Height = Left + Z
Right Height = Right + Z
2. A stereo pair of Z microphones is used to generate stereo height channels in conjunction with a spaced stereo microphone system. Using a spaced MZ stereo pair, left and right height can be generated as follows:

Left Height = Left + Left Z
Right Height = Right + Right Z

With this spaced MZ stereo pair (2nd option), time of arrival information between height channels is captured, while still maintaining phase coherence between sound arriving from the horizontal and median planes at each individual MZ pair. This option is more consistent with the way sound arrives at each of our ears.

9.9.3 With Height Systems Using Z Microphone Techniques

To create stereo high channels with surround, we considered four options:

Double MS + Z

Double MS is a compact three microphone coincident array that consists of two cardioid microphones back to back and a single figure of 8 microphone facing side to side (see (Holman 2008), pp. 93–94). By adding a single vertically oriented Z microphone to this system, the equivalent of Ambisonic B-format can be obtained as follows:

W = Middle + Middle Surround
X = Middle − Middle Surround
Y = Side
Z = Z

Thus the matrix to obtain a 7.0 (5 + 2) output is as follows:

Left = Middle + Side
Right = Middle − Side
Center = Middle
Left Surround = Middle Surround + Side
Right Surround = Middle Surround − Side
Left Height = Middle + Side + Z
Right Height = Middle − Side + Z

XY + Z

XY is a stereo system that uses two matched coincident directional microphones. They are rotated symmetrically to adjust the horizontal recording angle (Dickreiter

1989). To generate height channels, a figure of eight Z microphone is placed coincident to the XY pair oriented upward. Thus the matrix to decode XY + Z into 7.0 (5.0 + 2) is:

Left = X
Middle = X + Y
Right = Y
Left Surround = X−Y
Right Surround = Y−X
Left Height = X + Z
Right Height = Y + Z

Blumlein + Z

A Blumlein stereo system consists of a pair of figure-of-8 microphones rotated 90° (Pulkki 2002). By adding a third figure of eight microphone to the system oriented vertically to the Blumlein pair, a 7.0 (5 + 2) mix can be obtained using the same matrix described above for XY + Z.

Spaced 5.0 Array + Z channels

A 5 + 2Z array can be created by pairing two bidirectional Z microphones with the left and right microphones of virtually any 5.0 spaced microphone surround array. Height channels to complete a 7.0 (5 + 2) mix can be obtained by summing the channels as follows:

Left Height = Left + Left Z
Right Height = Right + Right Z

It should be noted that it is possible to add more than 2 height channels to a spaced surround microphone array using the Z microphone method, the number of Z channels possible is only limited to the number of horizontally oriented microphones used in the spaced array…" (from Geluso 2012).

Geluso summarizes: "…Preliminary evaluations of several 7.0 (5 + 2) recordings indicate that the 3D sound images of height channels made by the MZ technique are perceived as better than the up-mixed height channel derived from the 5.0 channels. Thus the addition of vertically oriented bidirectional Z microphones to stereo and surround microphone systems is an effective way to capture height sound. The MZ microphone technique is compact and does not require the use of specialty microphones. Moreover, being able to integrate the MZ microphone technique with familiar surround microphone techniques makes this method a practical solution for recordists who are interested in 3D recording. Using MS decoding techniques, height channels can be created and fine-tuned in post-production. Future research will test these techniques with more participants and expend these techniques to more height channels (e.g. 4Z and 5Z systems instead of 2Z)…" (for details see Geluso 2012).

9.10 The Zhang-Geluso '3DCC' Technique: A Native B-Format Approach to Recording

In (Zhang and Geluso 2019) the authors present a recording technique which has its roots in a proposal by Benjamin and Chen, who published their native B-format microphone technique in 2005 (see Benjamin and Chen 2005), introducing a microphone array that directly captures an Ambisonic B-format signal without having to be derived from cardioids. "...With an omnidirectional W and three figure-of-eight microphones for X, Y and Z capture, they created and evaluated two versions of native B-format: one using traditional bar microphones and one using smaller lavalier microphones. While their version native B-format microphone array used physical analogues of B-format's virtual capsules, the double MS-Z (DMS-Z) system introduced by Geluso in 2012 instead used two cardioid and two figure of eight microphones microphones arranged on B-format's X, Y and Z planes (see Geluso 2012). Unlike the original native B-format microphone technique, the omnidirectional W pattern is not physically present. However, the pattern can be easily derived by summing the two cardioid capsules, making it an implied W signal.

9.10.1 Dual –Capsule Technology

The 3DCC can be seen as building on the work of the DMS-Z array with the important introduction of dual-capsule microphones. While multi-pattern microphones use the summation of two cardioid signals to create a single polar pattern, dual-capsule microphones use two highly coincident capsules as discrete outputs (see Torio and Segota 2000a, b). One application of dual-capsule microphones that utilizes both capsules is surround sound, where one microphone can be used to capture a coincident left/right or front/back signal. The 3DCC is able to take advantage of these capsule relationships to create primary and secondary polar pattern outputs, moving beyond the figure-of-eight decoding of the B-format microphone technique.

9.10.2 3DCC Configuration

The 3DCC microphone technique is based on using three coincident dual-capsule microphones: microphone X, microphone Y, and microphone Z. The X microphone is oriented forward in the horizontal plane; we will consider this orientation to be at zero degrees (0°). The Y-microphone is also oriented in the horizontal plane so that its front capsule is facing left (from audience perspective) at 270°. The Z-microphone is oriented so that the front capsule is directly facing upward at 90° in the vertical plane. The balance and phase relationship between the front and rear signals define the orientation and directional sensitivity of each microphone output" (Zhang and Geluso 2019) (Fig. 9.22).

"Using the two discrete outputs of each dual-capsule microphone, the 3DCC recording technique can move beyond B-format to produce a number of primary and secondary outputs in different matched and mixed polar patterns.

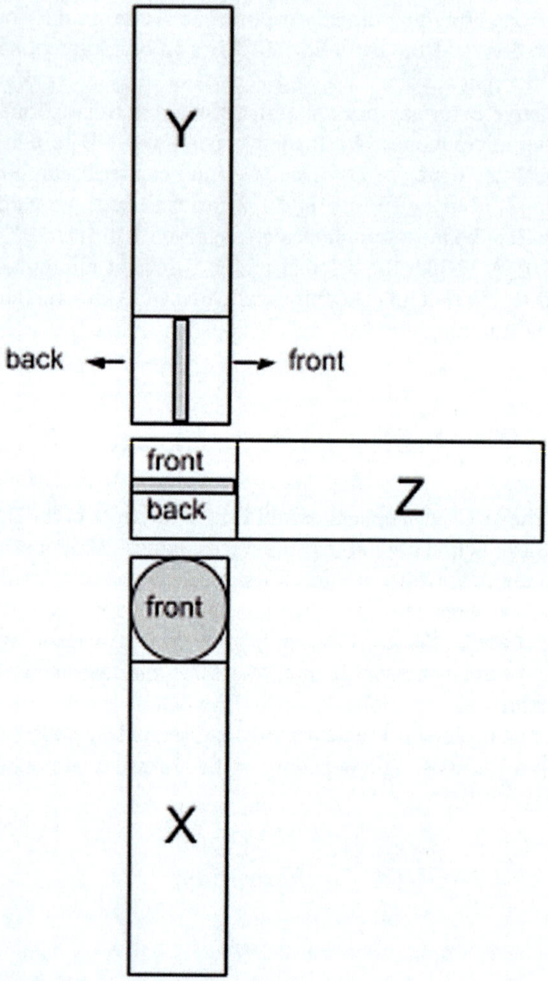

Fig. 9.22 Front view (performer perspective) of the 3DCC microphone configuration (DMS-Z mode) (from (Zhang and Geluso 2019))

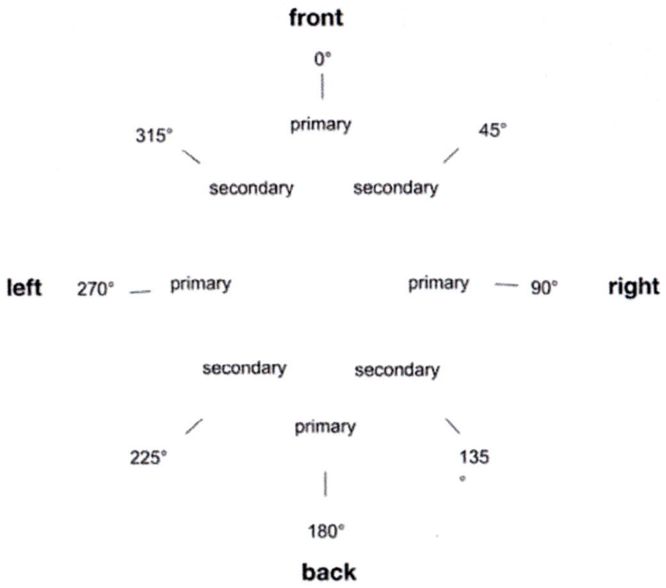

Fig. 9.23 Diagram showing the horizontal orientation of eight equally spaced directional signals derived from the 3DCC microphone array signals (from (Zhang and Geluso 2019)

9.10.3 *Primary Signals*

The six raw directional signals captured using the 3DCC microphone technique can be used for immersive sound applications without the conversion from A-format to B-format typically required when working with tetra-based sound field microphones as each microphone can individually produce a front and a back oriented signal. These six directional primary signals face 0°, 90°, 180°, and 270° in the horizontal plane and 90° (up) and 270° (down) in the vertical plane. These raw signals provide adequate coverage to capture and reproduce a fully immersive experience (see Fig. 9.23). All directional signals can be varied independently from omnidirectional, to cardioid, to figure-of-eight for added control over directionality in post-production. Using a figure-of-eight signal for the Z-microphone may be advantageous because of its excellent side rejection whereas hyper-cardioids may be used for the surround layer to enhance image clarity and width.

9.10.4 *Secondary Signals*

By combining neighbouring primary directional signals, we can create a set of high quality secondary directional signals offset by 45° (see Fig. 9.24). Creating secondary signals requires careful summing and weighting to steer the resulting polar pattern's

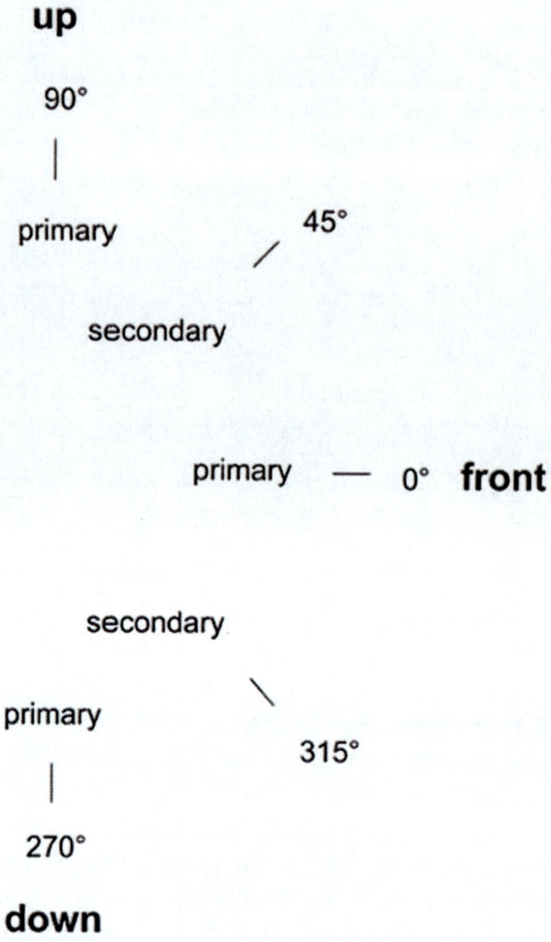

Fig. 9.24 Diagram showing the vertical orientation of five equally spaced directional signals that may be derived from the 3DCC microphone technique. (from (Zhang and Geluso 2019)

directionality and to compensate for a boost in gain due to polar pattern overlaps" (from Zhang and Geluso 2019).

9.10.5 *Practical Application of the 3DCC Microphone Array*

The 3DCC Microphone technique was used in a case study type of recording with a jazz quartet (see Zhang and Geluso 2019): "… In order to examine the robustness and understand the characteristics of the 3DCC's different outputs using musical source material, we examined the following recording using five different primary polar pattern sets derived from the same source material. This case study was recorded during New York University's Stephen F. Temmer Tonmeister course in June of

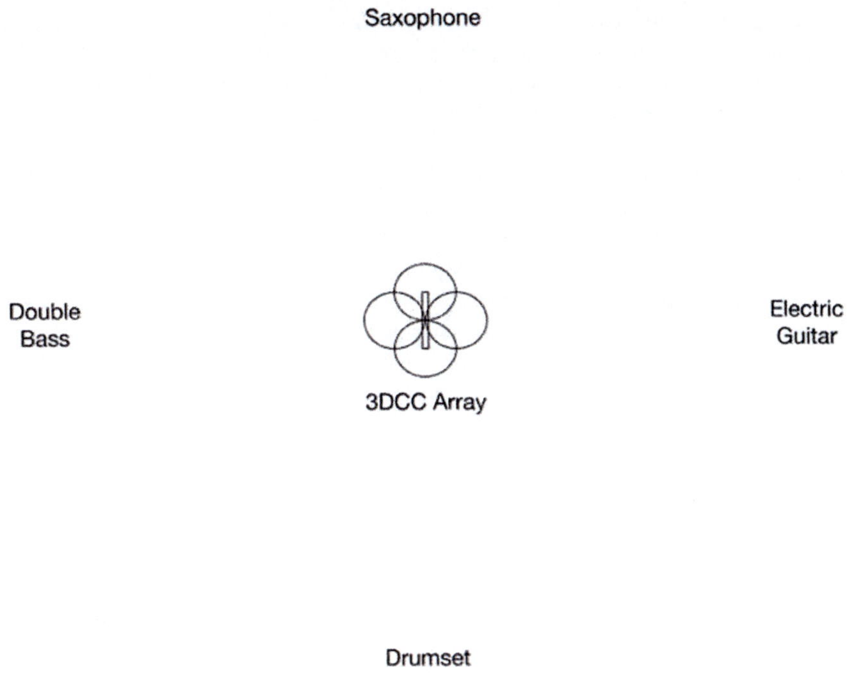

Fig. 9.25 Diagram of in-studio setup and instrument/3DCC relationships for the John Hadfield jazz quintet in Dimenna Center (from Zhang and Geluso 2019)

2017. It was chosen for having an even distribution of sources around the microphone system, which was positioned in the center of the four players who performed an improvisational jazz piece at Dimenna Center, a studio-like rehearsal space in Manhattan." (Fig. 9.25).

"Sennheiser MKH800 Twin microphones were used for the X, Y, and Z-microphones with their discrete outputs matrixed in Pro Tools for multichannel loudspeaker playback. During initial matrixing and evaluation, adjacent signals were compared using iZotope's Ozone Imager plugin. While the plugin is meant for examining the 'phantom center' in stereo sound, the imager can actually be used to compare the vector relationship between any two signals at a sample-by-sample basis, making it a useful tool to examine the correlation between adjacent signals. In order to reproduce the graphical representation of the analyzed signals in a multichannel environment as a whole (rather than pairwise), we chose to make polar vectorgraphs to represent the relationships between our primary signals. Evaluation took place in the Music and Audio Research Lab at NYU. For each case, the signal was decoded for a 11.0 loudspeaker playback system with seven channels (L, C, R, Ls, Rs, Lss, Rss) in the surround layer and four channels (LH, RH, LsH, and RsH) in the height layer" (from Zhang and Geluso 2019).

To evaluate the effectiveness of this method, Figs. 9.26 and 9.27 contain an analysis of different polar vectographs representing surround layer speaker signal relationship for two different polar patterns (omni and hypercardioid). Vectorgraphs were generated in Matlab using methods from Setsu Komiyama (see Komiyama 1997). Note that the signal shown in the vectorgraphs is not the direct signal coming from the speaker. Instead it shows the relationship between speakers in the surround layer. With left surround positioned at 225° and right surround at 315°, the scatter in-between displays the correlation of the two signals at 180°. The vector relationship falls in the center between each adjacent signal with the width of the scatter *inversely proportional* to the directionality of the primary polar pattern displayed. This demonstrates the system's ability to reproduce immersive sound in a multichannel setting.

Under laboratory listening conditions, we found that in this case using the 3DCC with a hypercardioid pickup pattern represented the best balance between signal width and image clarity, which is reinforced by its vectorgraph's visual appearance as the scatter is relatively diffused while still retaining the general shape of its reproduction pattern" (from Zhang and Geluso 2019).

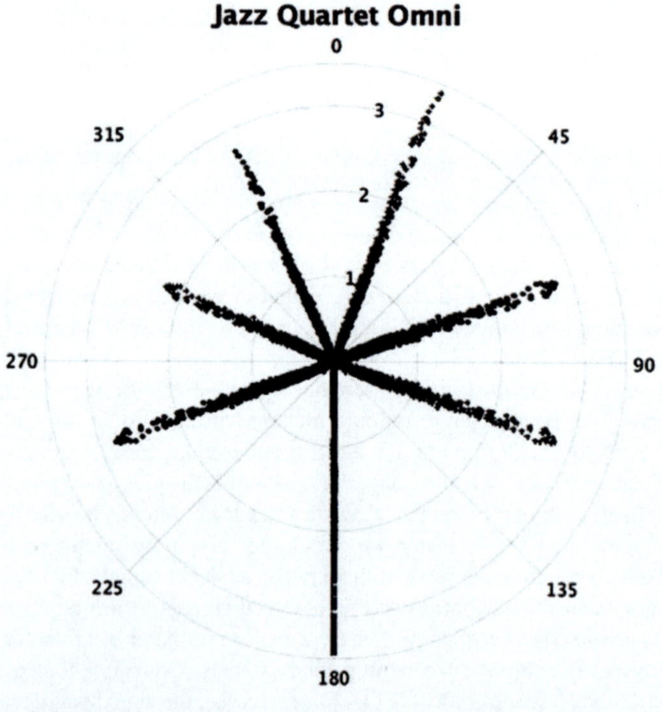

Fig. 9.26 Vectorgraph representing the correlation between adjacent loudspeakers signals for *omni* directional pattern of 3DCC microphone (horizontal surround layer).The scatter in these vectorgraphs demonstrates the relationship between adjacent signals and not the signals themselves. (from Zhang and Geluso 2019)

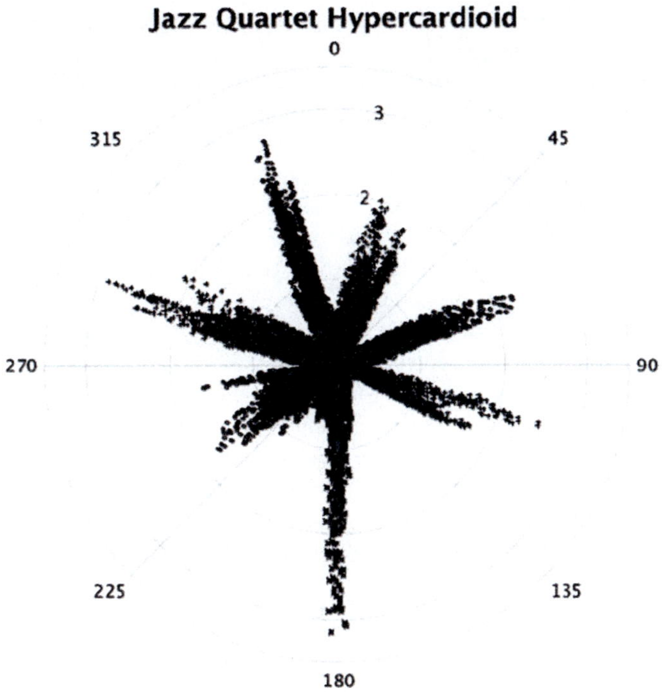

Fig. 9.27 Vectorgraph representing the correlation between adjacent loudspeakers signals for *hypercardioid* directional pattern of 3DCC microphone (horizontal surround layer).The scatter in these vectorgraphs demonstrates the relationship between adjacent signals and not the signals themselves. (from Zhang and Geluso 2019)

In Chap. 1 of this book another method is proposed, which represents the similarity (or dissimilarity) of two audio signals (e.g. the L and R channel of a 2-channel stereo recording, or two adjacent channels of surround sound signals), based on measuring the Frequency dependent Cross-correlation Coefficient (FCC) and is therefore somewhat more detailed than the above vectorgraph display method (see Figs. 9.26 and 9.27), as it takes into account also the possible change of signal correlation over frequency (see Figs. 1.6 and 1.13). This is relevant, for example, in the context of the spaciousness of a recording, which depends on sufficient signal de-correlation in the frequency band below 500 Hz (more detailed info on this can be found in Chap. 1 of this book as well as Sect. 1.4 in Pfanzagl-Cardone 2020). Previously the author of this book has used MATLAB for non-realtime analysis of the FCC, the code of which was published in the appendix of (Pfanzagl-Cardone 2020) and can be accessed freely on the website of the publisher (see https://link.springer.com/content/pdf/bbm%3A978-3-7091-4891-4%2F1). A more in-depth analysis on the differences between signal cross-correlation and signal coherence can be found in Chap. 2 of (Pfanzagl-Cardone 2020), and on the importance of adequate frequency dependent signal de-correlation for surround- and stereo-microphone techniques in Chaps. 7 and 8 of the same book.

Fortunately, since a few years, there is also the possibility of a real-time FCC analysis for 2-channel signals in the form of the "2BCmultiCORR" VST-Software plugin by MAATdigital (see MAATdigital, www.maat.digital/2bcmulticorr/), which offers sound-engineers the possibility to obtain information about signal cross-correlation over frequency in real-time in a very convenient fashion (see Fig. 9.28).

Also, Steinberg's 'Cubase' audio software contains similar tools for signal correlation measurement over frequency from Version 11 upwards (see Fig. 9.29).

In Fig. 9.29 the frequency dependent signal cross-correlation (FCC) of a small AB stereo recording (with 50 cm capsule spacing) is displayed: it is interesting to note that below approx. 200 Hz the correlation of the L and R signals rises above a value of +0.5 and—with falling frequency—continues to converge towards a value of + 1, which implies 'monophonic' sound (Compare with Fig. 2.4 in Pfanzagl-Cardone 2020). As very low signal-correlation, especially at frequencies below 200 Hz, was

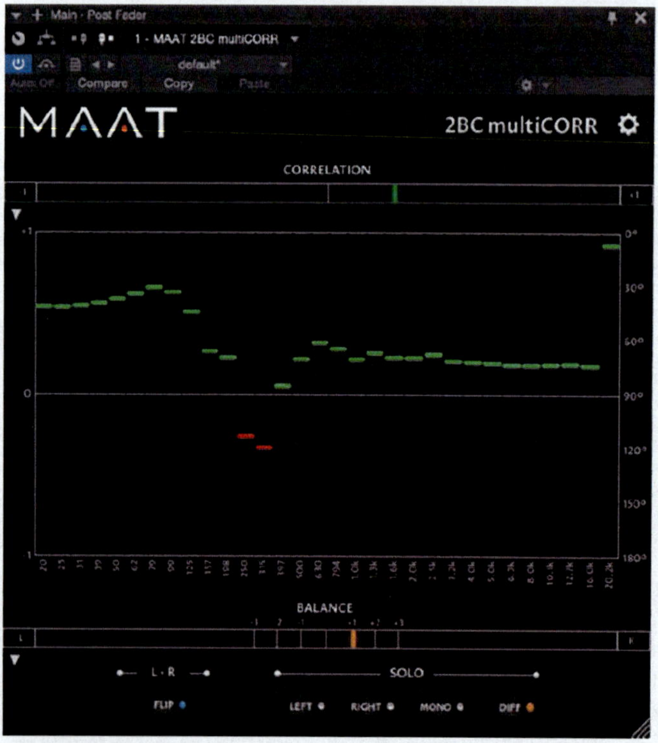

Fig. 9.28 The MAAT "2BC multiCORR" plug-in, offering Top-to-bottom: (1) a correlation meter (overall-signal 20 Hz–20 kHz), (2) frequency dependent cross-correlation readout of the stereo signal split into 31 frequency bands, (3) a "Stereo-Balance" Meter (graphic from: https://www.maat.digital/2bcmulticorr/ Accessed July 16th, 2022)

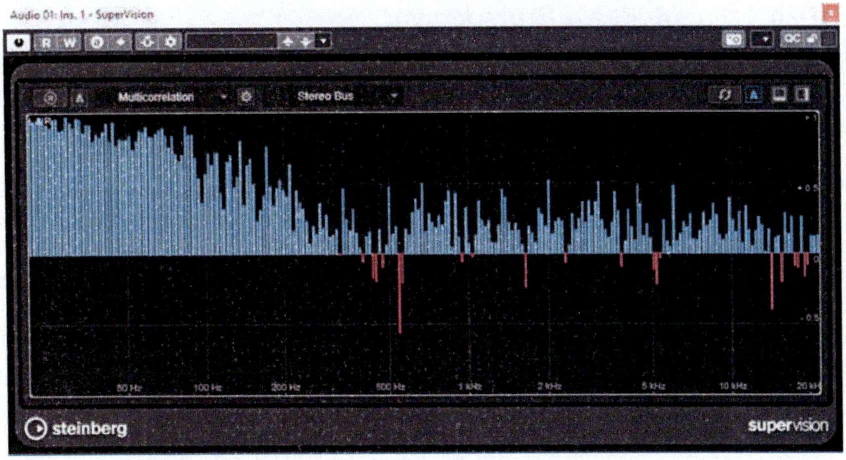

Fig. 9.29 A screenshot from Steinberg's 'Cubase' (V.12) 'SuperVision' metering section in Multicorrelation mode, displaying the cross-correlation of the L and R signal of the stereo bus with a resolution of 24 bands/octave

shown to be of paramount importance for good spatial reproduction (see—among others—Hidaka et al. 1995, Hidaka et al. 1997 as well as Beranek 2004) it can be deducted from Fig. 9.29 that small-AB recordings with a capsule spacing in the order of 50 cm are not suited to provide convincing spatial impression.

While the realtime FCC-analysis through MAAT's '2BC multiCORR' (see MAAT digital (2019)) or Steinberg's 'SuperVision' Multicorrelation Plug-In has the advantage of immediate signal relation display, the non-realtime analysis via the above MATLAB function is more 'objective' or balanced in the sense that the final outcome represents an analysis over the entire length of the audio sample which was used for the analysis and can therefore be regarded to be 'more complete' or 'more correct'.

More details on FCC analysis in relation to surround- and stereo- microphone techniques can be found in Chaps. 7 and 8 of (Pfanzagl-Cardone 2020). Also, the calculation for f_{crit}, the 'critical frequency' in a (small) AB pair of omni-directional microphones, below which the signal becomes more and more monophonic, can also be found in Chap. 8 of (Pfanzagl-Cardone 2020).

A short series of educational videoclips concerning microphone technique analysis (mainly for 2-channel stereo microphone techniques, but also including 3-channel techniques like DECCA and BPT), based on FCC measurements using the "2BCmultiCORR" frequency dependent signal cross-correlation plug-in can be found on the author's Youtube channel "futuresonic100", when searching for 'mic tech analysis' or 'Nevaton BPT'.

9.10.6 Height Signal Reproduction

"In the case of multichannel reproduction, height channels are commonly set to figure-of-eight in order to keep the surround layer source material in their null. The vector relationship is described in the polar plot in Fig. 9.28, which uses the example of a cardioid main signal and a figure-of-eight height signal. Since this is the result of a mixed polar pattern, the matrixed signal must be reduced by −1.66 dB. The resulting signal is source heavy and points downwards as the information received by the height layer contains mostly room ambience" (Zhang and Geluso 2019). One additional note by Paul Geluso is that they found that the weighting of the M and Z signals is very important, as is when working with any MS pair, to optimize a natural height effect when using the MZ system (see Fig. 9.30 in respect to this).

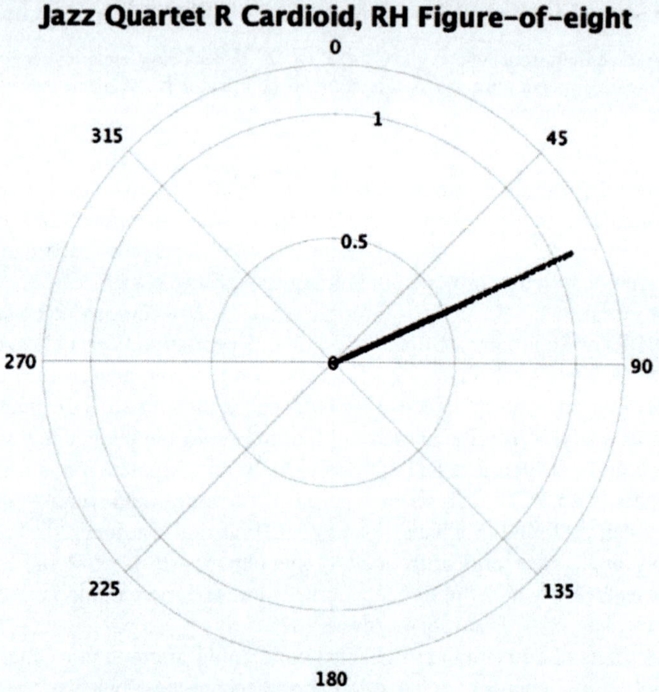

Fig. 9.30 Relationship between cardioid right channel at 90° and figure-of-eight right height at 0° (from (Zhang and Geluso 2019)

9.10.7 *Conclusion*

"Not only can the 3DCC microphone system be used to create an Ambisonic or stereo compatible signal, it can also produce an immersive six-channel or expanded eighteen-channel output with minimal signal processing. Various polar pattern configurations can be obtained for all directional signals in post-production as a result of using dual-capsule microphone technology. These signals are demonstrably stable and well-defined as illustrated by vectorgraphs created for multiple primary polar patterns" (from Zhang and Geluso 2019).

9.11 The Morton Lindberg "2L" Technique

Grammy Award winning sound-engineer and music producer Morten Lindberg states (from Lindberg 2015): "…Our '2L-cube' is only remotely inspired by Decca- or Mercury-tree. The microphone array is really a direct consequence of the speaker configuration in the AURO 3D playback system. Time of arrival, SPL and on-axis HF texture is directly preserved in this 5.1.4 or 7.1.4 microphone configuration. Proportions are cubical and the dimensions could vary from 150 cm (see Inglis 2022) for a large orchestral array down to 40 cm in an intimate chamber musical context. I always use omnis in the main array. But depending on the room, the music and the instruments I alternate between the DPA 4003 and the 4041 with the larger membrane, the latter providing a more focused on-axis texture.

I do appreciate the predictable channel based production flow in the AURO 3D configuration. Our recordings play back well in the Dolby ATMOS and the DTS:X speaker layout, but the concept of objects and local rendering introduce a few unpredictable factors to us as content producers.

The beauty of the recording arts is that there is no fixed formula and no blueprint. It all comes out of the music. Every project starts out by digging into the score and talking with the composer, if contemporary, and the musicians. It is not our task as producers and engineers to try to re-create a concert situation with all its commercial limitations.

On the contrary; we should make the ideal out of the recording medium and create the strongest illusion, the sonic experience that emotionally moves the listener to a better place.

Our approach is that the image is not created in the mix. It is made in the recording with dedicated microphone techniques. The composers and musicians should perform to the extended multi-dimensional sonic sculpture, allowing more details and broader strokes. Then immersive audio and surround sound is just a matter of opening up the faders. When the music is created, performed and recorded in immersive audio then stereo is our most time-consuming challenge; figuring out how to preserve the total impact and level of details from this sculpture into a flat canvas.

Recorded music is no longer a matter of a fixed one-dimensional setting, but rather a three-dimensional enveloping situation. Stereo can be described as a flat canvas, while surround sound is a sculpture that you can literally move around and relate to spatially; surrounded by music you can move about in the aural space and choose angles, vantage points and positions.

2L record in spacious acoustic venues: large concert halls, churches and cathedrals. This is actually where we can make the most intimate recordings. The qualities we seek in large rooms are not necessarily a big reverb, but openness due to the absence of close reflecting walls.

Making an ambient and beautiful recording is the way of least resistance.

Searching the fine edge between direct contact and openness—that's the real challenge! A really good recording should be able to bodily move the listener. This core quality of audio production is made by choosing the right venue for the repertoire, and by balancing the image in the placement of microphones and musicians relative to each other in that venue" (from Lindberg 2015).

In a typical recording situation, Lindberg will arrange the musical ensemble in a circular layout around his 2L-Cube which is at the center position to achieve 360° source imaging. Also, he may readjust the distance of individual musicians from the 2L-array in order to achieve an optimal level balance for each different musical work. "…Then all the musicians or voices in the ensemble are arranged physically around that microphone array. This is basically how balance is done with these recordings. If we need more second violins, then the whole second line in the group take their chairs forward half a meter. If we need less trombones—one meter back. That's how we balance this" (from Carlsson and Jopson 2020).The center microphone of the 2L-cube is placed slightly in front of the base point between the left and right microphones.

In Lee 2021 we also find the following notes concerning the 2L-cube, based on verbal communication exchange with the inventor Morten Lindberg: " … 2L-Cube is a microphone array developed by Lindberg. It employs nine omni microphones in a cube arrangement for a 4+5+0 reproduction. The width and depth dimensions of the cube could vary from 0.4 m to 1.2 m depending on the size of the ensemble, while the height dimension is kept constant as 1 m. The choice of omni microphones for 2L-Cube is more for their tonality rather than the polar response itself; an omni microphone would typically offer a more extended low-end in the frequency response compared to a unidirectional or bidirectional microphone. Furthermore, the exact vertical orientations of the upper layer microphones depend on the desired tonal characteristics. He often utilizes acoustic pressure equalizers to increase the directionalities of the microphones at high frequencies. This would produce some ICLDs ('Inter Channel Level Differences') vertically, which might be useful for avoiding the upward-shifting of source image in the vertical plane" (cited from Lee 2021) (see the photo, Fig. 9.31).

Morten Lindberg is rather purist in his approach to recording: "…My key to a good recording is to keep the signal path clean and short, with as few components as possible. A simple and pure signal path is the means to true high-resolution audio, not only at recording, but also preserving the purity all through editing, mix

Fig. 9.31 TrondheimSolistene with '2L-Cube', recording Reflections (2L-125-SABD) at Selbu church (photo courtesy Morten Lindberg, from Carlsson and Jopson 2020))

and mastering" (from Inglis 2022). Morten Lindberg avoids using any sort of post-production processing that alters the sound: "The most important aspect of post-production is to not destroy the fine qualities captured at recording. Instead of EQ in post, I much prefer to rather shift the angle or the distance of a microphone beforehand in recording. It just takes planning. With a good recording, I never use any EQ or dynamic processing at all." Instead, the focus of his post-production efforts is on sympathetic editing to create what he feels is the most compelling realisation of the composition. "Editing is an important tool. It makes it possible to combine the highest level of energy and details into an intense performance. To do that you need to use a sonically transparent workstation. For the past decade we've worked with Merging Technologies on their Pyramix system. Their latest development allows for an immersive workflow totally independent of destination format. I don't mix for a codec. I record for a playback environment. Any codec should then ideally provide for a transparent transport to the consumer" (from Inglis 2022).

9.11.1 Use of Center Speaker and LFE-Channel

"...From the sidelines, I would encourage all friends and colleagues to make considerate choices between phantom and hard centre [i.e. panning sources directly to the centre speaker as opposed to generating a phantom centre image from the left and right speakers]. There's a beauty to that lead vocal anchored into the centre speaker.

I also observe a tendency to use the LFE as a general bass-booster, replicating LF from main channels. This way of pre-mixing what should be left to bass management in the end-user environment is bound for trouble. Using LFE for dedicated musical components is way more effective to the extended experience" (from Inglis 2022).

From (Inglis 2022): "…In theory, one of the key design features of modern immersive audio formats is scalability. The engineer should be able to supply a single master recording which gets intelligently adapted to suit any distribution format and listening system. However, Morten Lindberg deliberately does not do this, and feels that the fidelity of his recordings is fully preserved only by making dedicated masters that map each microphone directly onto a loudspeaker in the surround playback system. To achieve this in Atmos, for example, he places all the mics on the lower level of the 2L Cube directly into the 7.1 bed, and uses the objects only to route the upper layer of mics to the overhead loudspeakers." "…Our philosophy is simple: one microphone straight to one speaker. For Auro-3D or Dolby Atmos, all 7.1.4 microphones go directly to their according loudspeaker." (from Inglis 2022). [Rem.: this finds its parallel in the univalent MMAD (multichannel microphone array design) of Michael Williams, see Chap. 6]. "…With diminishing numbers of loudspeakers, we do not sum or fold down. We take away sources. So for 5.1 only the lower bed of microphones are active, without the side-fills. Then it is usually only front left and right microphone playing in stereo, possibly with a slight texture added from the rear microphones. Pure, clean and minimalistic. The important aspect is to configure the array so time of arrival is captured and released in natural order" (from Inglis 2022).

9.11.2 Coincident and Ambisonic Versus Spaced AB

"…When making a straight-to-stereo recording, we have the option of using a coincident mic array or a spaced pair. Both approaches have their own technical strengths, but fundamentally, the choice is a subjective one. Some people prefer the pin-sharp stereo imaging of Blumlein, others the lush and expansive quality of spaced omnis. Similar options are available for surround recording, where the 360° sound field can be captured at a single point using an Ambisonic mic, or using a three-dimensional spaced array. Morten Lindberg is a firm adherent of the latter approach. The Ambisonic concept is a beautiful theory, and its application in VR is obvious. Early on, I experimented with a Soundfield microphone [see SoundField ST450 MKII (2018)]. The spatiality came out fine, but I never managed to capture that solid core of the instruments I was searching for" (from Inglis 2022).

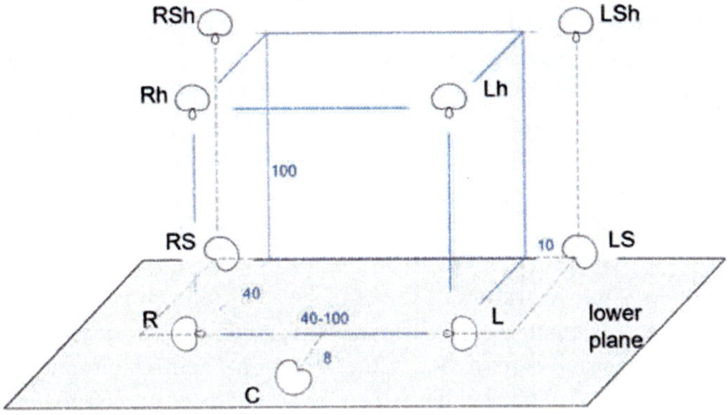

Fig. 9.32 OCT-3D Technique Microphone Layout schematic, using cardioid and super-cardioid mic patterns (from Wittek and Theile 2017)

9.12 The OCT-3D Technique

This technique was designed for recordings intended to be played back via the Auro 9.1 loudspeaker arrangement. The base section consists of the 'Optimized Cardioid Triangle' Surround (OCT-Surround) microphone arrangement (see Theile 2000), designed to capture sound for 5.1 systems (see Theile 2001). For the recording of 3D-Audio, four upward-facing super-cardioid microphones are added directly above the L, R, Left Surround and Right Surround microphones of the 'base layer' to create the OCT-3D array. While the front microphones capture mainly direct sound, the rear and upward-facing microphones are oriented to capture the diffuse sound and ambience of the environment (see Riaz et al. 2017 and Ryaboy 2015) (see Fig. 9.32).

9.13 The ORTF-3D Technique

Since its inception in the 1970s at the French national broadcasting agency ORTF ("Office de Radiodiffusion Télévision Française"), the ORTF microphone technique has experienced an extension of its sonic features from two to three-channel ("ORTF-Triple" or ORTF-T) for use as front-mic in 5.1 surround applications (see (Pfanzagl 2002) and (Pfanzagl-Cardone 2002) as well as full (quadraphonic) surround, intended mainly for ambience recording.

In (Wittek and Theile 2017) the ORTF-Surround system is described, as follows: "...One optimal solution for ambient recordings in multichannel stereophony is the 'ORTF surround' system, in which four super-cardioids are arranged in a rectangle with 10 × 20 cm side lengths. Here the distances between microphones help with decorrelation, and thereby lend the sonic impression its spatial openness. The microphone signals are routed discretely to the L, R, LS and RS channels. The signal separation in terms of level is ca. 10 dB; thus, the sonic image during playback is stable even in off-axis listening positions (see Fig. 9.33).

With eight or nine channels, the arrangement of the microphones becomes very difficult if the above-mentioned requirements are to be met. The simplest method for maintaining signal separation is to set up eight or nine microphones far apart from one another. Thus, a large nine-channel 'Decca Tree' arrangement is very well suited for certain applications, although it has severe disadvantages that limit its practical usability. For one, the sheer size of the arrangement is greater than 2 m in width and height. And the signal separation in terms of level difference is nearly zero; every signal is more or less available in all loudspeakers. Thus, this array can represent a beautiful, diffuse spaciousness, but stable directional reproduction isn't achieved beyond the 'sweet spot'. This can be helped by adding spot microphones.

An optimal ambience arrangement for eight channels is offered by the new 'ORTF-3D' system developed by Wittek and Theile. It is more or less a doubling of the 'ORTF Surround' system onto two planes, i.e. there are four super-cardioids on each level (upper and lower), forming rectangles with 10 and 20 cm side lengths. The two 'ORTF Surround' arrangements are placed directly on top of one another.

The microphones are furthermore tilted upward or downward in order to create signal separation in the vertical plane. Thus an 8-channel arrangement is formed, with imaging in the horizontal plane that somewhat corresponds to the 'ORTF Surround'

Fig. 9.33 ORTF-Surround outdoor microphone array in basket (courtesy Wittek, Schoeps Microphones)

system. The microphone signals are discretely routed to four channels for the lower level (L, R, LS, RS), and four for the upper level (Lh, Rh, LSh and RSh).

In VR applications, virtual loudspeaker positions forming an equal-sided cube are binauralized.

Lee et al. (Lee and Cribben 2014) found that the decorrelation of the diffuse field is less important in the vertical domain than in the horizontal domain. This means whereas it is clearly audible that an A/B microphone pair sounds wider than an X/Y pair when reproduced between L/R, there is only a little audible difference when reproduced between L/Lh. This helps a lot in the design of compact 3D ambience microphone. Imaging in the vertical dimension is produced by angling the microphones into 90° X/Y pairs of super-cardioids. Such a two-channel coincident arrangement is possible due to the high directivity of the super-cardioids, and the imaging quality and diffuse-field decorrelation are both good. This results in an eight-channel array with high signal separation, optimal diffuse-field correlation, and high stability within the playback space. All requirements are optimally fulfilled, yet the array is no larger than the compact ORTF Surround system—a decisive practical advantage (see Figs. 9.34 and 9.35).

Numerous test recordings have shown that the ORTF-3D approach produces very beautiful, spatially open and stable 3D recordings.

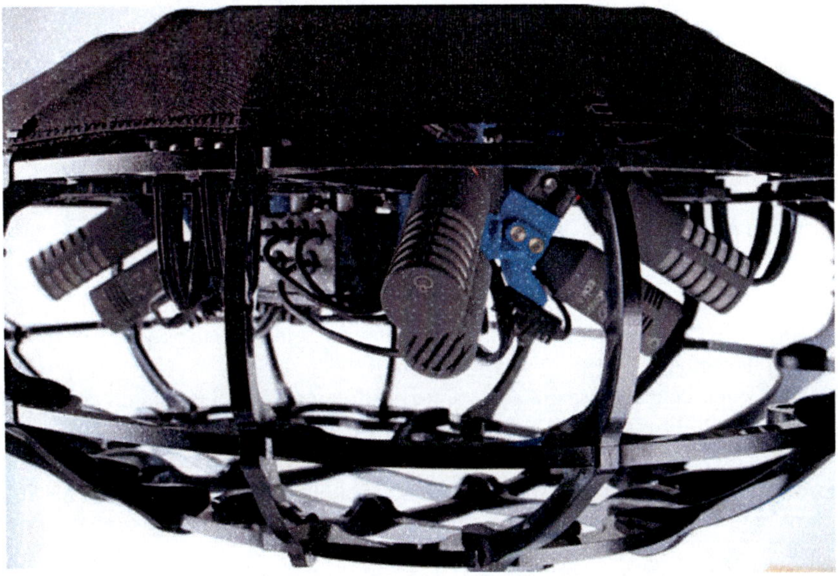

Fig. 9.34 The ORTF-3D outdoor microphone array (courtesy Wittek, Schoeps Microphones)

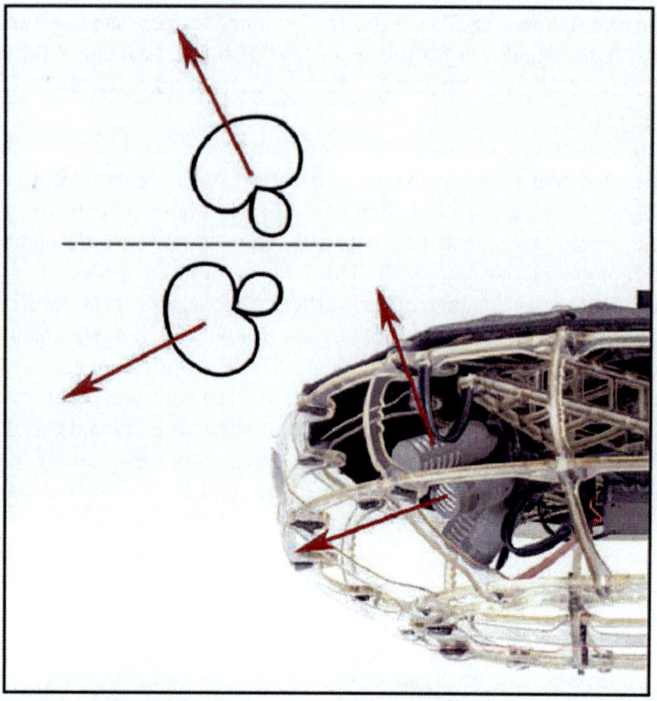

Fig. 9.35 Detail of capsule orientation for the ORTF-3D microphone array (from Wittek and Theile 2017)

9.13.1 Conversion of the ORTF-3D Setup for Dolby Atmos and Auro3D

The eight channels of the ORTF-3D are L, R, LS, RS for the lower level, and Lh, Rh, LSh and RSh for the upper level. They are routed to eight discrete playback channels without matrixing.

The Center channel remains unoccupied. A Center channel is seldom desired in ambience recording; it would distort the energy balance between front and rear, and require significantly greater distances among microphones in order to maintain the necessary signal separation. If a Center signal should be necessary for a specific reason, e.g. to cover the shutoff of a reporter's microphone, a simple downmix of the L and R signals at low level is sufficient.

In Auro3D the loudspeaker channels L, R, LS, RS, HL, HR, HLS and HRS are fed.

With Dolby, the integration in the Atmos production environment is equally simple; the channels L, R, LS, RS are simply laid down in the corresponding channels of the surround level, the so-called 'Atmos bed', whereas the four upper channels are placed as static objects in the four upper corners of the Cartesian space in the Atmos panning tool. These are then rendered in playback through the corresponding front or rear loudspeakers" (from Wittek and Theile 2017).

9.14 The 'Minotaur 3D Array' (Olabe and Lagatta)

The Minotaur 3D Array consists of the evolution of an LCR main system updated to Dolby Atmos and other immersive audio delivery formats.

As can be observed in Fig. 9.36: in the horizontal plane we can see the Left, Centre and Right microphones forming a triangle. Between the left and right capsules there are 2.2 m. The center capsule is at 0.8 m forward from the midpoint between L and R. These 3 microphones should have omnidirectional capsules and only if the room was very reflective L and R could be changed to wide cardioids.

On the same plane (but outside of the photo) the surround microphones can be found 2 m behind the front array, Ls and Rs are cardioid mics, angled 30° each side (see Fig. 9.36). The two mics are separated by 2.5 m from one another. The Lrs and Rrs channel signals are also captured by cardioid capsules and arranged with the same angle and capsule distances as the surround microphones but at 4 m from the main array rather than 2 m.

The top (or height) plane can be found 0.5 m above the main plane. Two NOS systems (Cardioid capsules) are mounted on this level the L Tops and the R Tops: Ltf, Ltr/Rtf, Rtr. [Rem.: the NOS microphone system is named after the Dutch National

Fig. 9.36 Minotaur 3D microphone array in the Auditorium of the Opéra National de Bordeaux (surround channels not shown in image) (photo © Iker Olabe)

Table 9.1 Showing channel inputs of the Minotaur 3D array (right) and equivalent Dolby Atmos corresponding master outputs (left). Compare with Fig. 4.15 for signal-layout and corresponding speaker placement in a 7.1.4 Dolby Atmos mix

Dolby atmos master	Minotaur 3D array
L	Main L (Omni)
R	Main R (Omni)
C	Main C (Omni)
LFE	None
Ls (Left Surround)	Ls (Card.) 2 m 30º
Rs (Right Surround)	Rs (Card.) 2 m −30º
Lrs (Left Rear Surround)	Lrs (Card.) 4 m 30º
Rrs (Right Rear Surround)	Rrs (Card.) 4 m −30º
Ltf (Left Top Front)	Ltf NOS (Card. or Hyper-Card.)
Ltr (Left Top Rear)	Ltr NOS (Card. or Hyper-Card.)
Rtf (Right Top Front)	Rtf NOS (Card. or Hyper-Card.)
Rtr (Right Top Rear)	Rtr NOS (Card. or Hyper-Card.)

Broadcasting Corporation (Nederlandse Omroep Stichting). It consists of 2 cardioid microphones, the capsules of which are spaced by 25 cm and angled at ±45° towards the sound source in a regular 2-channel stereo-recording application.] In case of a hall with a strong diffuse sound field, when looking for a strong '3D effect' the cardioid capsules could be exchanged for hyper-cardioids (see Table 9.1).

The 'Minotaur 3D Array' was named after a piece by the renowned French composer Thomas Bangalter and the visual association or resemblance of this array to the Greek mythological figure. The concept was created by sound engineers Iker Olabe and Florian Lagatta.

9.15 '6DOF' Mic System for VR-Applications (Rivaz-Mendes et al.)

In Rivaz-Mendes et al. 2018 a case study is presented "…on practical spatial audio recording techniques for capturing live music performances for reproduction in a six-degrees of freedom (6DOF) virtual reality (VR) framework.

The end-goal is to give the listener the ability to move close to or even around musical sources with a high degree of plausibility to match the visuals. The recording workflow facilitates three major rendering schemes—object-based using spot microphones and diffuse field capture microphone arrays, Ambisonics with multiple-placed sound-field microphones and hybrid approaches that utilise the prior two methods. The work is presented as a case-study where a jazz ensemble is recorded at Studio 3 of Abbey Road Studios London using the proposed techniques.

9.15.1 '6DOF' Versus '3DOF' Recording

Three-Degrees of Freedom (3DOF) denotes the ability of a VR user to perform a set of rotational movements i.e. yaw, pitch and roll, within a virtual environment although users are still confined to a fixed point in space. 3DOF+ adds an extra degree of flexibility to the head movements (e.g. parallax) but still possesses much of the limitations of a 3DOF system in that the movement is bounded within a local region about the head. Effective ways of recording music for its presentation in a 3DOF VR context were explored and evaluated by the authors by means of multichannel microphone arrays for 360 VR music production (Riaz et al. 2017). Binaural rendering of the signals is the commonly used delivery method.

Unlike 3DOF and 3DOF+, a 6DOF VR system permits free-movement within the VR space using both rotational and translational movements (up-down, left–right, and front–back). Audio recorded and reproduced in a 6DOF VR format presents the user with the possibility to move close and around any given sound sources with matching visuals.

Prior approaches that were explored include the use of sound field recordings that react to head movement (see (Frank M et al. 2015) and (Schultz and Spors 2013)) real-time auralization, and the interpolation of signals to cover the recording area and account for movement within such space (see Kearney 2010, Southern et al 2009, Tyla and Choueiri 2016, Kearney et al. 2011). The methods presented in this paper derive from a combination of traditional recording techniques, diffuse field capture and sound field recording through Ambisonics.

9.15.2 Recording Setup

The recording for the research presented here was conducted at Studio 3 of Abbey Road Studios, UK. The ensemble were The Vicente Magalhães jazz band which comprised of a vocalist, double bass, drums and piano. The reasoning behind choosing this setup was to present timbrally contrasting elements that would be easily distinguishable from one another, as well as to record a completely acoustic, naturally balanced performance.

The backline consisted of a Ludwig jazz drum kit (kick drum, snare drum, high and low toms, hi hat, ride and cymbal), double bass, and a Yamaha grand piano. A total of 13 spot microphones were used to capture the entire band. A Neumann U 47 FET was used as the kick drum out microphone, along with a pair of AKG C 414 B-ULS for the snare top and bottom positions. Furthermore, two Sennheiser MD 421 were used for the high and low toms, and a pair of Schoeps MK 4 as stereo overheads. For the double bass, a Neumann U 47 FET was setup at the bridge and a Neumann KM 84 was positioned at the neck. The piano was captured with a spaced stereo pair of DPA 4011, whilst the vocals were recorded through a Chandler Ltd

REDD microphone. Every spot microphone was then sent into the preamps of the SSL 9000 J console in the control room.

As a reference point for the proposed Ambisonic recording setup, musicians were placed around a centrally positioned MH Acoustics Eigenmike® microphone for optimal natural acoustic balance, as illustrated in Fig. 9.37. The Eigenmike® is a higher-order Ambisonic microphone comprised of 32 omnidirectional capsules based on a pentakis-dodecahedron array (see MH-Acoustics 2018).

Of all musicians, it was necessary to move the singer closest to the microphone, based on auditioning the vocal level relative to the other instruments. The Eigenmike® utilizes a proprietary software and hardware interface. As such, it was connected via

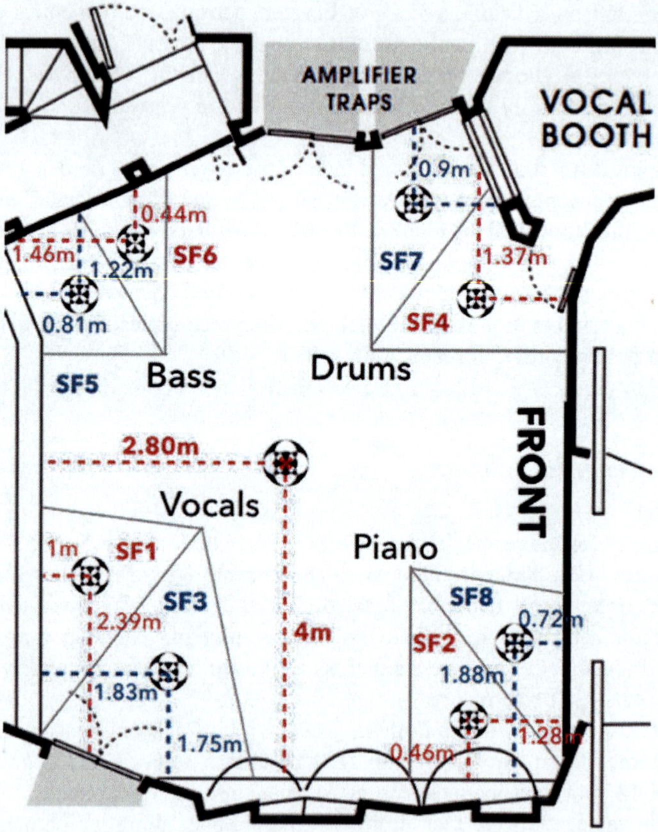

Fig. 9.37 SoundFieled microphone positions; Eigenmike® is shown in the centre (from Rivaz-Mendez et al. 2018)

Firewire to its own laptop as a standalone solution, independent from the rest of the session. There, the performance was recorded into a 32-channel track session in the Digital Audio Workstation (DAW) by means of software 'Reaper'.

The recording space was then divided into 9 zones relative to the band setup. Each musician was located in the convergence point of three zones: a main zone, shared by all instruments in the acoustic line-of-sight of the Eigenmike®, and two individual zones behind them, each within the range of a 1st order sound-field microphone (see SoundField), as shown in Fig. 9.37. Each zone acted as a self-contained 'acoustic space' around the musician, providing the possibility to move around and behind the sources in a 6DOF VR system. With the zones defined, a total of 8 sound-field microphones were positioned across them. It is worth mentioning that, although different models of sound-field microphones were used, the same models within adjacent zone pairs were used where possible. All the microphones were facing in the same direction as the Eigenmike®, towards the front wall, with their capsules set at a height of 1.60 m.

Five SoundField ST450 MKII (SF1, SF2, SF3, SF4, and SF8), one ST422 (SF5) one ST250 (SF6), and one ST350 (SF7) were used. The microphone positions within each of the zones is shown in detail in Fig. 9.37. Each of these microphones has its own preamplifier and outputs a 4-channel B-Format signal. These outputs went directly via the tielines into the SSL 9000 J console with a total of 32 tracks.

For diffuse field capture, a modified Hamasaki Cube (see Riaz et al 2017) was arranged. Due to the band setup and recording space restrictions, the microphones were configured in a cross-like fashion. The cross was symmetric, with a spacing of 1 m between each opposing capsule (see Fig. 9.38), with a height layer at 4 m, as shown in Fig. 9.39. Two Neumann KU 100 binaural dummy-heads were also used for binaural reference to obtain two contrasting perspectives of the performance (see Figs. 9.38 and 9.39).

9.15.3 Recording Process

Once the set up was completed, and a line check of the instruments conducted, calibration of the diffuse and sound-field microphones was carried out using an EMI RS145 noise generator to ensure relative levels could be matched in post-processing.

The spot and sound-field microphones used the preamplifiers and lines of the SSL 9000 J console, whilst the two Neumann KU100 and the 8 Neumann U87 that formed the modified Hamasaki Cube were amplified by stepped Neve 1081 preamps. The Eigenmike® used its own self-contained recording setup into Reaper.

A total of 89 channels were recorded for each take of the session. 32 for the Eigenmike® and 57 for the rest of the microphones into Pro Tools. The complete channel count included: 13 channels for the spot microphones; 32 channels for the Eigenmike®; 32 channels for the 8 sound-field microphones (4 per microphone); 8 channels for the modified Hamasaki Cube; 4 channels for the Neumann KU100s.

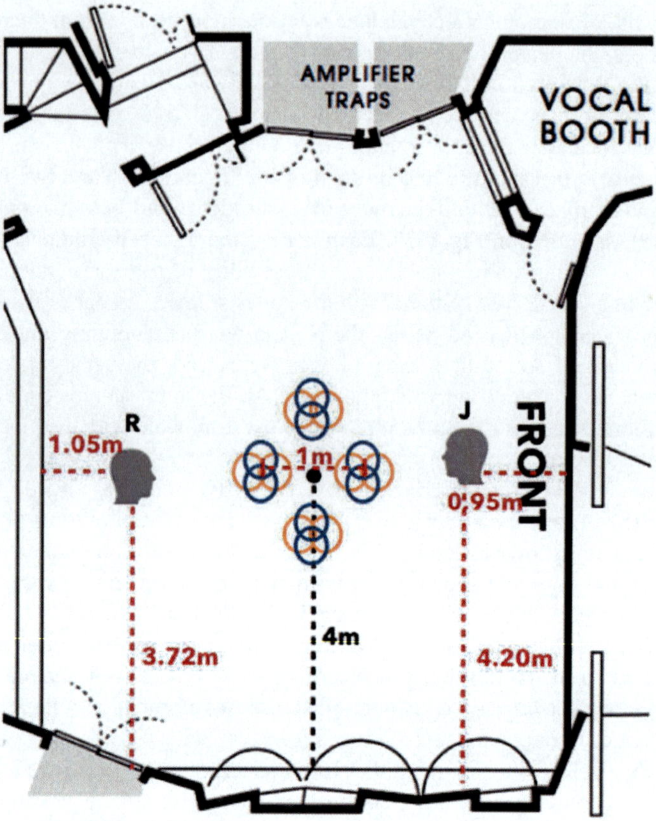

Fig. 9.38 Microphone positions of modified Hamasaki Cube and Neumann KU100 artificial human heads (from Rivaz-Mendez et al. 2018)

The SSL 9000 J in the control room of Studio 3 is a 96-channel console, readily accommodating the 57 tracks required.

In terms of practicality, the most complex and time-consuming setup was the modified Hamasaki Cube. The session was recorded into Pro-Tools at 24-bit, 48 kHz sample rate. A clapperboard was used as a way of syncing each take with the parallel Eigenmike® recording session in Reaper.

9.15.4 Rendering Approaches

Real-time update of the final binaural signals is dependent on the workflow used. There are two main workflows accommodated with the recordings:

Fig. 9.39 Final physical setup at Abbey Road Studio 3 (from Rivaz-Mendez et al. 2018)

Object-Based Rendering: For the object-based approach, the spot microphones can be panned relative to the user's positional and rotational data. The amplitude at which the microphone is encoded is based on the distance between the user and the source. Diffuse field microphone recordings are also encoded as spot sources, but spatially distributed about the sphere and kept at a constant level. The advantage of the object-based approach is that it can be rendered using any format e.g. Binaural-based Ambisonics, direct HRTF convolution etc. and does not require the Ambisonic recordings from the different zones. The downside is that the sources need to be encoded with specific directivity patterns, which may not match that of the actual instrument.

Sound-field Rendering: For single Ambisonic microphone recordings plane wave expansion of the sound-field is possible, as shown for example in (SADIE-II Binaural Database 2018). This assumes that there is only one source coming from a particular direction relative to the microphone, which may not always be the case, and also does not consider the directivity of the sources for 6DOF movement. Another approach is interpolation across multiple Ambisonic microphones. In the Studio 3 recordings, zones were defined such that within the boundaries of each zone, the order of musicians spread across the sound-field does not change. Interpolation can be achieved linearly or using weighted interpolation (Southern et al. 2009, Tylka and Choueiri 2016) alongside matrix operations to distort the angular perspective of the sources within a zone using Lorentz transformations. Ideally, at each boundary, identical Ambisonic signals may be produced by the transformation of either zone's recording—requiring no crossfade. However, due to non-point sources and room reflections a small transition is needed to facilitate a smooth changeover.

In the following section, we demonstrate a hybrid approach using practical tools available to content creators. The workflow utilizes Reaper for Ambisonic en/transcoding and *Unity-3D* (see Unity 3D 2018) for the 6DOF environment.

9.15.5 Example Implementation

The first step in the post-production phase was processing of the Ambisonic recordings. Within Reaper, a 4-channel 1st-order decoder track was created using the AmbiX suite of plug-ins (SADIE-II Binaural Database 2018). The use of 1st-order Ambisonics over higher-order for the Eigenmike® recording was a result of current channel count limitations within *Unity 3D*. The raw 32-channel Eigemike recording was converted to 1st-order using the MH-Acoustics EigenUnits encoder (MH-Acoustics 2018). A 1st-order MaxRe binaural decoder was used for monitoring with KU100 HRTFs from the SADIE-II database (SADIE-II Database 2018). All files were converted to Ambix format and exported for use in Unity.

A computer model of Studio 3 was designed in Blender (Blender 2018) and imported into Unity for the creation of the 6DOF VR environment. The model was based on physical measurements and photographic documentation of the studio. Virtual avatars of the musicians/instruments were included to be used as visual cues for the sound sources.

Google's Resonance Audio SDK (Resonance Audio 2018) was utilised, and the following elements were added as components to the game objects in the Unity project:

- *Resonance Audio Listener*: Allows to set up a "listener" for all of the audio sources in the project.
- *Resonance Audio Source*: When this component is applied to an object, various parameters like directionality, occlusion, gain and spread can be modified. This was applied to the spot microphones. Object rendering is at 3rd order Ambisonics.
- *Resonance Audio Sound Field*: This component allows playback of 1st Order Ambisonic files and reacts to the Audio Listener component movement for weighted interpolation based on listener position. This was applied to the Ambisonic microphone recordings. A balance between 1st and 3rd order components can readily be implemented for optimal sounding results.

Once the audio implementation stage was completed, it was possible to test the experience through the HTC Vive (HTC Vive 2018) and a pair of headphones. Two laser-emitting sensors (part of the HTC Vive system) were positioned in opposite corners of a 3.7 m × 3.7 m room. The sensors tracked and processed the movement of users wearing the headset within the physical space, accounting for such movement to be experienced in the virtual model of Studio 3. Users were then able to move in 6DOF within the up-to-scale virtual representation of the studio whilst experiencing corresponding changes in sound when moving close to and around the virtual avatars of the musicians.

9.15.6 Conclusions

This case study of a recording at Abbey Road Studios has presented a workflow for the practical recording of live music production for presentation in a 6DOF VR framework. The recording workflow is adaptive to multiple rendering approaches and is scalable if the rendering approach is known prior to the recording. Further work will include an evaluation of the recordings using different rendering approaches to see which methods gives the most plausible results and to see whether a more streamlined recording setup might be devised, based on the practicality considerations and conclusions obtained from such tests. Furthermore, replacing the 1st-order microphones with higher order microphones or upscaled versions might create a higher quality experience" (from Rivaz-Mendez et al. 2018).

9.16 Binaurally Based 3D-Audio Approaches

In Chap. 4 of (Pfanzagl-Cardone 2020), an examination of two 'enhanced' binaurally based recording techniques for 3D audio can be found: the Pan-Ambiophonic 2D/3D System by Miller (see Miller 2004a; b), as well as the BACCH (TM) 3D Sound system by Prof. Edgar Choueiri, head of the '3D Audio and Applied Acoustics (3D3A) Lab' of Princeton University (see Choueiri 2010) (Figs. 9.40 and 9.41).

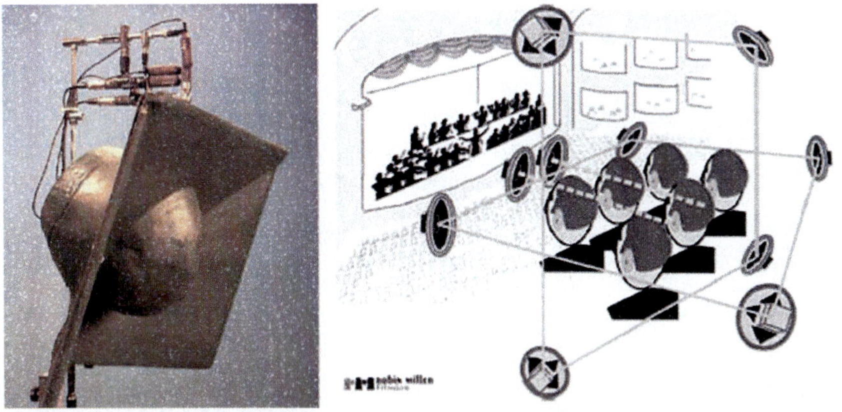

Fig. 9.40 Left—The 'PanAmbiophone'—a total of 8 recording channels for 'PerAmbio 3D' (from (Miller 2004a)); Right—'PerAmbio 3D/2D' System (Pat. Pend.) plays both Ambiophonic and 3D (with height) recordings using 10 speakers (from (Miller 2004a))

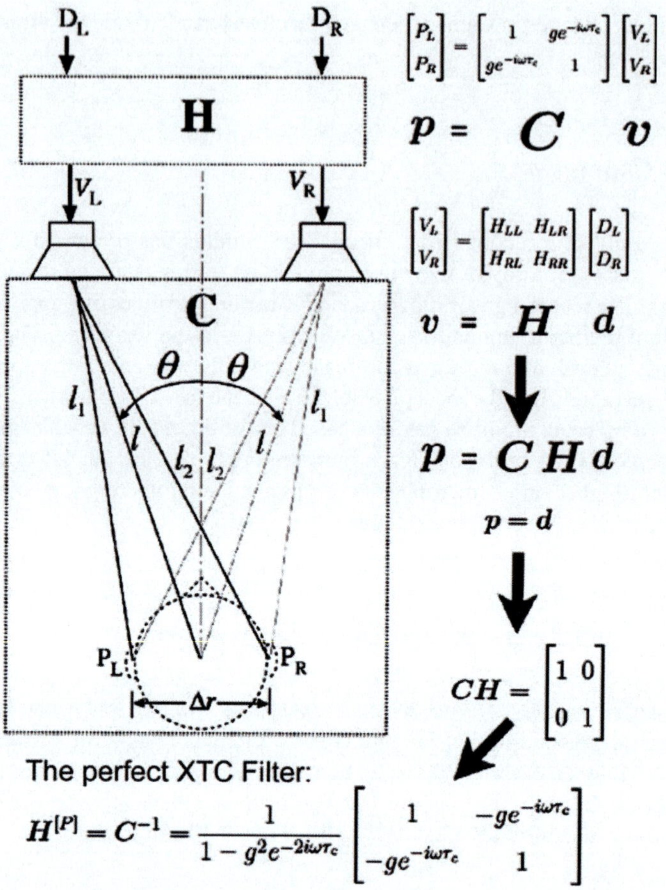

Fig. 9.41 Schematic of the BACCH(TM) Crosstalk-Cancellation filter XTC (from Choueiri 2010)

References

Barbour JL (2003) Elevation perception: phantom images in the vertical hemisphere. In: Proceedings to the 24th Audio Engineering Society International Conference: Multichannel Audio—The New Reality

Barrett N (2012) The perception, evaluation and creative application of higher order ambisonics in contemporary music practice. Ircam Composer in Research Report 2012, pp 1–6

Benjamin E, Chen T (2005) The native b-format microphone. Paper presented at the 119th Audio Engineering Society Convention

Berg J, Rumsey F (2001a) Verification and correlation of attributes used for describing the spatial quality of reproduced sound. Paper presented at the Audio Engineering Society 19th International Conference

Beranek L (2004) Concert halls and opera houses: music, acoustics and architecture, 2nd edn. Springer, New York

Blauert J (1974) Räumliches Hören. S. Hirzel Verlag, Stuttgart

Blauert J (1997) Spatial hearing. The MIT Press
Blender (2018) Available at https://www.blender.org/ Accessed 17 Jun 2018
Bowles D (2015) A microphone array for recording music in surround-sound with height channels. Paper 9430 presented at the 139th Audio Engineering Society Convention, New York, Oct 2015
Camerer F, Sodl C (2001) Classical music in radio and TV—a multichannel challenge. The IRT/ORF Surround Listening Test. http://www.hauptmikrofon.de/stereo-surround/orf-surround-techniques. Accessed 31 May 2019
Carlsson P, Jopson N (2020) Morten Lindberg: High resolution. Resolution V19.6 Winter 2020. pp 17–22
Chapman M et al. (2009) A standard for interchange of Ambisonic signal sets. In: Ambisonics symposium 2000, Graz, Austria, 2009
Choueiri E (2010) Optimal crosstalk cancellation for binaural audio with two loudspeakers. http://www.princeton.edu/3D3A/Publications/BACCHPosterV4.pdf Accessed: 27 Feb 2018
Daniel J (2009) Evolving views on HOA: from technological to pragmatic concerns. In: Ambisonics symposium 2009, Graz, Austria, June 2009, pp 1–18
Dickreiter M (1989) Tonmeister technology: recording environment, sound sources, microphone techniques. Temmer Enterprises Inc
Dickreiter M, Dittel V, Hoeg W, Wöhr M (2008) Handbuch der Tonstudiotechnik. Band 1, 7th edn. K. G. Saur Verlag, München
Dietel M, Hildenbrand S (2012) Untersuchungen zu den Höhenkanälen in Auro 3D. In: Proceedings to the 27. Tonmeistertagung des VDT, Nov 2012, pp 701–714
Ellis-Geiger J (2016) Music production for Dolby Atmos and Auro 3D. Paper 9675 presented at the 14st Audio Engineering Society Convention, Los Angeles
Farina A (2007) Advancements in impulse response measurements by sine sweeps
Rank M, Zotter F and Sontacchi, A (2015) Producing 3D audio in ambisonics. Paper presented at the 57th International Conference of the Audio Engineering Society: The Future of Audio Entertainment Technology
Fukada A, Tsujimoto K, Akita S (1997) Microphone techniques for ambient sound on a music recording. Paper 4550 presented at the 103rd Audio Engineering Society Convention, New York, Sept 1997
Geluso P (2012). Capturing height: the addition of Z microphones to stereo and surround microphone arrays. Paper 8595 presented at the 132nd Audio Engineering Society Convention
Griesinger D (1997) Spatial impression and envelopment in small rooms. Paper 4638 presented at the 103rd Audio Engineering Society Convention
Griesinger D (1998) General overview of spatial impression, envelopment, localization and externalization. In: Proceedings to the Audio Engineering Society 15th International Conference on Small Room Acoustics, Denmark Oct/Nov 1998, pp 136–149
Haigh C, Dunkerley J, Rogers M (2021) Classical recording: a practical guide in the Decca tradition (Audio Engineering Society). Routledge
Hamasaki K (2003) Multichannel recording techniques for reproducing adequate spatial impression. In: Proceedings to the Audio Engineering Society 24th International Conference on Multichannel Audio—The New Reality, Banff , Canada, 2003
Hamasaki K, Hiyama K (2003) Reproducing spatial impression with multichannel audio. Paper presented at the Audio Engineering Society 24th International Conference on Multichannel Audio—The New Reality, Banff, Canada, 2003
Hamasaki K et al. (2004) Advanced multichannel audio systems with Superior Impressions of Presence and Reality. Paper presented at the 116th Audio Engineering Society Convention, Berlin, May 2004
Hamasaki K, Van Baelen W (2015) Natural sound recording of an orchestra with three-dimensional sound. Paper presented at the 138th Audio Engineering Society Convention, Warsaw, Poland
Heller A, Lee R, Benjamin EM (2012) Is my decoder ambisonic? Paper presented at the 125th Audio Engineering Society Convention, San Francisco, USA

Heller A, Benjamin EM (2014) The ambisonic decoder toolbox: extensions for partial-coverage loudspeaker arrays. In: Linux audio conference 2014, Karlsruhe

Hibbing M (1989) XY and MS microphone techniques in comparison. Paper presented at the 86th Audio Engineering Society Convention, Hamburg, March 1989

Hidaka T, Beranek L, Okano T (1995) Interaural cross-correlation, lateral fraction, and low- and high-frequency sound levels as measures of acoustical quality in concert halls. J Acoust Soc Am 98(2), Aug 1995

Hidaka T, Beranek L, Okano T (1997) Some considerations of interaural cross correlation and lateral faction as measures of spaciousness in concert halls. In: Ando Y, Noson D (eds) Music and concert hall acoustics. Academic Press, London

Holman T (2008) Surround sound up and running. Focal Press, Burlington, MA, USA

Howie W, King R (2015) Exploratory microphone techniques for three-dimensional classical music recording. Convention E-Brief presented at the 138th Audio Engineering Society Convention, Warsaw, Poland, May 2015

Howie W, King R, Martin D (2016) A three-dimensional orchestral music recording technique, optimized for 22.2 multichannel sound. Paper 9612 presented at the 141st Audio Engineering Society Covention, Los Angeles 2016

Howie W, King R, Martin D, Grond F (2017) Subjective evaluation of orchestral music recording techniques for three-dimensional audio. Paper 9797 presented at the 142nd Audio Engineering Society Convention, Berlin, 20–23 May 2017

Howie W, Martin D, Benson DH, Kelly J, King R (2018) Subjective and objective evaluation of 9ch three-dimensional acoustic music recording techniques. Paper presented at the Audio Engineering Society International Conference on Spatial Reproduction—Aesthetics and Science, Tokyo, 7–9 Aug 2018

HTC Vive (2018) Available at https://www.vive.com/uk/ Accessed 15 Jun 2018

Inglis S (2022) Mixing atmos: Morten Lindberg. Sound-On-Sound, July 2020 pp 112–117

Kearney G et al. (2011) Real-time walkthrough auralisation of the acoustics of Christ Church Cathedral, Dublin. In: Proceedings of the Institute of Acoustics 34

Kearney G, Doyle T (2015) An HRTF database for virtual loudspeaker rendering. Paper 9424 Presented at the 139th Audio Engineering Society Convention

Kearney G (2010) Auditory scene synthesis using virtual acoustic recording and reproduction. PhD Thesis, Trinity College Dublin, 2010

Komiyama S (1997) Visual monitoring of multichannel stereophonic signals. J Audio Eng Soc 45(11):944–948

Lee H (2011 a) A new microphone technique for effective perspective control. Paper 8349 Presented at the 130th Audio Engineering Society Convention, May 2011

Lee H (2016a) Capturing and rendering 360° VR audio using cardioid microphones. Paper presented at the conference on audio for virtual and augmented reality, Los Angeles, 30 Sept–1 Oct 2016a

Lee H (2016b) Towards the development of intelligent microphone array designer. Paper Presented at the 2nd Audio Engineering Society Intelligent Music Production Workshop, Sept 2016b

Lee H, Gribben C (2014) Effect of vertical microphone layer spacing for a 3D microphone array. J Audio Eng Soc 62(12):870–884

Lee H, Rumsey F (2013) Level and time panning of phantom images for musical sources. J Audio Eng Soc 61(12):753–767

Lee H (2021) Multichannel 3D microphone arrays. J Audio Eng Soc 69(1/2):5–26

Lindberg M (n.a.) 3D Recording with the '2L-cube'. http://www.2l.no/artikler/2L-VDT.pdf (Accessed 4 Aug 2020)

MAATdigital (2019) 2BCmultiCORR. (Software Plug-In) http://www.maat.digital/2bcmulticorr/ Accessed 24 May 2019

Mason R, Rumsey F (2000) An assessment of the spatial performance of virtual home theatre algorithms by subjective and objective methods. Paper 5149 to the 108th Audio Engineering Society Convention, Paris, May 2000

Meyer J (1999) Akustik und musikalische Aufführungspraxis. 4th edn. Verlag Erwin Bochinsky (PPVMEDIEN Gmbh)

MH-Acoustics Eigenmike® (Online). Available at https://mhacoustics.com/ Accessed 18 June 2018

Miller R (2004a) Spatial definition and the PanAmbiophone microphone array for 2D surround & 3D fully periphonic recording. Paper presented at the 117th Audio Engineering Society Convention, San Francisco 2004

Miller RE (2004b) System and method for compatible 2D/3D (full sphere with height) surround sound reproduction. US Patent 2004b/0247134

Nipkow L (2010) Angewandte Psychoakustik bei 3D Surround Sound Aufnahmen. In: Proceedings to the 26. Tonmeistertagung des VDT, Leipzig, 2010, pp 786–795

Nipkow L (2012) Eigenschaften von mikrofonierten Raumsignalen bei 3D Audio/Auro 3D. In: Proceedings to the 27. Tonmeistertagung des VDT, Cologne, Nov 2012

Pfanzagl E (2002) Über die Wichtigkeit ausreichender Dekorrelation bei 5.1 Surround-Mikrofonsignalen zur Erzielung besserer Räumlichkeit. In: Proceedings to the 21. Tonmeistertagung des VDT, Hannover, 2002

Pfanzagl-Cardone E (2002) In the light of 5.1 Surround: Why AB-PC is Superior for Symphony-Orchestra Recording. Paper 5565 presented at the 112th Audio Eng Soc Convention, Munich, 2002

Pfanzagl-Cardone E (2020) The art and science of surround and stereo recording. Springer-Verlag, GmbH Austria. https://doi.org/10.1007/978-3-7091-4891-4

Pfanzagl-Cardone E, Höldrich R (2008) Frequency-dependent Signal-Correlation in Surround- and Stereo-Microphone Systems and the Blumlein-Pfanzagl-Triple (BPT). Paper 7476 presented at the 124th Audio Eng Soc Convention, Amsterdam, 2008

Power PJ (2015) Future spatial audio: subjective evaluation of 3D surround systems. Dissertation, University of Salford, UK

Pulkki V (2002) Microphone techniques and directional quality of sound reproduction. Paper presented at the 112th Audio Engineering Society Convention, Munich, May 2002

Resonance audio (2018) Available at https://developers.google.com/resonance-audio/ Accessed 5 Aug 2018

Riaz H, Stiles M, Armstrong C, Chadwick A, Lee H and Kearney G (2017) Multichannel microphone array recording for popular music production in virtual reality. e-brief presented at the 134rd Audio Engineering Society Convention, New York, Oct 2017

Rivas-Méndez D, Armstrong C, Stubbs J, Stiles M and Kearney G (2018) Practical recording techniques for music production with six-degrees of freedom virtual reality. Paper presented at the 145th Audio Engineering Society Convention, New York

Rumsey F (2002) Spatial quality evaluation for reproduced sound: terminology, meaning, and a scene-based paradigm. J Audio Eng Soc 50(9):651–666

Rumsey F, Lewis W (2002) Effect of rear microphone spacing on spatial impression for omnidirectional surround sound microphone arrays. Paper 5563 presented at the112th Audio Engineering Society Convention, April 2002

Ryaboy A (2015) Exploring 3D: A subjective evaluation of surround microphone arrays catered for Auro-3D reproduction system. Paper 9431 presented at the 139th Convention of the Audio Eng Soc, New York, Oct 2015

SADIE-II Binaural Database (2018) Available at http://www.sadie-project.co.uk Accessed 19 Sept 2018

Sengpiel E (2016) www.sengpielaudio.com/SRAflash.swf. 2016

Schultz F, and Spors S (2013) Data-based binaural synthesis including rotational and translatory head-movements. Paper presented at the 52nd International Conference of the Audio Engineering Society: Sound Field Control-Engineering and Perception

SoundField ST450 MKII (2018) Available at http://www.soundfield.com/products/st450mk2 Accessed 18 June 2018

Southern A, Wells J and Murphy D (2009) Rendering walk-through auralisations using wave-based acoustical models. Paper presented at the 17th European signal processing conference IEEE (pp 715–719)

Stiles M (2018) Recording spatial audio. Resolution 17(2):49–51

Theile G (1980) Über die Lokalisation im überlagerten Schallfeld. Dissertation, Technische Universität Berlin, Germany

Theile G (1990) On the performance of two- channel and multi-channel stereophony. Paper 2932 presented at the 88th Audio Engineering Society Convention

Theile G (1991) On the naturalness of two-channel stereo sound. Paper presented at Audio Engineering Society 9th Int Conference, Detroit, 1–2 Feb 1991

Theile G (2000) Mikrofon- und Mischungskonzepte für 5.1 Mehrkanal-Musikaufnahmen. In: Proceedings to the 21. Tonmeistertagung des VDT, Hannover, 2000, pp 348

Theile G (2001) Multichannel natural music recording based on psychoacoustic principles. In: Proceedings to the 19th International Conference of the Audio Engineering Society, 2001, pp 201–229

Theile G, Wittek H (2011) Principles in surround recordings with height. Paper 8403 presented at the 130th Audio Engineering Society Convention, May 2011

Theile G, Wittek H (2012) 3D audio natural recording. In: Proceedings to the 27.Tonmeistertagung des VDT, Cologne, Nov 2012

Pesch P (2010) Die Lokalisation von Phantomschallquellen im oberen Halbraum: Untersuchungen zur Erweiterung der Binauralen Raumsynthese. Verlag Dr. Möler

Torio G, Segota J (2000a) Unique directional properties of dual-diaphragm microphones. Paper presented at the 109th Audio Engineering Society Convention, 2000a

Torio G, Segota J (2000b) Unique directional properties of dual-diaphragm microphones. Preprint presented at the 109th Int. Audio Engineering Society Convention

Tylka JG, and Choueiri E (2016) Soundfield navigation using an array of higher-order ambisonics microphones. Paper presented at International Conference on Audio for Virtual and Augmented Reality of the Audio Engineering Society, 2016.

Unity 3D (2018) Available at https://unity3d.com/ Accessed 17 June 2018

Van Daele B, Van Baelen W (2012) Productions in auro-3D: professional workflow and costs. White paper by Auro-Technologies, Feb 2012

Williams M (2013a) The psychoacoustic testing of the 3D multiformat microphone array design, and the basic isosceles triangle structure of the array and the loudspeaker reproduction configuration. Paper 8839 presented at the 134th Audio Engineering Society Convention, May 2013a

Williams M (1987) Unified theory of microphone systems for stereophonic sound recording. Paper 2466 presented at the 82nd Audio Engineering Society Convention, 1987

Williams M (2005) The whys and wherefores of microphone array crosstalk in multichannel microphone array design. Paper 6493 presented at the 118th Audio Engineering Society Convention, Barcelona, May 2005

Williams M (2007) Magic arrays, multichannel microphone array design applied to multi-format compatibility. Paper 7057 presented at the 122nd Audio Engineering Society Convention, Vienna, 2007

Williams M (2008) Migration of 5.0 multichannel microphone array design to higher order MMAD (6.0, 7.0 & 8.0) with or without the Inter-format Compatibility Criteria. Paper 7480 presented at the 124th Audio Engineering Society Convention, Amsterdam, May 2008

Williams M (2012a) Microphone array design for localization with elevation cues. Paper 8601 presented at the 132nd Audio Engineering Society Convention, Budapest, April 2012a

Williams M (2012b) 3D and multiformat microphone array design for the GOArt project. In: Proceedings to the 27. Tonmeistertagung des VDT, Cologne, Nov 2012b, p 739

Williams M (2013b) The psychoacoustic testing of the 3D multiformat microphone array design, and the basic isosceles triangle structure of the array and the loudspeaker reproduction configuration. Paper 8839 presented at the 134th Audio Engineering Society Convention, Rome, May 2013b

Williams M (2014) Downward Compatibility configurations when using a univalent 12 Channel 3D microphone array design as a master recording array. Paper 9186 presented at the 149th Audio Engineering Society Convention, Los Angeles, Oct 2014

Williams M (2016a) Microphone array design applied to complete hemispherical sound reproduction—from integral 3D to comfort 3D. Paper presented at the 140th Audio Engineering Society Convention, Paris, June 2016a

Williams M (2016b) The basic philosophy of the 3D microphone array for recording and reproduction. Paper presented at the 29th Convention of the Verein Deutscher Tonmeister, Nov 2016b

Williams M, Le Du G (1999) Microphone array analysis for multichannel sound recording. Paper 4997 presented at the 107th Audio Engineering Society Convention

Wittek H (2000) Masters thesis, Institut für Rundfunktechnik (IRT)

Wittek H (2002) Image assistant V2.0. http://www.hauptmikrofon.de. Accessed 24 June 2008

Wittek H, Theile G (2002) The recording angle—based on localisation curves. Paper 5568 presented at the 112th Audio Engineering Society Convention, Munich

Wittek H, Theile G (2017) Development and application of a stereophonic multichannel recording technique for 3D Audio and VR. Paper presented at the 143rd Audio Engineering Society Convention, New York, Oct 2017

Wittek H, Haut C, Keinath D (2006) Doppel-MS—eine Surround-Aufnahmetechnik unter der Lupe. In: Proceedings to the 24. Tonmeistertagung des VDT, Leipzig, 2006

Wittek H, Rumsey F, Theile G (2007) Perceptual enhancement of wavefield synthesis by stereophonic means. J Audio Eng Soc 55:723–751

Zacharov N et al. (2016) Next generation audio system assessment using the multiple stimulus ideal profile method. In: Proceedings to the 8th international conference on quality of multimedia experience, Lisbon, Portugal, 2016, pp 1–6

Zhang K, Geluso R (2019): The 3DCC microphone technique: a native B-format approach to recording musical performance. Paper 10295 presented at the 147th Audio Engineering Society Convention, Oct 2019

Zielinsky G (2011) More reality mit Auro 3D. VDT-Magazin 2:24–27

Zotter F, Frank M (2012) All-round ambisonic panning and decoding. J Audio Eng Soc 60(10):807–820

Chapter 10
Comparative 3D Audio Microphone Array Tests

Abstract This chapter presents and summarizes the results of a comparison of almost 30 different 3D audio microphone techniques, gained through a number of individual test sessions by different researchers. The first presented test by Luthar and Maltezos contrasts a Decca-derived recording approach (omni microphones) with a Fukada/OCT use of directional microphones for an orchestral recording. The sonic results were evaluated in an informal manner by the authors of the study. An interesting comparison between Twins-Square versus Double-MSZ was made in a study by Ryaboy for the recording of a pop music ensemble in 2015. The technical details and results of subjective evaluation are described. Informal results are presented for a rather extensive comparative 3D audio microphone array test, that has taken place at Abbey Road Studios in 2017, led by Riaz and Stiles. It involved a total of 11 different microphone systems: mh acoustics EM32 'Eigenmike®'(HOA), SoundField ST450 MKII, ESMA (Equal Segment Microphone Array), ORTF-3D, Sennheiser AMBEO (FOA), OCT-3D, PCMA 'Perspective Control Microphone Array', Stereo XY-pair, IRT-Cross, Hamasaki Cube, SoundField ST450 MKII, and a Neumann KU100 dummy head. The most extensive comparative 3D audio microphone array recording so far has been conducted by Lee at the University of Huddersfield, including a total of 17 different 3D mic arrangements, the results of which have been partially evaluated by means of objective acoustic measurements (spectrum, frequency dependent signal correlation and direct-/diffuse sound ratio). Results and evaluation of an informal listening test on 2L-Cube, OCT-3D, ORTF-3D, PCMA-3D and Sennheiser 'Ambeo' by Gericke and Mielke are presented. Conclusions are dawn by summarizing also the results of other subjective listening test (e.g. by Kamekawa and Marui) and an attempt at a qualitative ranking of 3D microphone techniques is undertaken. In the outlook the importance of overall signal-decorrelation in the low frequency band for good spatial impression is revealed, based on previous measurements of the author. The relevance of the BQI_{rep} ('Binaural Quality Index for reproduced music') in relation also to 3D audio is argued.

Keywords 3D audio · Immersive audio · Mic arrays · Microphone technique · Subjective evaluation · Objective measurement · Spatial hearing · Abbey Road

E. Pfanzagl-Cardone, *The Art and Science of 3D Audio Recording*,
https://doi.org/10.1007/978-3-031-23046-2_10

10.1 The Luthar-Maltezos 9.1 Experiment (Decca vs. Fukada/OCT Tree)

In Luthar and Maltezos (2015) the results of an investigation in the effect of the polar pattern of the microphones used for the L, C, R channels in a 9.1 immersive music recording setup on the effect of perceived spatial impression and cohesiveness of the resulting sound-image are presented. Even though this project lacks a subjective listening test comparison and is solely based on the impressions of the two authors of the paper, it seems nevertheless of interest as it is strictly oriented at the real-world requirements (and restrictions) of an orchestra (with soloists) recording in a traditional concert hall (with apparently very good acoustics) and aims for an optimization of microphone patterns (set in locations, chosen according to architectural–mechanic possibilities/restrictions).

Even though the evaluation of the authors is only of informal nature the outcome may nevertheless be considered meaningful for the sound-engineer in working practice: the exclusive use of 'omni' microphones (as in one version of the Luthar-Maltezos 9.1 Technique) may indeed sound sonically superior to the 'cardioids only' Fukada/OCT array, as described in Sect. 10.1.1, which might be more accurate in terms of localization, but probably sound less 'nice', as cardioids usually are not able to provide the same degree of 'naturalness', which can be achieved with pure pressure transducers ('omnis') or with pure velocity microphones in an appropriate configuration.

From Luthar and Maltezos (2015): "... The authors wanted to expand on research already begun on the polar characteristics of microphones used in a 3-D recording setup within a classical music context (Howie and King 2015; Geluso 2012) along with more 'set' arrays designed with commercial releases in mind. The focus of the research was to address the impact that the LCR image has on the overall image stability. The center channel in an LCR array is utilized to increase directional stability (Theile 2001). Microphone polar characteristics of the center channel and its subsequent influence on the lateral LCR image is an important consideration when choosing a surround microphone setup; it affects the width, depth, and realism of the sound stage when reproduced for playback. The authors chose to focus efforts on expanding already existing microphone arrays for a sense of continuity and familiarity and to easily compare how the added height channels influence the 5.1 setup.

For this experiment, the authors were able to utilize the rehearsal time of the Stavanger Symphony Orchestra and record the ensemble, along with any soloists. The orchestra was of standard size and spread across the stage in conventional form. One of the challenges presented in the experiment was the limitation of choices for microphone placement due to the technical restrictions of the hall and the rehearsal requirements of the orchestra. Microphones were hung from the balconies for the height channels, as well as the lateral layer surrounds, and the LCR tree was placed off the stage but raised at an adequate height to compensate. Unlike other multichannel recordings that place the musicians around a microphone array, this experiment tests

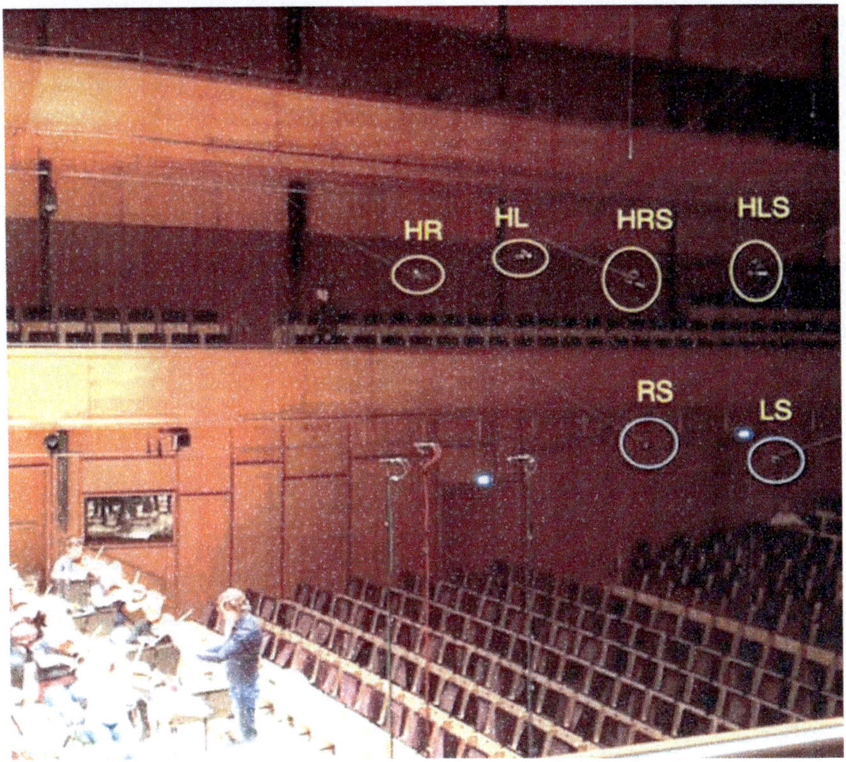

Fig. 10.1 Microphones in the Stavenger Konserthus hall—suspended microphones are circled (from Luthar and Maltezos 2015)

a more 'traditional' concert recording setup where the microphones are placed in front of the musicians (Fig. 10.1).

10.1.1 Recording Methodology

The desire was to test different microphones for the LCR channels and find a compromise so this could be done successfully with minimal complications within the confines of available gear. Based on these requirements, a compromise was found between the measurements for a Decca Tree formation, with 3 omnidirectional microphones (DPA 4006s), and a Fukada Tree/OCT adaptation of 3 cardioid microphones (KM140s) (Fukada 2001). Outriggers were not used. The microphones were coincidently placed on a dual microphone array bar, with each pair of left, center, and right microphones on three stands. The cardioid microphones were on the outer side of the L and R stereo bars, at an angle of approximately 45°. A compromise in height and width was reached so as to benefit both configurations equally. The L, C, and R microphones were approximately 3.25 m above the stage. The LS and RS omnidirectional microphones (Schoeps MK2, facing the rear) were hung from the balconies so taking measurement of them proved to be quite difficult. They were approximately 3 m above the stage. No inaccuracy was anticipated from being marginally lower than the main array (Fig. 10.2).

The height microphones were hung about 1.5 m above the main layer, based on best practices, as well as reference to research done prior on the height preference between the main array and height array (Lee and Gribben 2014). The front height L (FHL) and front height R (FHR) microphones were MKH 8020s facing front and the height surround L and height surround R were 2 DPA 4006s facing backward (with

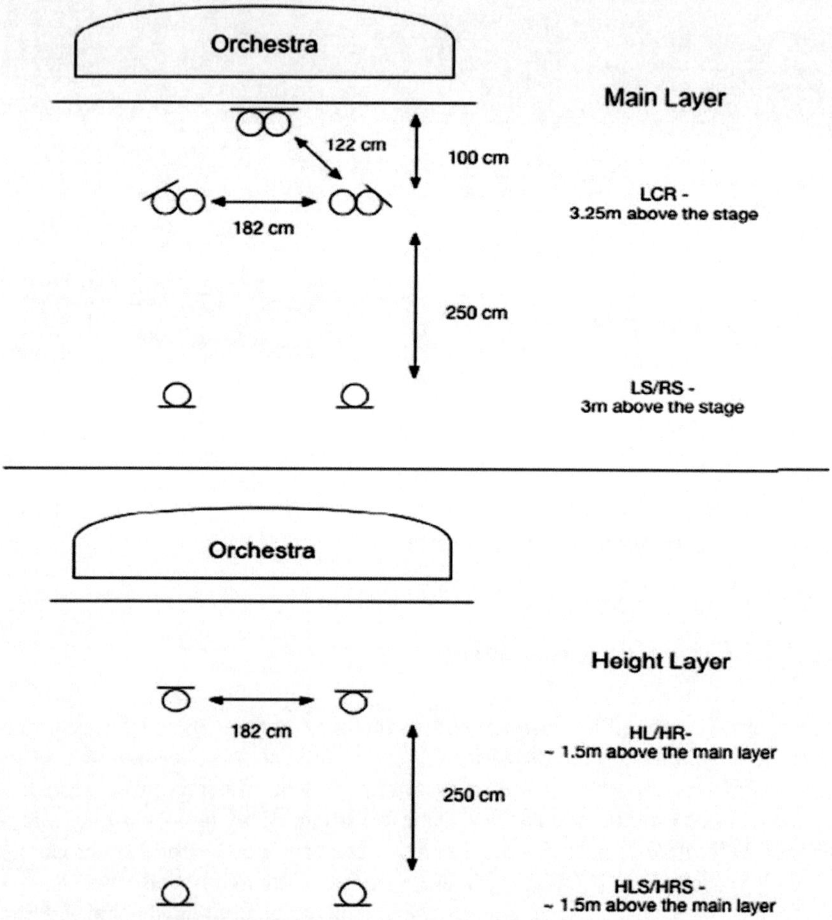

Fig. 10.2 Schematic of microphone placement for the Luthar-Matezos 9.1 recording session (from Luthar and Maltezos 2015)

diffuse field grids on). The microphones were chosen from an available selection and paired according to how the authors felt each microphone's technical specifications would benefit the experiment. The 0.1 channel was not represented by a microphone, but rather was derived from the crossover points in the Genelec subwoofer, which the L, C, and R channels were connected to during playback. All microphones were captured at 24 bits/48 k sampling rate, into 2 Fireface UFXs connected via ADAT, and recorded into Pro Tools.

10.1.2 Recording Reproduction and Playback

The facility in which listening occurred was not equipped with a standardized multi-channel system beyond 5.1 so the authors were able to construct a 9.1 system based on professional references. The 5.1 system was calibrated according to the ITU-R BS 775 standard for 5.1 reproduction (ITU-R BS.775-3 2012). The authors' personal speakers were used for the height surrounds. The speakers were all Genelec 1031A, except the rear height channels, which were Genelec 8020B. Output was calibrated for each speaker based on mixing procedures for music (using full range pink noise at 78dBSPL). The height channels were positioned about a meter above the 5.1 layer and were angled as close as possible to 30° towards the listener, which is a commonly specified technical standard (see 'Auro3D' and 'Dolby Atmos'). The channels were mixed according to the authors' aesthetic choices, and common mixing practices. No processing (compression, reverb, etc.) was implemented." (from Luthar and Maltezos 2015)

10.1.3 Choice of Microphone Polar Pattern in Relation to Program Material

"...The three examples of music that were analyzed were:

1. Janáček—"*Jealousy*" overture—a loud, brassy excerpt,
2. Elgar—Cello Concerto, Mvt. 2—soloist excerpt,
3. Rachmaninov—Symphonic Dances, Mvt. 1—sparse orchestration, woodwind-centric.

For this recording, the authors' opinion was that the set of microphones producing the most natural image in the space were the 9 omnidirectional microphones. The acoustics of the concert hall are ideal, and the omnidirectional microphones were the best at recreating that space. While there was quite a large amount of diffuse sound, this wasn't perceived as objectively negative, but rather, a matter of personal taste.

In the cello excerpt a pleasant result was found by using the 2 DPA [omni] microphones for the L and R, and the KM140 for the center (– 3 dB from the L and R channels)." (from Luthar and Maltezos 2015) [Rem.: some similarity can be found in the CHAB-technique by Olabe and Alvarez (for reference see Olabe 2014, as well as Sect. 3.3.9 in Pfanzagl-Cardone 2020) which uses a center hemi-cardiod in addition to L and R omnis; with the difference, that the center-capsule is on the same line as the omnis and not in a triangular configuration. However the sonic advantages reported for the use of a more directional pattern for the center capsule are the same].

"... This helped stabilize the image and bring a bit more detail out of the cello. For similar situations, this might be a good tactic to use for helping bring out detail in a soloist. This combination also proved successful for the other selections, helping anchor the orchestral image, or to help bring the woodwinds a bit more forward. Using only the KM140s as the LCR microphones caused an auditory disconnect between the height channels and the horizontal sound stage, which was not found when the LCR were either all omnidirectional or with a cardioid in the center. The use of only cardioids had a tendency to flatten the frontal image and the sound was perceived as more aggressive and lacking any sense of 'air'. In general, the height channels helped provide a natural, but improved, spatial impression with a greater sense of envelopment. The radiation patterns of the instruments were reproduced more accurately with the addition of the height channels. For example, the bassoon in the 'woodwind' excerpt was more present and full with the height channels engaged. It was also easier to localize lower frequencies (from the double basses, for example). The positive effect of the height channels was minimized, however, when the LCR configuration was all KM140s.

10.1.4 Conclusions

The authors felt that the experiment was a good showcase of the possibilities for concert recording with an added height element. As it was noted that a more directional center channel was useful for helping stabilize the image and provide extra clarity when soloists were in front of the orchestra, it might be of interest to experiment with different directional microphones for the center channel. It would also be interesting to utilize different polar patterns in the height channels, and see if/how they affect the center image. The authors are confident that future recordings will yield useful and creative results. Recording and reproducing music in 9.1 (or greater) is still in its infancy and leaves much to be explored and discovered." (from Luthar and Maltezos 2015).

10.2 Twins-Square Versus Double-MSZ—A Comparative Test

An interesting comparison of two 3D audio microphone-techniques, used for the recording of a pop-style music ensemble may be cited from (Ryaboy 2015): "…Currently, many microphone arrays that are used for multi-channel audio with height channels use traditional surround sound arrays aided by an additional microphone array placed in the upper plane for capturing height information. OCT-9 [now termed: OCT-3D] is a recommended system for Auro-3D that uses OCT-Surround array with four additional spaced super-cardioid microphones placed directly above the Left, Right, Surround Left, and Right channels. Wide A/B is another system recommended for Auro-3D, which uses nine omnidirectional microphones, a microphone for each channel, positioned in the same fashion as the Auro-3D speaker configuration. Both of these microphone arrays are using spaced microphone approach. Spaced, or time-of-arrival techniques, are based on the time delay differences caused by the different paths from the sound source to the microphones, while in coincident microphone techniques, the microphones are positioned as close to each other as possible and the stereo image is created by differences in sound pressure levels. Although the aforementioned arrays may be suitable for large concert halls, they may not work in smaller spaces and recording studios that don't have large live rooms. Consequently, the two microphone techniques examined in the study were a fully coincident Double-MSZ array and a semi-coincident array Twins Square.

Double-MSZ technique is compatible with B-Format Ambisonics, and captures four signals WXYZ, also called 'B-Format', which represent omnidirectional pressure and figure of eight pressure gradient in horizontal and vertical planes, creating a sphere that can be manipulated, using similar principles as the MS decoding.

Introduced by Gregor Zielinsky, Twins Square array is based around a unique Sennheiser MKH800 Twin multi-pattern microphone, and combines coincident as well as time-of-arrival approaches. Unlike traditional multi-pattern microphones that combine signals from two transducers into a single output, the Twins' transducers can each be accessed separately, which not only provides the ability to route each output to different channels, but it gives the mixing engineer an ability to decide during post production, which pick-up pattern is better suited for mixing. The microphone array utilizes four Twin microphones arranged in a rectangular fashion, with two microphones mounted at different heights but on the same depth plane at a wide AB distance. Consequently, the front-facing capsules provide signals for the front facing speakers, while the capsules facing the back provide information for the rear channels. An additional cardioid microphone is placed in between of the bottom two microphones for the center channel. What makes this microphone technique unique is that it is coincident from front to rear, and spaced in the vertical and horizontal planes.

In concert realism perspective, the aim of the surround microphone array in addition to properly capturing direct sound is to capture correct spatial attributes of the recording space. Direct sound influences the correct directionality of the source while early reflections are one of the main attributes that determine the perception of depth, distance, and special impression (Theile and Wittek 2011). Although spaced and coincident approaches are aimed at achieving these goals, the resulting sound image is somewhat different. One of the more studied aspects on this topic is the addition of height channels, and correct placement of microphones to capture information feeding those channels. It has been observed in studies researching spatial hearing that the horizontal stereo principles, such as time and amplitude differences may not be applicable in the vertical plane. For example, panning of phantom images in the vertical plane is unstable (Barbour 2003) and localization in the vertical plane relies more on spectral cues rather than inter-aural time and amplitude cues (Lee and Gribben 2014). Similarly, some of the principles guiding horizontal stereo recording techniques may not be applicable when microphones are spaced vertically.

A recent study, examining overall preference and spatial impression of a spaced/coincident microphone approaches by comparing four different levels of the vertical spacings, showed that microphones configured in coincidental fashion were not only preferred, but produced greater spatial impression for transient rich material, and there were no significant differences between spaced layer configurations. In his paper 'Channel-Z' (Geluso 2012), Geluso introduces a coincident technique that uses a figure-of-eight microphone oriented upwards combined with traditional stereo arrays. His preliminary research demonstrated that Channel-Z was rated better for 3D imaging criteria compared to recordings where height channel signals contained up-mixed height signal.

10.2.1 Spatial Audio Evaluation

Subjective evaluation of audio is common practice in the field of music technology where some of characteristics that are evaluated are timbral quality, width and depth of soundstage, image stabilization, and localization. When it comes to multichannel audio, new characteristics arise, such as envelopment, spaciousness and spatial impression. Evaluating these characteristics is a somewhat challenging task, as these qualities are subtle, and difficult to quantify. Current perceptive studies that investigate these elements tend to conflate several of the aforementioned characteristics into a single rating. For example, in a study that tested effects of vertical microphone spacing subjects evaluated 'Overall Spatial Impression', a term predefined by the researchers, 'as encompassing all possible spatial percepts from three dimensions for the environment as well as the source components' (Lee and Gribben 2013). Although such studies may yield conclusive results about preference and perceived differences of recordings, they produce little information about specific characteristics of perceived audio experience by the subject. For example, when a recording is being evaluated for envelopment, envelopment may be perceived as a single immersive sound source, or an enveloping environment. Similarly, a recording where the perspective is such where one is surrounded by musicians, may also be considered enveloping (Rumsey 2002). Francis Rumsey defines Spatial Impression as 'an auditory perception of the location, dimension and parameters of a sound source and acoustic environment in which the source is located'. In his paper 'Spatial Quality Evaluation for Reproduced Sound' Rumsey, proposes an approach to evaluating spatial impression by dividing the listening experience into hierarchical scenes ranging from individual instrument sources (micro-scenes) to spaces and environments (macroscenes) (Rumsey 2002). The present study uses Rumsey's scene based recommendations, and focuses on evaluating envelopment, localization and spatial impression based on three dimensions, width, depth and height.

10.2.2 Recording Conditions

The recordings took place at NYU's James L. Dolan Studio. The musical program of choice was a five-piece Klezmer band consisting of Violin, Accordion, Clarinet, Voice, Upright Bass, and marching band style Drum. The band was chosen for its distinct instrumentation, which can be easily identified in a recording. A multicapsule Josephson C700s microphone with an addition of Sennheiser MKH800 was used for the Double-MSZ array. C700s microphone features three capsules, omnidirectional (W), front-facing figure-of-eight (X), and a side-facing figure-of-eight (Y) each having its own separate output. MKH800 provided the Z capsule by operating in figure-of-eight mode and facing upwards. The array was placed in the center of the room at about 3.5 ft. (approx. 1 m) off the floor (Fig. 10.3).

Twins Square array was implemented using four MKH800 Twin mics, with both arrays sharing the same center channel. The microphones were spaced about 10 ft. (approx 3 m) apart; the spacing between the lower and the upper plane mics was 5 ft. (1.5 m). The lower plane microphone capsules were on the same plane as the Josephson middle capsule, with Double-MSZ array centered between left and right Twin microphones. The band was arranged in a wide semi-circle, and the arrays were placed about 6 ft. in front of the band.

10.2.3 Reproduction System

The listening test took place at the NYU Music and Audio Research Laboratory. The lab is a dry listening room, equipped with sixteen identical Genelec 8030A active speakers that can be arranged for a number of different speaker configurations. The speaker arrangement of choice was a nine-channel Auro-3D layout. Subwoofer was not used in the evaluation. Playback channels from the Double-MSZ array were derived using the arithmetic in Table 10.1.

Although only nine of the sixteen speakers were used in the experiment all sixteen were turned on, and the subjects were not aware of which speakers were operating. Subjects were seated in a marked sweet spot of the room in front of a small table, with the questionnaire sheets and a computer mouse to control the graphic user interface.

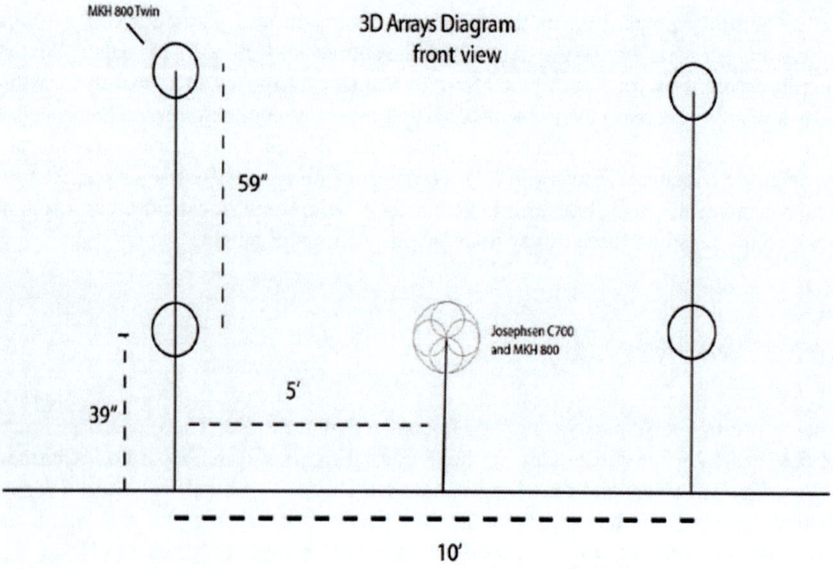

Fig. 10.3 Microphone positions for the twins-square and double-MSZ mic arrays (from Ryaboy 2015)

Table 10.1 Ambisonic-style processing applied to the Double-MSZ microphone signals (from Ryaboy 2015)

Left	W + X + Y
Right	W + X − Y
Center	W + X
Left-surround	W − X + Y
Right-surround	W − X − Y
Left-height	W + X + Y + Z
Right-height	W + X − Y + Z
Left-height-surround	W − X + Y + Z
Right-height-surround	W − X − Y + Z

Head movement was not restricted, but the subjects were instructed to not make any drastic changes to the listening position. The gain levels for each speaker were calibrated using pink noise and SPL meter measuring 76 dB at listening position. The 17 test listeners were presented with a total of 20 clips of music at random that they were asked to evaluate." (from Ryaboy 2015).

More details on the test stimuli, test method, statistical evaluation etc. can be found in (Ryaboy 2015), the essence of which is summarized below:

"… The present study investigated two different 3D recording techniques, in a small studio environment, for Auro-3D speaker configuration. Using a perception test, the study evaluated the recordings on three-dimensional qualities, and whether the recording method influenced the end result. The study found that there were significant differences between techniques as well as sources themselves in terms of localization in the horizontal as well as vertical planes, and perceived room size. The Twins Square technique showed a wider soundstage and wider overall ensemble width, which corresponds with previous research suggesting that spaced techniques exhibit greater spatial impression due to larger microphone spacing which result in lower degree of correlation between signals (Lee and Gribben 2014). Both techniques placed all sources correctly to where the players were positioned during the recording except, Double-MSZ technique incorrectly placed the Cymbal closer to the center. The Bass and Accordion were rated as being the widest sources. The higher Bass width may be due to the fact that lower frequencies are less directional. The width of Bass was slightly narrower for Tune 2 for both techniques, the reason for this may be that the bass in Tune 1 was bowed, while it was plucked in Tune 2, which may give it a slightly percussive sound and perhaps make it more easily locatable. Accordion may have been picked as the widest simply due to its size, and a similarly large frequency range which in the case of both tunes provided lower and mid-register harmony. Also, both the Bass and Accordion did not have an easily distinguishable melody, which could catch the listener's ear, while more melodic sources were localized easier and were rated narrower.

Double-MSZ showed higher elevation for all sources. The explanation for this may be the position of the array. Since it was positioned about 3.5 ft. off the floor, at about musicians' waist level, it is possible that the resulting angle elevated sources, due to direct sound entering the Z channel. During post-test interviews several subjects revealed that they thought they heard different perspectives, one of sitting in the front row with musicians on stage, and another being on stage with musicians. Another conclusion that could be drawn is that Double-MSZ showed stronger phantom imagery in the vertical plane as the technique is able to maintain a stable elevation angle. This is also supported by the lower Blur values for Double-MSZ. It could be concluded that Double-MSZ is much more sensitive to the position than a spaced array but has more stable imagery. Bass and Accordion showed highest 'Blur' values, which may indicate that these sources may also be wide in the vertical plane (Fig. 10.4).

Height graphs also revealed that clarinet values for the Twin array were evenly dispersed between first three levels and may be caused by high directivity of the instrument, since many clarinet players move their instrument up and down as they play.

The study did not find any strong links between depth and microphone arrays. One possibility for this may be simply that the musicians were all positioned relatively on the same plane with only slight deviations in depth, which may not have been picked up by the arrays or the listener. Correlation between height and depth is definitely something that should be looked into further as there were no immediate reasons for this phenomenon. Many subjects expressed difficulties gauging depth in the post-test interviews. The perceived room size values show that there were noticeable differences between techniques, with Twin technique showing much higher values. These results don't necessarily show that one technique is better than another at reproducing

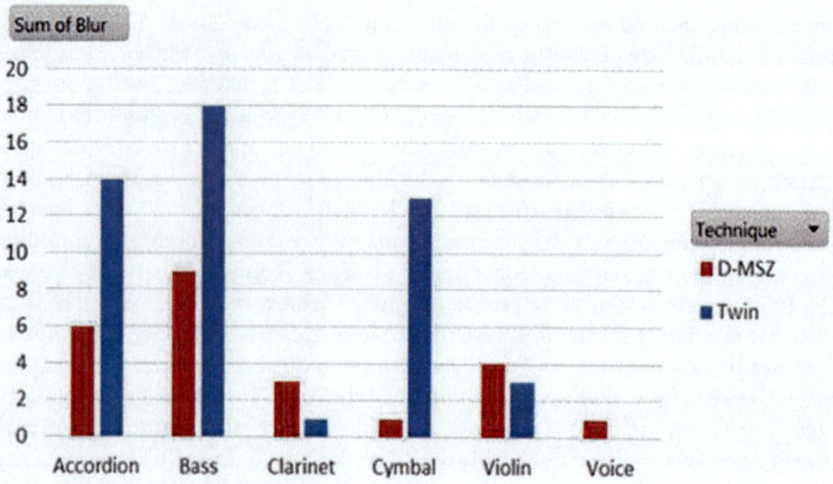

Fig. 10.4 Comparison of 'blur' ('source width'/'localization accuracy') relative to sound-source and mic technique (Double-MSZ vs. Twins-Square)

the room size or that one technique has better spaciousness than the other, however, it clearly shows that there's a difference in the way the arrays capture the ambience of the room. Since Dolan Studio is bigger than the research lab where the test took place and a bit smaller than a club it would be a 2 on the given scale. Therefore Twin Squares produced a slightly exaggerated room size, while Double-MSZ a smaller, more intimate room size. The high standard deviation shows that subjects did not have a clear consensus on the perceived room size for either technique. Lastly the results show that elevated sources for Double-MSZ did not have an impact on the room size. However, widths and depths of sources correlated slightly with room size for both techniques, possibly showing that subjects interpreted room size in terms of source or ensemble envelopment, and depth.

10.2.4 Conclusion

The Twins Square technique demonstrated a more enveloping soundstage, wider ensemble width, and a more spacious environment compared to the Double-MSZ technique, which showed better localization in vertical and horizontal planes, stable imagery, and a more intimate perspective. The test did not find any strong linking between microphone technique and perception of depth; however, both techniques showed a correlation between depth and height. Additionally, the test showed that the source type has a significant influence on its localization, where sources with lower frequency range demonstrated poorer localization than those in the higher frequency spectrum; however, sources that exhibit stronger directionality are more sensitive to source movement. While these findings solely describe this specific recording situation, I believe they could possibly be observed with other material and different recording environments, which is certainly something to explore in the future." (cited from Ryaboy 2015).

10.3 The "Howie 3D-Tree" Versus "Hamaski-3D" Versus "HOA" for 22.2 Orchestra Recording

In a paper presented at the 142nd Audio Engineering Society Convention (Howie et al. 2017) Howie analyzed the results of a subjective listening test, concerning three different 3D-audio orchestra recording techniques for 22.2 immersive audio, which—according to the author—was the first comparative study on that subject (Figs. 10.5 and 10.6).

Technique 1 (Howie 3D-Tree): One of the explored techniques under test was proposed by the authors in (Howie et al. 2016); following is a rough description: "... Primarily direct orchestral sound is captured by a modified 'Decca Tree' of five omnidirectional microphones, the middle three of which are outfitted with acoustic

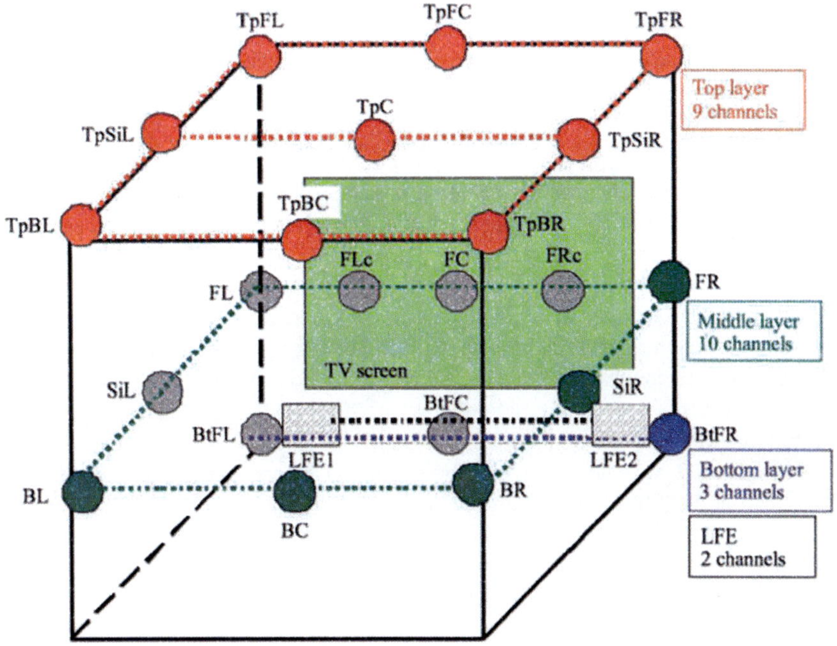

Fig. 10.5 22.2 multichannel loudspeaker configuration (from Howie et al. 2016), adapted from ITU-R BS.2159-4 (2012)

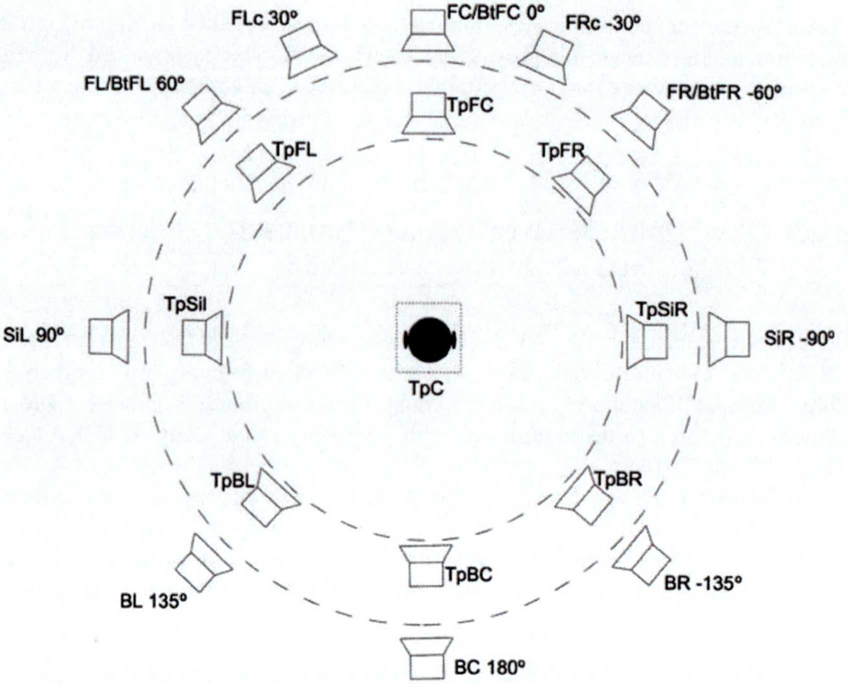

Fig. 10.6 Loudspeaker layout schematics for 22.2, including angle information (Bottom Layer loudspeakers BtFL, BtFC, BtFR and Subwoofers LFE1, LFE2 are not shown) (graphic courtesy and © Will Howie)

pressure equalizers. Three directional microphones placed 1 m above the stage floor provide signal for the bottom channels, vertically extending and anchoring the orchestral image. Widely spaced directional microphones capture decorrelated, spatially diffuse ambience, and are assigned to the remaining main layer and height channels. The technique is designed to retain the traditional 'concert perspective' that is typical of most multichannel classical music recordings. Microphone orientation typically mirrors assigned playback channel orientation: for example, the TpFL microphone would have a horizontal orientation of around 60°, and a vertical orientation of approximately 45°." (from Howie et al. 2017).

Technique 2 was designed by Hamasaki and his coauthors, as described in Hamasaki et al. (2004) and (Hamasaki and Van Baelen 2015). "… The technique is a logical extension of Hamasaki's earlier publications on multichannel music recording, particularly 'Reproducing Spatial Impression With Multichannel Audio', co-authored with Hiyama (Hamasaki and Hiyama 2003). Direct sound from the orchestra is captured by an array of 5 super-cardioid microphones, placed at equal intervals across the sound stage. In Hamasaki et al. (2004), ambient sound is captured with an array of laterally oriented bi-directional microphones—an extension of the well-known 'Hamasaki Square' (Hamasaki and Hiyama 2003). The placement and spacing of the bi-directional microphones ensures minimal capture of direct and rear wall sound, and that the ambient sound field is decorrelated across the audible frequency spectrum. Several of these ambience microphones are assigned to the front channels, to be mixed in if the recording engineer feels the orchestral sound is too "dry". In Hamasaki and Van Baelen (2015) this approach is updated, using vertically oriented super-cardioid microphones as height channels. This version of the technique is representative of current 3D orchestral music recordings being made by NHK recording engineers, and thus can be considered a de-facto production standard for 22.2. Neither Hamasaki et al. (2004) nor Hamasaki and Van Baelen (2015) specify microphones for the bottom channels. For this study, three Sanken CUB-01 miniature boundary microphones have been added to the technique, each placed as far down-stage as possible (see Fig. 10.7 and Table 10.2). These microphones were chosen for their minimal visual impact, an important factor in broadcast sound recording, as well as to contrast with the bottom channel microphones used in Technique 1. Placement of the front five microphones for Technique 2 was based on available hanging points, and the increased 'reach' of hyper-cardioid microphones as compared with omnidirectional microphones. Like the 'Hamasaki Square', Technique 2 included 3 frontal ambience (FrAmb) microphones to be mixed in with the direct orchestral sound as necessary. Microphones for all three techniques were either hung or placed on telescopic stands in the hall, depending on their desired height and location. adjustments were made based on monitoring the orchestra's rehearsals. ..." (from Howie et al. 2017)

Technique 3: "... When considering the complexity, cost, and time associated with setting up either Techniques 1 or 2, the potential advantages to using a single-point, Ambisonics-based capture system become obvious. As such, for this study, the Eigenmike (em32) was chosen as a 3rd recording technique. The em32 from company 'Mh acoustics' is a spherical microphone array where each of the 32 capsules is calibrated for magnitude and phase response. The accompanying software 'Eigenstudio' converts the microphone signals into 3rd order Ambisonics B-format. 16 channels were recorded following the ACN channel order convention with N3D normalization (see Chapman et al. 2009). The em32 was placed approximately 1 m above the conductor. Professional recording engineers tend to place microphones based on previous experience, known best practices, and most importantly, what they hear in the monitoring environment. Recording with the Eigenmike, as such, presents a unique set of challenges. There is little published information detailing placement and optimization strategies for music recording using spherical HOA microphones, especially where the desired sound scene utilizes the traditional 'ensemble in front, ambience surrounding' perspective. Daniels discusses several experimental recordings done with spherical HOA microphones, mixed for two-dimensional playback (Daniel 2009). For a large ensemble recording where the goal was to keep the ensemble imaged in front of the listener, Daniels placed a 20 capsule HOA sphere near several other (unidentified) 5.1 microphone arrays. Barrett (2012) and Power (2015) both used the Eigenmike for music recording as part of their respective studies, but provided no methodology for placement and/or optimization. The 32-channel output from the Eigenmike is recorded to a computer running Eigenstudio

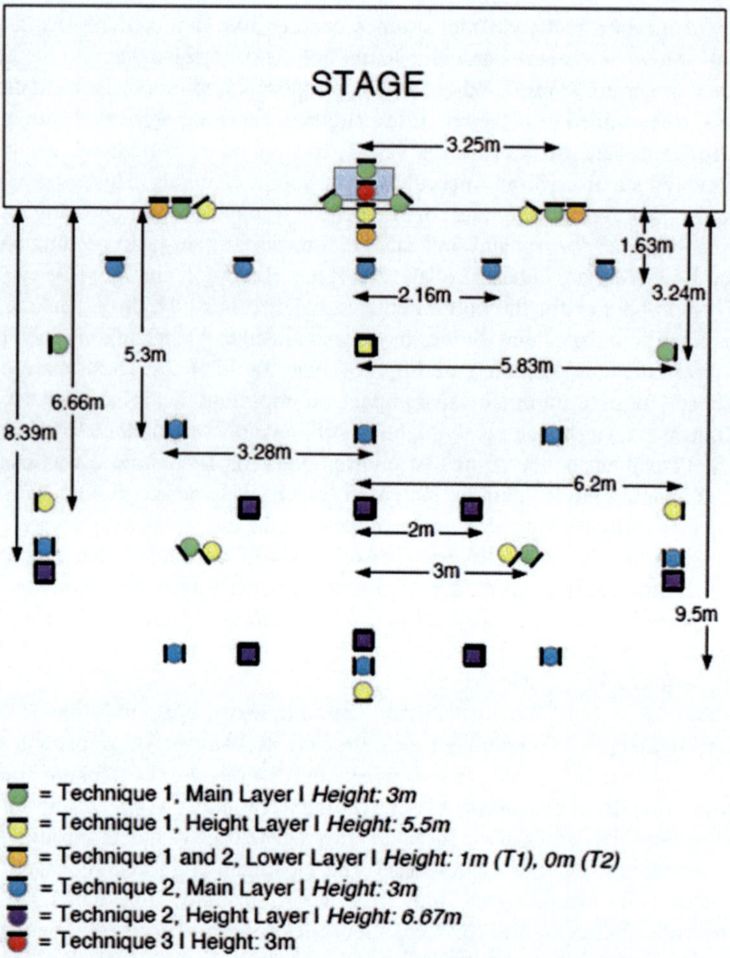

Fig. 10.7 Microphone placement for techniques 1–3, overhead view; height is referenced to stage floor (from Howie et al. 2017)

software via firewire output from an mh acoustics EMIB termination box. There is no effective way to monitor a 22.2 rendering of these signals in real time. For this study, the beam-pattern of an omnidirectional microphone was sent from the Eigenstudio recording software to Studio 22 for monitoring. Though not ideal, this gave the recording team some degree of information (distance and balance of instrumental groups) for microphone placement optimization. The result was the Eigenmike being placed in the centre of Technique 1's 'Decca Tree' (Fig. 10.7) …" (from Howie et al. 2017).

The orchestral recordings took place at Pollack Hall, a medium sized concert hall with a seating capacity of 590. The hall measures 36 m long by 18 m wide by 12 m

Table 10.2 Listing of microphones used for techniques 1 and 2 (from Howie et al. 2017)

Channel	Technique 1	Technique 2
FL	Schoeps MK2S	Neumann KM185
FLc	Schoeps MK2H	Neumann KM185
FC	Schoeps MK2H	Sennheiser MKH 8050
FRc	Schoeps MK2H	Neumann KM185
FR	Schoeps MK2S	Neumann KM185
BL	Schoeps MK21	Schoeps MK8
BC	Neumann KM120	Neumann KM120
BR	Schoeps MK21	Schoeps MK8
SiL	Neumann KM184	Sennheiser MKH30
SiR	Neumann KM184	Sennheiser MKH30
TpFL	Schoeps MK4	Neumann KM185
TpFC	Schoeps MK4	Neumann KM185
TpFR	Schoeps MK4	Neumann KM185
TpC	Schoeps MK41	Schoeps MK41
TpBL	Schoeps MK4	Neumann KM185
TpBC	Schoeps MK4	Sennheiser MKH 8050
TpBR	Schoeps MK4	Neumann KM185
TpSiL	Schoeps MK4	Sennheiser MKH50
TpSiR	Schoeps MK4	Sennheiser MKH50
BtFL	DPA 4011	Sanken CUB-01
BtFC	DPA 4011	Sanken CUB-01
BtFR	DPA 4011	Sanken CUB-01
FrAmb L	N/A	Sennheiser MKH 800
FrAmb C	N/A	Sennheiser MKH 800
FrAmb R	N/A	Sennheiser MKH 800

high, which results in an overall volume of 7776 m^3. According to the RT60 data provided by the authors (which are between 2.3 s at 63 Hz and 1.4 s at 4 kHz) the corresponding 'critical distance' (or reverberation radius) values are roughly between 3.3 and 4.3 m.

10.3.1 Listening Test Conditions and Creation of the Stimuli

"... All three techniques simultaneously captured the final orchestral dress rehearsal: Techniques 1 and 2 were recorded to Pro Tools 10 at 96 kHz/24bit resolution. Spot microphones for the woodwinds, bass and tympani were also recorded. Technique 3 was recorded to a separate laptop computer, whose audio interface was locked to the RME Micstacys' master clock. A single, 30-s musical excerpt was chosen as stimuli—the passage contains dense orchestration representative of the piece it was derived from (Tchaikovsky's 5th Symphony), and has a fairly even dynamic envelope. The techniques under investigation were balanced by a team of three recording engineers with extensive professional experience recording and mixing orchestral music. It was observed that Technique 2 did not contain enough low frequency content for a satisfying mix, largely due to the low frequency roll-off typical of highly directional microphones. In accordance with Hamasaki and Hiyama (2003), the FL and FR omnidirectional channels from Technique 1 were added to Technique 2's mix, low-passed at 200 Hz. Once ideal balances were achieved, 24-channel mixes of the musical excerpt were made for each technique. To create an optimal 22.2 mix of the Eigenmike recording, a custom-made decoder for the speaker positions in Studio 22 was built by using the Ambisonic Decoder Toolbox by Heller and Benjamin (2014) the decoder matrix for a dual band All-Round decoder (Zotter and Frank 2012) was calculated, which allowed for adjustment of balance between high and low frequencies with phase matched filters per (Heller et al. 2008). The crossover frequency (400 Hz) and the gain for the balance (+ 1 dB HF) were chosen to perceptually match the mixes from Techniques 1 and 2. The three resultant stimuli were level matched by ear. These results were then confirmed by objective means. A Neumann KU-100 Dummy Head microphone was placed in the listening position at ear level, and used to record the playback of each stimulus. Integrated loudness measures (LUFS 9) were then performed for each recording. All stimuli were found to be within 0.5 dB of each other." (from Howie et al. 2017).

10.3.2 The Test Results

"… Based on previous research into spatial audio evaluation (see Berg and Rumsey 2001a, b; Zacharov et al. 2016), the attributes Clarity, Scene Depth, Naturalness, Environmental Envelopment (termed 'Room envelopment' in Fig. 10.8) and Sound Source Envelopment were chosen and had to be rated by 23 subjects (students or faculty within the Graduate Program in Sound Recording at McGill University). A new term, 'Quality of Orchestral Image', was included based on the sonic imaging goals of Technique 1.

10.3.3 Overall Performance of Recording Techniques

Figure 10.8 shows a clear similarity of ratings between Techniques 1 and 2 for all the subjective attributes under investigation: clarity, scene depth, naturalness, environmental envelopment (room envelopment), sound source envelopment, and quality of orchestral image. Given this, and its consistently high mean scores across all attributes, the three-dimensional recording technique proposed in Howie et al.

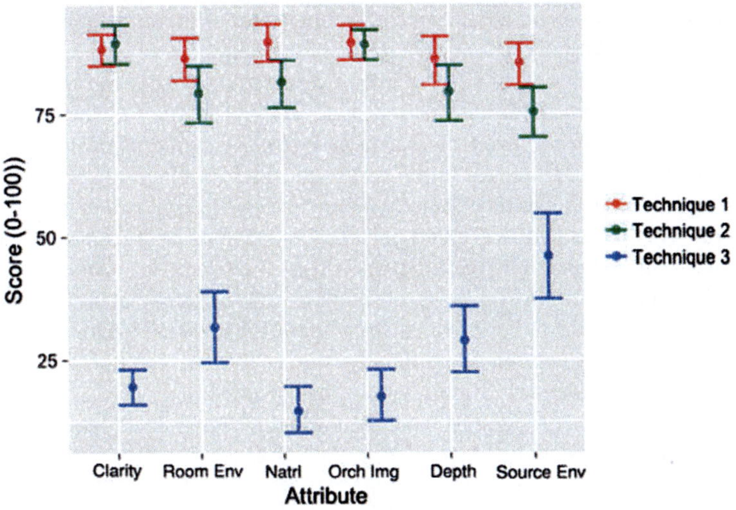

Fig. 10.8 Average rating for each attribute of microphone techniques 1–3 (adapted from Howie et al. (2017) (Technique 1 = Howie 3D-Tree; Technique 2 = Hamasaki 3D; Technique 3 = HOA)

(2016) should be considered a well performing, valid production technique for three-dimensional orchestral music recording. Concepts from both Technique 1 and 2 could also easily be combined to form any number of hybrid techniques. For example, broadcast recordings involving picture would benefit from the bottom channel microphone design from Technique 2, which is more visually transparent. Also very clear are the consistently low scores across all perceptual attributes for Technique 3. This matches well with the trend observed in previous research comparing two-dimensional recording techniques. For example, Camerer and Sodl (2001) observed a lack of depth and adequate spatial impression for the Soundfield MKV, a 1st order Ambisonics recording system. These observations are echoed in the current study, with the Eigenmike performing poorly for Scene Depth, Environmental Envelopment and Sound Source although a convenient alternative to large spaced microphone arrays, may not yet be suited to professional 3D music recording, especially given the monitoring difficulties discussed above.

10.3.4 Naturalness and Sound Source Envelopment

'Naturalness' appears frequently as a subjective attribute in multichannel audio evaluation, and has been shown to correlate strongly with the impression of 'presence' (Berg and Rumsey 2001a, b) and overall preference of sound quality (Mason and Rumsey 2000). Frequently observed by both subjects and researchers were unpleasant and unnatural 'out of phase' sonic artefacts present in Technique 3. This may explain why amongst all attributes, Technique 3s mean rating was lowest for 'naturalness'. In this study, a lack of perceived 'naturalness' may also be an issue of perspective bias. Techniques 1 and 2 deliver a 'cinematic' perspective for reproduced orchestral music, with the orchestra appearing entirely in front of the listener—a perspective most listeners have grown accustomed to. Technique 3, however, presents a much wider orchestral image, with direct sound sources covering almost 180° of frontal sound, likely due to the spherical nature of the Eigenmike. It is possible that the more 'wrap-around' direct sound perspective delivered by Technique 3 is also perceived as being 'unnatural'. Rumsey has written, "Envelopment, on the other hand, must be subdivided into environmental envelopment and source-related envelopment, the former being similar to LEV in concert halls and the latter to envelopment by one or more dry or direct foreground sound sources." (see Rumsey 2002). It was assumed that Technique 3s wider orchestral image would be rated highly for Sound Source Envelopment. However, although that attribute represented Technique 3s highest rated mean, it still scored well below Techniques 1 and 2. Clearly, listeners did not find a wider 'wrap around' orchestral image to be more enveloping. In this study, the listeners' impression of Sound Source Envelopment may be closer to Griesinger's concept of Continual Spatial Impression (Griesinger 1997)—a fusion of continuous direct sound and reflected energy that results in a sense of envelopment connected to the sound source. Technique 1 seems to best represent this type of spatial impression. ..." (from Howie et al. 2017).

10.4 Comparison of 9-Channel 3D Recording for Solo Piano

An interesting example on how to mike a solo piano for 3D audio can be found in Howie et al. (2018):

"... As with traditional stereo and 5.1 surround sound, acoustic music recording techniques for 3D audio can often be divided into one of three categories: spaced, near-coincident, or coincident.

10.4.1 Spaced Recording Techniques

Spaced 3D microphone techniques aim to capture and reproduce spatial sound information through time of arrival differences between microphone signals. A linear, one-to-one microphone signal to loudspeaker relationship is typically maintained. Another common feature in many proposed techniques is an emphasis on distant spacing between rear and height microphones to prioritize decorrelation between microphone signals. Several authors have commented on the importance of minimizing direct sound capture in height channel signals in order to ensure instrument or ensemble image stability at ear level (see Bowles 2015; Theile and Wittek 2011; King et al. 2016; Howie et al. 2016) while maintaining a traditional 'concert' perspective.

King et al. (2016) suggest the use of acoustic pressure equalizers when using omnidirectional microphones for rear and height channels, ensuring increased microphone directivity and channel separation at frequencies above 1 kHz, but maintaining an efficient capture of low frequency information (see King et al. 2016; Woszczyk 1990). Bowles (2015) on the other hand, suggests that to minimize direct sound in the height channels, hyper-cardioid microphones should be used, angled such that the nulls of the microphones are facing the soundstage. Hamasaki and Van Baelen (2015) describe a similar approach, suggesting upward facing hyper-cardioid microphones for height channel capture, placed very high above the 'main' microphones.

10.4.2 Near-Coincident Recording Techniques

Near-coincident 3D recording techniques use smaller spacing between microphone capsules: typically less then 1 m. The sound scene is captured using a combination of timing and level differences between microphone signals. Michael Williams has written extensively on his '3D Multiformat Microphone Array' (see Williams 2012, 2013, 2016), which is designed to prioritize localization of direct sounds in the horizontal and vertical plane, while minimizing interaction effects between the two loudspeaker layers (Williams 2012). Wittek and Theile have expanded their 'OCT Surround' (Optimized Cardioid Triangle Surround) technique for 3D audio, adding four upward facing hypercardioid microphones, placed 1 m above the main layer microphones (Theile and Wittek 2011). Main layer microphones use a mixture of cardioid (Centre, Rear Left, and Rear Right) and hypercardioid (Left and Right) polar patterns. Wittek and Theile also introduced '3D ORTF' [now termed: ORTF-3D] (see Wittek and Theile 2017) an ambience capture system for 3D audio and VR applications that, as the name implies, is comprised of four closely spaced ORTF pairs. For height channel microphones spaced less than 2 m above main layer microphones, Lee (2011) and Wallis and Lee (2017) both suggest the use of directional polar patterns to minimize vertical inter-channel crosstalk, set at angles of at least 90° or 105° respectively.

10.4.3 Coincident Recording Techniques

Most publications addressing coincident microphone techniques for three-dimensional acoustic music recording have focused on Ambisonics-based recording techniques (see Geluso 2012; Ryaboy 2015; Bates et al. 2017). A notable exception is Martin et al.'s single instrument capture arrays for 3D audio (Martin et al. 2016a) 'Double-XY', for example, combines a traditional XY cardioid pair with a 2nd, vertically oriented coincident cardioid pair. Though not designed to capture a complete sound scene, as there is no information captured for the rear channels, Martin et al. have shown these techniques create sonic images with well-defined horizontal *and* vertical extent, which is highly valuable for achieving realistic or hyper-realistic recreations of acoustic instruments (Martin et al. 2016b).

Geluso (2012) and Ryaboy (2015) both discuss a native B-format capture approach for acoustic music recording: 'Double MS + Z'. As the name implies, a vertically oriented coincident bi-directional microphone is added to the Double MS array. For ease of coincident spacing, both authors suggest the use of a Sennheiser MHK800 Twin microphone to capture both front and rear M components. As with coincident surround techniques, 1st order A-format capture systems such as a Soundfield microphone, or higher order spherical microphones such as the Eigenmike can also be used for coincident 3D sound capture (see Howie et al. 2017; Bates et al. 2017; Ikeda et al. 2016).

10.4.4 *Objective Measures for Multichannel Audio Evaluation*

Several authors have investigated objective measures for multichannel audio that may act as predictors of subjective listener evaluations for spatial sound attributes (see Power et al. 2014; George et al. 2010; Choisel and Wickelmaier 2006; Mason and Rumsey 2002). Interaural cross correlation (IACC) has been used in concert hall acoustics as an objective measurement for aspects of spatial impression, and is meant to quantify the dissimilarity of signals at the two ears (see Rossing 2007). Investigating the impact of 3D audio on 'envelopment', Power et al. found a strong negative correlation between mean listener envelopment scores for the various reproduction systems under investigation and measured IACC values for binaural dummy-head recordings made of the testing stimuli (Power et al. 2014). Choisel and Wickelmaier (2006) reported a strong negative correlation between $IACC_f$ and perceived 'spaciousness', comparing $IACC_f$ measurements of binaural recordings of the stimuli with listener evaluations. Masson and Rumsey (2002) found perceptually grouped IACC measurements (PGIACC) on experimental stimuli correlated highly with listener subjective data. George et al. (2010) showed that the measures 'Area of sound distribution' and 'extent of sound distribution' were successful in predicting listener scores for 'envelopment' in surround sound recordings. Both measures were designed to 'model the extent of sound distribution'." (from Howie et al. 2018).

More detailed information on signal correlation, IACC and spatial impression can also be found in Chap. 2 of this book.

10.4.5 The Recording Techniques Under Investigation

"… Four techniques were selected for investigation, drawn from the current literature on three-dimensional acoustic music recording. For this study, all techniques were optimized for reproduction using ITU 4 + 5 + 0 [(see ITU-R 2014)]. This 9-channel 3D audio standard calls for five loudspeakers at ear level at 0°, ± 30°, and ± 110°, and two pairs of elevated height channels at ± 30° and ± 110°. A "concert" perspective sound scene (i.e., direct sound in front, ambience above and surrounding) was maintained with all techniques.

Techniques 1 and 2 are both spaced techniques, based on designs described in Howie et al. (2016) and King et al. (2016). Microphone type and placement strategy for the Left, Centre, and Right channels are identical for both techniques: spaced omnidirectional microphones. Both techniques also utilize large spacing between rear and height channel microphones. This ensures a high degree of decorrelation between signals that contain primarily ambient information. Technique 1, a reduction of a recording method originally optimized for 22.2 (see Sect. 10.3), uses wide-cardioid microphones for the rear channels, and cardioid microphones for the height channels. This ensures a lack of direct sound information being captured by the ambience microphones, resulting in a more stable frontal sound image. Technique 2 uses omni-directional microphones for rear and height channels, all fitted with acoustic pressure equalizers.

Technique 3, OCT 9 [i.e. OCT-3D], is described in detail in Theile and Wittek (2011) as well as in Chap. 8 of this book. The technique is designed to prioritize clear directional imaging and a high degree of channel separation. Adequate decorrelation between microphone signals ensures that a 'natural spatial impression' is reproduced.

Technique 4, Geluso's 'Double MS + Z', is described in detail in Geluso (2012) and Ryaboy (2015). Like other native B-format capture systems, the microphone signals require matrixing or post-processing to achieve the correct decoding for a given reproduction system. This is in contrast to Techniques 1–3, all of which maintain a linear relationship between microphone signals and respective loudspeaker channels.

All four recording techniques under investigation were setup for simultaneous recording of a solo piano. The recording venue was the Music Multimedia Room (MMR) at McGill University, a large scoring stage measuring 24.4 m × 18.3 m × 17 m. At the time of recording, no acoustical treatment was installed in the room, resulting in an RT60 of approximately 4.5 s. For all techniques, microphone choice and placement were based on the recommendations of the technique's creators (for details see Table 10.3; Fig. 10.9).

Microphone placement for all techniques was optimized by a team of two professional recording engineers, both of whom had previous experience recording 3D audio. Techniques 1 and 2 shared the same microphones for the Left, Centre, and Right channels. The remaining channels for Techniques 1 and 2 used different microphones (Table 10.3) that shared the same placement and capsule angles (see Fig. 10.9).

Table 10.3 Microphones as used in the comparative recording of solo piano by Howie et al. (2018); (APE means acoustic pressure equalizer)

Recording channel	Microphone
Tech 1 + 2 L	Schoeps MK 2H w/APE
Tech 1 + 2 C	Schoeps MK 2H w/APE
Tech 1 + 2 R	Schoeps MK 2H w/APE
Tech 1 Rear L	Schoeps MK 21
Tech 1 Rear R	Schoeps MK 21
Tech 1 Top Front L	Schoeps MK 4
Tech 1 Top Front R	Schoeps MK 4
Tech 1 Top Rear L	Schoeps MK 4
Tech 1 Top Rear R	Schoeps MK 4
Tech 2 Rear L	DPA 4006 w/APE
Tech 2 Rear R	DPA 4006 w/APE
Tech 2 Top Front L	DPA 4006 w/APE
Tech 2 Top Front R	DPA 4006 w/APE
Tech 2 Top Rear L	DPA 4006 w/APE
Tech 2 Top Rear R	DPA 4006 w/APE
Tech 3 L	Schoeps MK 41
Tech 3 C	Schoeps MK 4
Tech 3 R	Schoeps MK 41
Tech 3 Rear L	Schoeps MK 4
Tech 3 Rear R	Schoeps MK 4
Tech 3 Top Front L	Sennheiser MKH 8050
Tech 3 Top Front R	Sennheiser MKH 8050
Tech 3 Top Rear L	Sennheiser MKH 50
Tech 3 Top Rear R	Sennheiser MKH 50
Tech 4 M	Sennheiser MKH 800 Twin
Tech 4 S (horizontal)	Sennheiser MKH 800 P48
Tech 4 S (vertical)	Sennheiser MKH 800 P48

Recordings were made to a Pyramix workstation at 96 kHz/24bit resolution, monitored in McGill University's Studio 22. The studio is equipped with 28 full-range, two-way loudspeakers (ME Geithain, model M-25) powered by Flying Mole class D amplifiers, and an Eclipse TD725SWMK2 stereo sub-woofer. The loudspeakers are arranged for reproduction of both 22.2 (9 + 10 + 3) and 4 + 5 + 0. The room's dimension ratios and reverb time fulfill ITU-R BS.1116 requirements. Matrixed 9.1 monitoring of Technique 4 was made possible using Pyramix's internal mixer, following the matrixing scheme in Ryaboy (2015).

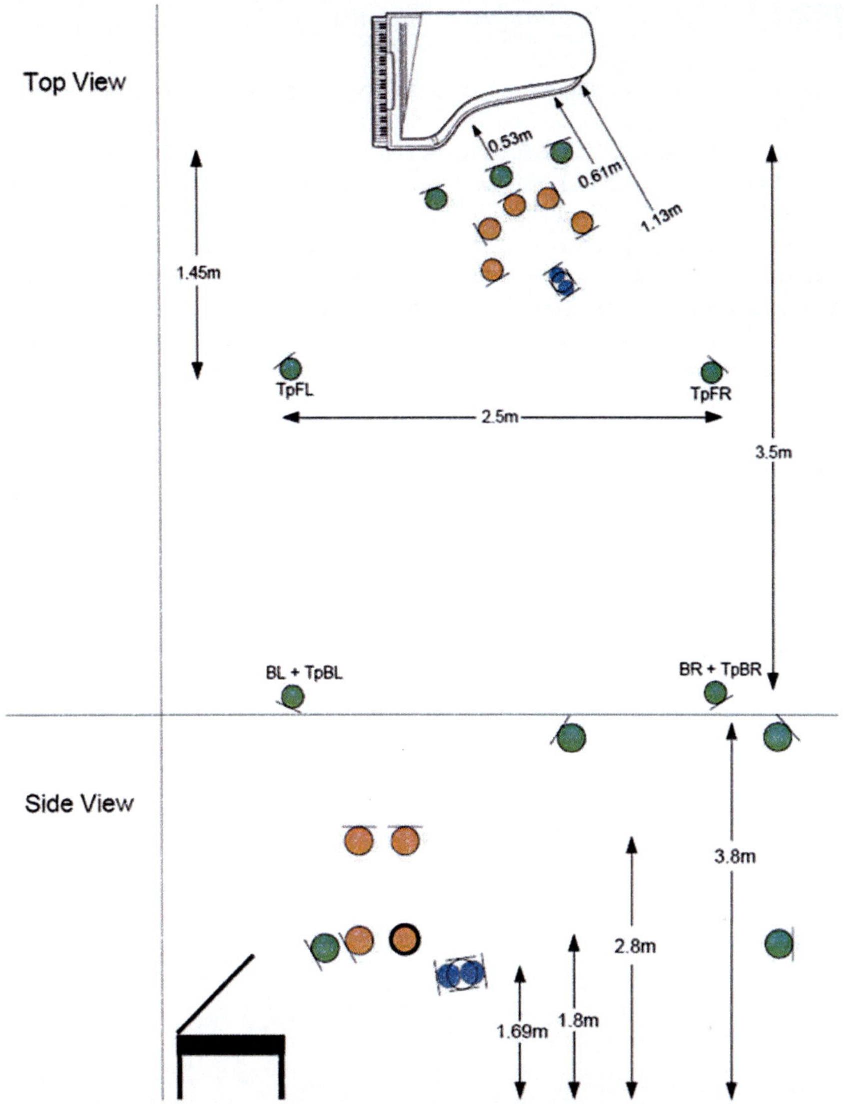

Fig. 10.9 Overhead and side view of all microphone techniques setup in MMR. Green = Tech 1 and 2, Yellow = Tech 3, Blue = Tech 4. Spacing between Tech 1 and 2 Left and Right microphones is ≈ 1.4 m. (graphic is Fig. 1 from Howie et al. 2018)

10.4.6 Subjective Evaluation of Stimuli

A double-blind listening test was designed to identify perceptual differences between the four recording techniques. Four perceptual attributes were chosen for investigation: 'envelopment', 'naturalness of sound scene', 'naturalness of timbre', and 'sound source image size'. 13 subjects participated in the listening test, all of whom were either students or faculty within the Graduate Program in Sound Recording at McGill University. All subjects had previous experience performing triad or pairwise comparison-style listening tests or ear training activities. Since the attribute ratings were relative and not absolute, each participant's data was normalized for mean and standard deviation. The purpose of this is to normalize each individual participant's use of the rating scale. Z-scores were computed for each participant, within each attribute. The attribute rating results averaged over all participants are visualized in Fig. 10.10.

In addition to the subjective listening comparison three sets of objective features were calculated on binaural dummy-head recordings of the stimuli. The first set of features were related to the inter-aural cross-correlation coefficient (IACC), the second set was derived from a binaural model designed to predict room acoustic attributes. The features of the second set have been shown to correlate with subjective assessments of 'reverberance', 'clarity', 'apparent source width' and 'listener envelopment'.

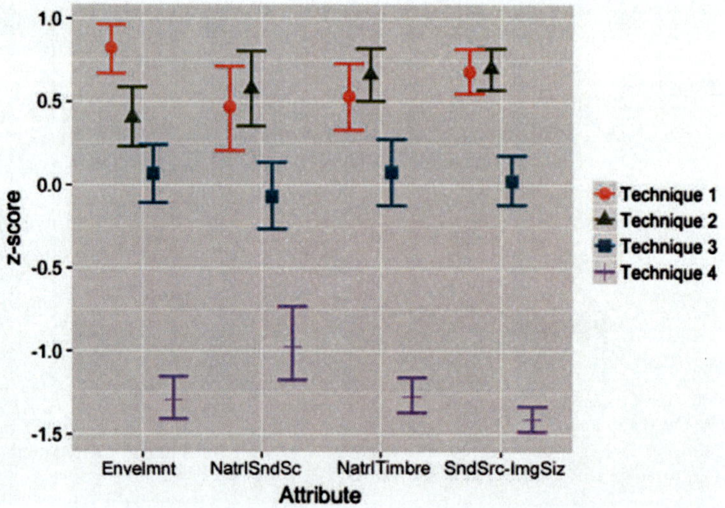

Fig. 10.10 Attribute ratings averaged across all participants for each recording technique. (attributes: envelopment, naturalness of sound scene, naturalness of timbre, sound source image size) (graphic has been adapted from Fig. 2 of Howie et al. 2018) [Tech. 1 = spaced omnis (wide-card.) after Howie et al. (2016) and King et al. (2016) Tech 2 = OCT-3D after Theile and Wittek (2011); Tech 3 = Double MS + Z after Geluso (2012) and Ryaboy (2015)]

The third set of features were designed to characterize signals' monaural frequency content. These features were the Spectral Centroid, Spectral Crest factor, Spectral Flatness, Spectral Kurtosis, Spectral Skew, Spectral Spread, and Spectral Variation. All were calculated using the open source 'Timbre Toolbox' (see Peeters et al. 2011).

10.4.7 Comparing Subjective Attribute Ratings with Objective Signal Features

A linear regression analysis was conducted to investigate relationships between the subjective attribute ratings discussed in the previous section and the objective signal features. Many features were found to correlate strongly with 'sound source image size'. Two of these features, $IACC_f$, and Spectral Variation, were also predictive of 'envelopment'. The strong relationship between $IACC_f$ and both 'envelopment' and 'sound source image size' was unsurprising as this feature was explicitly designed to predict spatial attributes of binaural signals" (from Howie et al. 2018).

10.4.8 Conclusion

This research stands out in that it is not only based on the commonly found subjective listening tests, but in addition three sets of objective features were calculated on binaural dummy-head recordings of the stimuli. These objective features were related to the inter-aural cross-correlation coefficient (IACC) of these binaural recordings, a binaural model designed to predict room acoustic features and a third set of features, designed to characterize signals' monaural frequency content. "A study was undertaken to compare four different 9-channel three-dimensional acoustic music recording techniques, all optimized for capturing a solo piano. The four techniques range in design philosophy: spaced, near-coincident, and coincident. Results of a subjective listening test showed the two spaced techniques as being equally highly rated for the subjective attributes 'naturalness of sound scene', 'naturalness of timbre', and 'sound source image size'." Technique 3 (OCT-3D) was rated lower than both spaced techniques, but higher than technique 4 (Double-MS + Z). "Listeners rated the coincident technique significantly lower than all other techniques under investigation for all perceptual attributes. Binaural recordings of the stimuli were analyzed using several different objective measures, some of which were found to be good predictors for the perceptual attributes 'envelopment' and 'sound source image size'." (from Howie et al. 2018).

10.5 Comparative Recording with Several 3D Mic-Arrays at Abbey Road Studios

Another case study shall be presented, in which a very interesting 3D mic-array comparison in the domain of pop-music has taken place at Abbey Road Studios London, led by Hashim Riaz and Mirek Stiles (see Riaz et al. 2017). The purpose of the experiment was to address the recording aspects and needs of VR (Virtual Reality) music production as well as to document a recording session that enables future subjective comparison of applicable recording methods: "... Recording of a London-based Indie-Pop ensemble called 'Nova Neon' was undertaken at Studio 3 of Abbey Road Studios. The band is a five-piece who play a sophisticated style of Indie-Pop using a mix of both synthetic sounds and acoustic instruments. Studio 3 comfortably accommodates a five-piece ensemble and boasts a high ceiling, allowing for diffuse soundfield capture using microphone arrays as shown in Fig. 10.11.

A GoPro Omni camera rig was set up in the middle of the room (Pos. A) with the musicians arranged in a pentagon around the camera as shown in Fig. 10.12. The drummer, bassist and vocalist were placed to the 'front', whilst the two guitars were set to the far left and right just behind the camera. The rationale for this arrangement was to over-exaggerate the positioning of sound sources over 360° creating an increased sense of envelopment. The bass guitar was recorded in isolation to avoid excessive spill in the lower bass frequencies.

Fig. 10.11 Overview of recording setup showing 3D mic-arrays at Studio 3, Abbey Road (Stiles 2018)

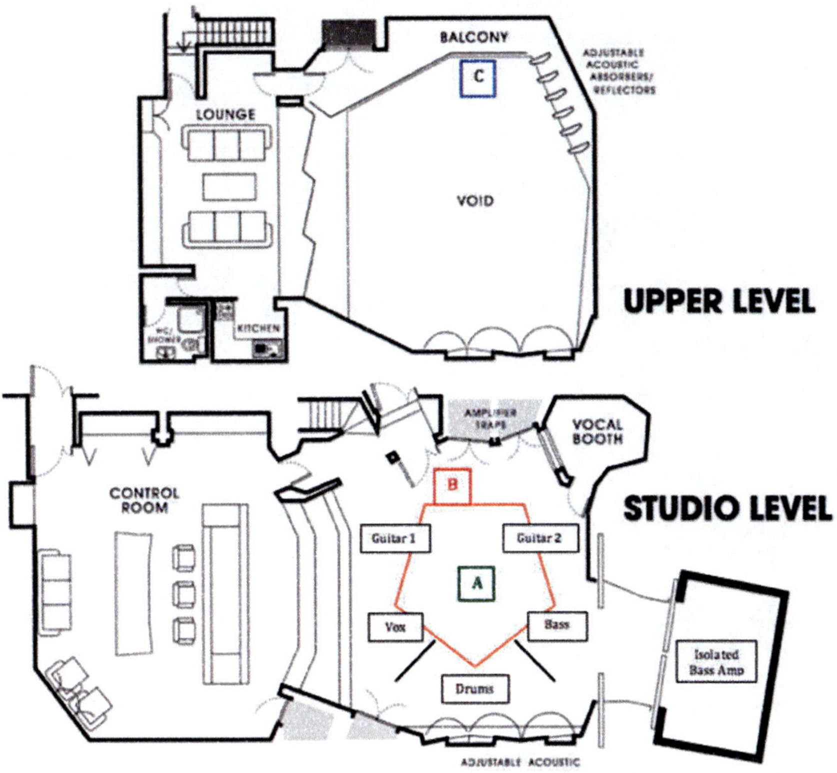

Fig. 10.12 Schematic view of recording setup and 3D-Audio microphone arrays of positions A, B, C at Studio 3, Abbey Road (from Stiles 2018)

10.5.1 Microphone Setup

The spot microphones were positioned on each instrument by Abbey Road engineers using their usual recording techniques and workflow. The bands backline consisted of a Spaun TL series drum kit with Paiste cymbals, an Ampeg Portaflex Bass Amp, an Orange Rockaverb 50 combo electric guitar amp and a Fender Blues Junior guitar amp. Further technical details on the band setup and spot microphones used can be found at http://www.sadie-project.co.uk\AbbeyRoadRecordings.html. The multi-channel microphone arrays found at Positions A, B, and C in Fig. 10.12 are now described (Figs. 10.13 and 10.14):

Microphone Arrays at Position A

Position A is located in the centre of recording space and in the middle of the musicians. Alongside the Omni rig, the following arrays were set up at this position:

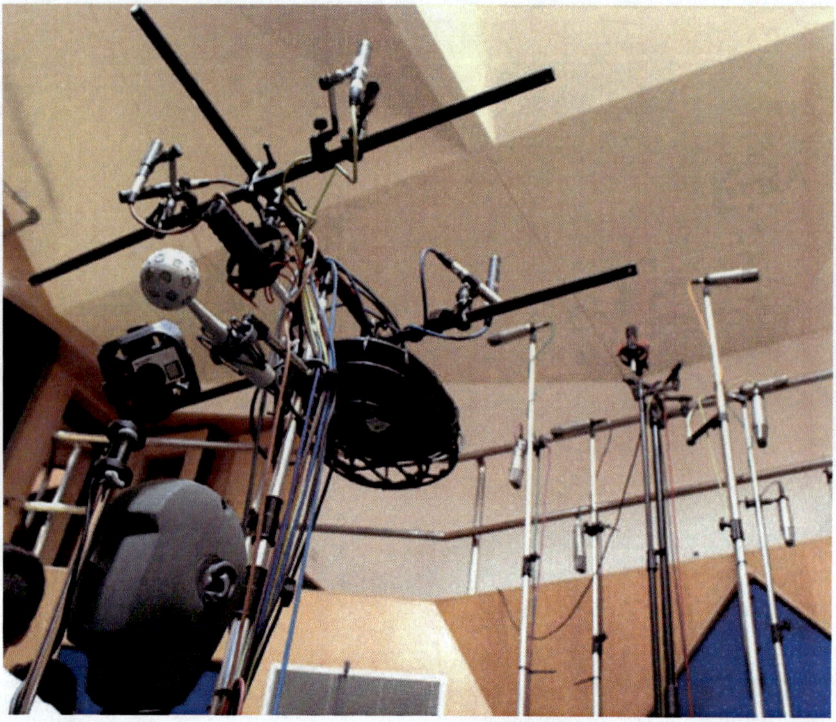

Fig. 10.13 Arrangement of several 3D audio microphone systems (and a Neumann KU100 artificial human head) as used in the comparative recording at Abbey Road Studios in London (from Riaz et al. 2017): front (bottom to top): Neumann KU100, mh acoustics M32 'Eigenmike'®, SoundField, ST450, ORTF-3D (in basket), ESMA; rear (bottom to top): Hamasaki-Cube, SoundField ST450 View 1 of Position A and C microphone arrays. The front of the arrays face the drum kit (Reproduced with permission of Abbey Road Studios and the University of York)

- **Neumann KU100:** The Neumann KU100 dummy head is a binaural microphone. It was positioned facing towards the reference point (drum kit).
- **MH Acoustics EM32 Eigenmike:** This is a 4th order Ambisonic microphone comprising of 32 omnidirectional electret capsules arranged in a pentakis dodecahedron. The Eigenmike was placed endfire just above the GoPro Omni rig at a height of 1.81 m.
- **Soundfield ST450 MKII:** This microphone uses four sub-cardioid capsules in a coincident tetrahedral arrangement. The outputs of the capsules (A-Format) are converted into 1st order Ambisonics (B-Format) (Gerzon 1973) signals using SoundField's own processor. The microphone was placed at a height of 2 m in end-fire to allow for space for the Omni rig and Eigenmike.
- **Equal Segment Microphone Array (ESMA):** Based on the 'four segment array' proposed by Williams (1991, 2003), the ESMA captures sound from 360 degree in the azimuth plane using four cardioid microphones positioned in a square.

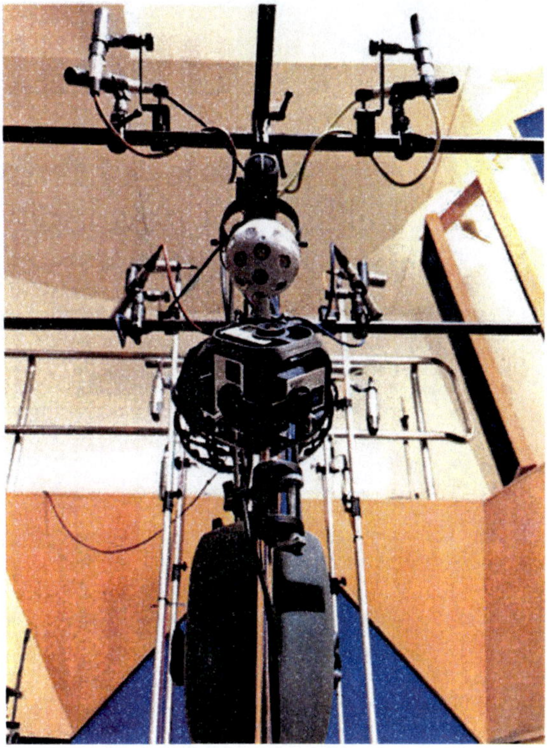

Fig. 10.14 View 2 of Position A and C microphone arrays. The front of the arrays face the drum kit (from Stiles 2018)

Changing the distance between each capsule from 25 to 50 cm has been shown to improve localisation accuracy (Lee 2017). Each of the four microphones were angled down by 45° to optimise direct sound capture of the instruments. The array height was 2.15 m. An additional four upward-facing cardioid microphones (with 50 cm separation) were added to capture diffuse sound within the recording environment. This vertical coincident configuration is based on experimental results reported in Lee and Gribben (2014).

- **ORTF-3D Surround:** This is a near-coincident microphone array based on the ORTF technique and adapted to capture sound from all directions. The bottom plane consists of 4 supercardioid microphones positioned in a 10 cm × 20 cm rectangle. The angles between the left and right capsules are 80°, and 100° between the front and rear capsules. The upper plane consists of four upwards facing supercardioid microphones to create coincident X–Y pairs at 90° between the upper and lower planes (Schoeps Mikrofone 2017a).

Microphone Arrays at Position B

Position B is located to the rear of Studio 3. The objective for this position was to capture a 180° view of the musicians and provide a different perspective for the VR

Fig. 10.15 Position B microphone arrays (from Riaz et al., and Kearney 2017)

experience with a Samsung Gear 360 camera. The following arrays were placed at this position (see Fig. 10.15):

- **Stereo X–Y Pair:** Two Neumann KM184 cardioid microphones were used in a coincident X–Y arrangement to produce a stereo recording angle of 115°. The stereo pair was positioned to face the drum kit at a height of 1.94 m to the coincident point.
- **Sennheiser AMBEO:** This is a B-format Ambisonic microphone that uses a tetrahedral arrangement of capsules (Sennheiser 2017). The AMBEO was placed just above the Samsung Gear 360 camera at a height of 1.59 m to the centre of the microphone grill.
- **OCT-9 Surround:** OCT-9 [now termed OCT-3D] was designed for the Auro 9.1 loudspeaker arrangement. The base section is the optimised cardioid triangle surround (OCT-Surround) array, designed to capture sound for 5.1 systems (see Theile and Wittek 2011; Theile 2001). Four upward-facing supercardioid microphones are added to create the OCT-9 array. The front microphones capture direct sound whilst the rear and upward-facing microphones capture the diffuse sound and ambience of the environment.
- **Perspective Control Microphone Array (PCMA):** The PCMA arrangement is similar to OCT-Surround but five widely spaced coincident pairs are used in place of single microphones. Each pair uses cardioid microphones creating virtual microphones through different mixing ratios (Lee 2012). This allows perceived source distance and source width to be changed post-recording (Lee 2012). Due

to space and equipment limitations the PCMA set-up for the recording session used the rear facing and upward facing microphones from the OCT-9 Surround. Coincident stereo pairs were set-up for only the front left and front right positions. The PCMA was set-up at a height of 1.82 m to the microphone capsules.

Microphone Arrays at Position C

Position C was located behind Position A at the rear of Studio 3. The Position C arrays were placed much higher in the room in order to capture more of the diffuse field and a greater sense of the recording space. The following arrays were used:

- **IRT Cross:** This array is generally used to capture reflected diffuse sound and direct environmental sounds such as applause and crowd noise. It can be created using four cardioid microphones positioned in a square with 20 cm - 25 cm spacing between each capsule (Schoeps Mikrofone 2017b). The IRT Cross can also be used in combination with other arrays to capture both direct and diffuse sound and provide a full spatial representation of a musical performance within an environment. The IRT Cross was positioned at a height of 3.5 m.
- **Hamasaki Cube**: Derived from the Hamasaki square, this array is designed to capture reverberation and diffuse sound. It consists of eight bi-directional microphones positioned in a cube arrangement with equal spacing of 1 m between each microphone. The nulls of each microphone are pointed in the direction of direct sound (Theile and Wittek 2011). Due to space limitations, the dimension of the Hamasaki Cube was reduced from 1 m to 70 cm. Eight Neumann U87 condenser microphones were used to create the Hamasaki Cube, which was positioned at a height of 3 m to the microphone capsules on the lower layer of the cube (Figs. 10.16 and 10.17).

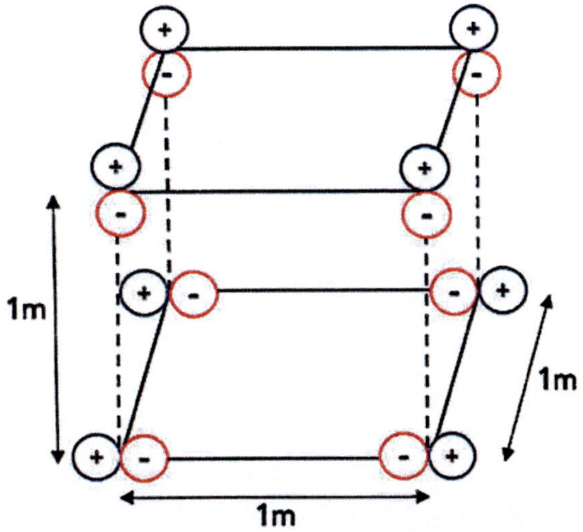

Fig. 10.16 Hamaski Cube schematic (from Stiles 2018)

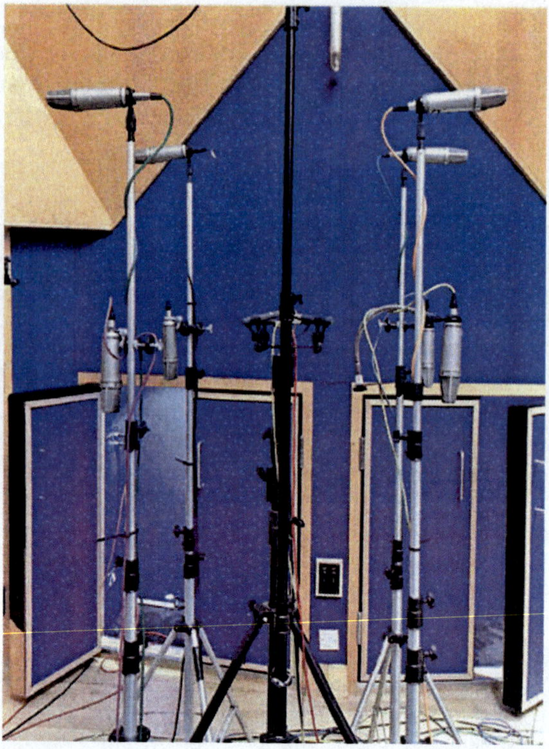

Fig. 10.17 Hamaski Cube, using Neumann large diaphragm microphones, setup at Abbey Road, Studio 3 (from Stiles 2018)

- **Soundfield ST450 MKII**: This was positioned between the front microphones of the IRT Cross at a height of 3.45 m. This allowed for an accurate soundfield recording higher up in the recording space to capture room ambience.

10.5.2 The Recording Process

With the exception of the Eigenmike, each microphone was plugged into tie lines in the studio and routed to a channel on the SSL 9000 J 96-channel mixing console. The spot microphones used the pre-amps on the mixing console as did each channel of the Hamasaki Cube. Other microphones utilised stepped preamplification from thirty six AMS Neve Montserrat pre-amplifier channels and twelve AMS Neve 1081 pre-amplifier channels to ensure matched gains on each channel. The session was recorded on a ProTools HD rig at a sampling rate of 48 kHz and with a bit-depth of 24 bits. This was the practical option as there were in excess of ninety channels being recorded in the session and the file sizes had to be taken into consideration. The Eigenmike was recorded onto a separate system synchronised to the Pro Tools HD session.

A 5.1 surround system comprising of five Bowers and Wilkins 800D speakers was utilised. Where possible arrays were 'folded down' to the 5.1 system for monitoring. Spot microphones were monitored in stereo as it would be in a standard recording session. Timbral qualities such as the brightness of the microphones could be identified whilst monitoring providing an early indication to how the different microphone arrays might perform. An 8-channel auxiliary feed was sent to a separate binaural monitoring system that utilised Reaper and the AmbiX plugins (Kronlachner 2014). Including the Eigenmike, 122 channels were recorded simultaneously in the session for each take. A session of this size required meticulous planning and attention to detail to ensure successful set-up, documentation, routing and recording. For each take the cameras needed to be checked, set to record and take-slated, the band given feedback and each input channel monitored to make sure record levels were optimal. Only as a team is it possible to undertake such an intensive audio and visual recording session in one day. In terms of practicality, it was difficult to set-up and position each microphone within the space available. Like the Ambisonic microphones, the ORTF-3D array was practical due to its size and housing, requiring minimal effort to set-up and position. This would be extremely useful in a field recording scenario or whenever video capture is involved and the recording equipment needs to be discrete. In contrast, the Hamasaki Cube had to be assembled using individual microphone stands for each microphone. This set-up would not be ideal for a performance space where floor space is limited and the microphones need to be discrete. Alternatively it would be better to rig some of the arrays from the ceiling or lighting rig, or to utilize specially constructed microphone trees, as was the case with the OCT-9. Another practical consideration is the battery life of the 360° cameras. Both the GoPro Omni and Samsung Gear 360 cameras do not allow the use of mains power whilst recording. This meant that the session was paused to allow for the cameras to be recharged.

10.5.3 Binaural Processing in Reaper

Co-ordination from the console aux feeds to the Reaper system meant that arrays could be separately auditioned in binaural-based Ambisonics over headphones. Ambisonic encoding and decoding of the spot mics and arrays was achieved using the AmbiX plug-ins (Kronlachner 2014). A 36-channel mix-bus was created for 5th order processing. A 50-point Lebedev Grid binaural decoder using MaxRe weighting from the SADIE database (Subject 002—A KU100 dummy head), was utilised (Kearney 2017). The Soundfield recordings and a B-Format reverb were sent to a separate 1st-order binaural decoder comprised of a cube using MaxRe weighting. Similarly a 3rd-order binaural decoder was created for the Eigenmike using a 26-point Lebedev grid with MaxRe weighting. The 360 video was previewed using Kolor's GoPro VR Player (Kolor.com 2017). SpookSyncVR software (spook.fm 2017) was employed to synchronise the video to the audio in Reaper. SpookSyncVR is built in Max MSP as a stand-alone application that allows data exchange between GoPro VR Player and Reaper using Open Sound Control (OSC).

By designating the GoPro VR Player as the 'master' and Reaper as the 'slave' it was possible to synchronise, play and stop the audio and video together. Using SpookSyncVR it is possible to gather the X (yaw) and Y (pitch) positional data from a headset such as an Oculus Rift and transfer the information to Reaper so that the soundfield can be rotated accordingly to create an enhanced listening experience. To achieve this, an AmbiX Rotator plug-in was inserted onto each of the three decoder tracks just before the AmbiX Binaural Decoder plug-in. A 5th-order Rotator was used for the HOA decoder track, a 3rd-order Rotator for the Eigenmike decoder and a final 1st-order Rotator was inserted onto the FOA decoder. The incoming pitch and yaw data from GoPro VR Player could then be assigned to the pitch and yaw values for each of the rotator plug-ins using SpookSyncVR's 'learn' mode.

10.5.4 Summary and Informal Evaluation

After listening to the recordings from each of the multichannel microphone arrays and Ambisonic microphones it was possible to predict how each recording technique might perform in future formal listening tests. Listening to just the spot microphones did not yield a strong sense of the recording space due to a lack of room ambience being captured as the microphones were placed so close to each source. However, localisation with just spot microphones is very good, albeit with compromised distance perception. The combination of spot microphones and individual position A arrays works well to aid localisation of individual sound sources and capture more of the room's ambience inducing a greater sense of the recording space. The arrays in Position C at the back of the recording space seem to further improve the perceived sense of the recording space at the cost of localisation accuracy. Future work will report on a subjective evaluation of combinations of the different recording techniques undertaken. The reader is directed to http://www.sadie-project.co.uk\AbbeyRoadRecordings.html to listen to example recordings from the different arrays and for further technical details on the session." (from Riaz et al. 2017).

In Stiles (2018) for some of the 3D-microphone techniques an individual evaluation of their sonic characteristics and operational practice is given:

Ambisonics

"Worth noting is that Ambisonics has two different channel order standards: Furse-Malham (FuMa) and Ambix. Without going into detail, my advice is to stick to Ambix whenever you can and always keep track of conversion workflow. I tend to find placing Ambisonic microphones too far or high in a room doesn't really work to great effect, but in the mid and near field, once decoded, they can sound stunning. For example, I recorded a solo acoustic guitar/vocal performance with the usual close mics, and a nice stereo pair to pick up some room. In between the close and distant mics, I placed an FOA microphone. Once decoded and blended into the final mix it created a beautiful sense of depth and space I would have found challenging—if not impossible—to create using standard mics. The same can be said for close miking:

for example, a FOA microphone on a harp or piano sounds stunning. The Ambisonic microphone has an ability to pick up a large spread of detail from all around in every direction. A midfield Ambisonic microphone can act as a sort of directional glue between closer mics and ambient microphones.

Multi-channel Microphone Arrays (MMAs)

The other side of the coin is spatial multi-channel microphone arrays (MMAs), which I think pick up much more useful information for reflections and room than ambisonic microphones. The session in question was a joint research collaboration between Abbey Road Studios and the University of York, conducted in Studio Three with then student, now Abbey Road employee, Hashim Riaz. The concept was capturing a live band for an immersive virtual reality experience. On this session, we had all sorts of spatial (MMA) arrays out on the floor (around 80 channels to be precise). There are a couple of favourites I would like to mention: the Hamasaki Cube and Equal Segment Microphone Array.

Hamasaki-Cube

The Hamasaki Square was designed to capture the ambient and diffused sounds of a concert hall with surround sound in mind. As we are now in the world of 3D sound, we add another layer to capture some height—hence the Hamasaki Cube. The cube consists of eight figure-of-8 microphones placed 1 m apart from each other. For the bottom layer, the null of the figure-of-8 pattern is pointing towards the sound source, so we can reduce the direct sound and increase the diffuse sound captured by the array. The top layer has the null pointing towards the sides, so we capture the diffused sound from above and below. The overall array should be placed as far and high from sound source as to capture some exciting room reflections.

Equal Segment Microphone Array (ESMA)

The Equal Segment Microphone Array (or ESMA as I like to call it) is another 8-channel configuration, this time consisting of cardioid pattern mics. The idea is to capture sound in a 360-degree azimuth plane positioned in a square arrangement (angle of 90° between each mic) with 50 cm between each capsule to minimise inter-channel cross talk. This is then augmented by another four upward facing cardioids to capture some height information. When the eight channels of each array are subsequently processed via an object or ambisonic panner, the result can be extremely lush, to my ears. Once blended in with close and ambisonic microphones they really help create an overall spatial audio scene." (from Stiles 2018) (Rem.: More info can be found on: abbeyroad.com/spatial-audio).

10.6 '3D-MARCo'—3D Microphone Array Recording Comparison (Lee and Johnson)

The comparative multichannel 3D audio microphone review (Lee 2021), as well as signal analysis (Lee and Johnson 2021), are currently among the most extensive ones, using a total of 71 microphones simultaneously, from which 17 different 3D-audio microphone setups can be derived. Conducted by Prof. Hyunkook Lee, head of the APL—Applied Psychoacoustics Lab at Huddersfield University, the resulting analysis published in Lee and Johnson (2021) concerns mainly the inter-channel signal-correlations between the respective microphones. As sound source for the correlation measurements, a loudspeaker array, slightly above floor level and aimed at the microphone arrays, was used in the recording venue, St. Paul's church with an average reverb time of 2.1 s.

In Fig. 10.18 some of the 3D-audio microphone arrays, which have been used in the MARCo ("3D Microphone Array Recording Comparison") recording are displayed, including their physical position and capsule directivity. Figures 10.19, 10.20 and 10.21 show the placement of all the microphone array systems used in relation to the recording venue, as well as the sound sources (loudspeaker or musicians).

The comparatively dry sound of the Ambisonic recording principle, as mentioned in previous chapters, can also be recognized in Fig. 10.22, in which the Direct-to-Reverberant Ratio (DRR) is documented for all microphone channels for the objective measurements which used a loudspeaker at + 45° left of the microphone

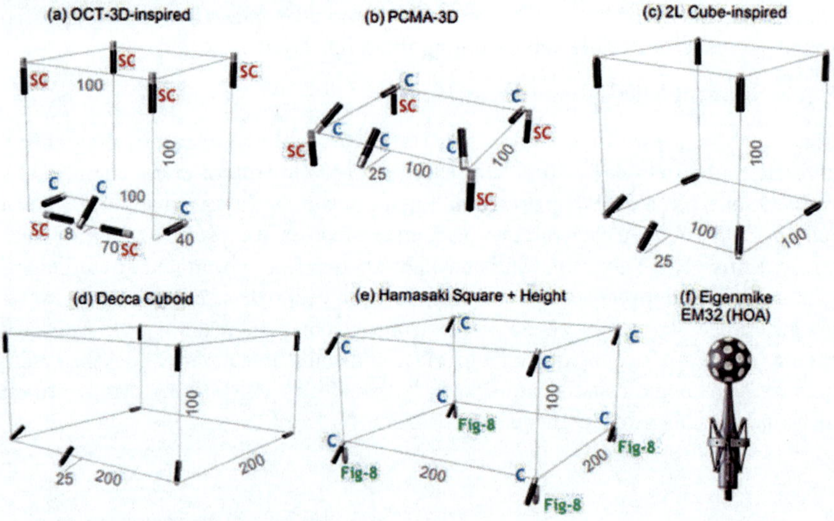

Fig. 10.18 Some of the 3D-Audio mic arrays as used in the MARCo recording. All microphones except for the Eigenmike® and the Hamasaki-Square (Schoeps CCM8) were from the DPA d:dicate series. Capsules were omnis, except if noted otherwise (C = cardioid, SC = super-cardioid, Fig-8 = Figure of eight) (Fig. 10.18 has been adapted from Fig. 1 from Lee and Johnson 2021)

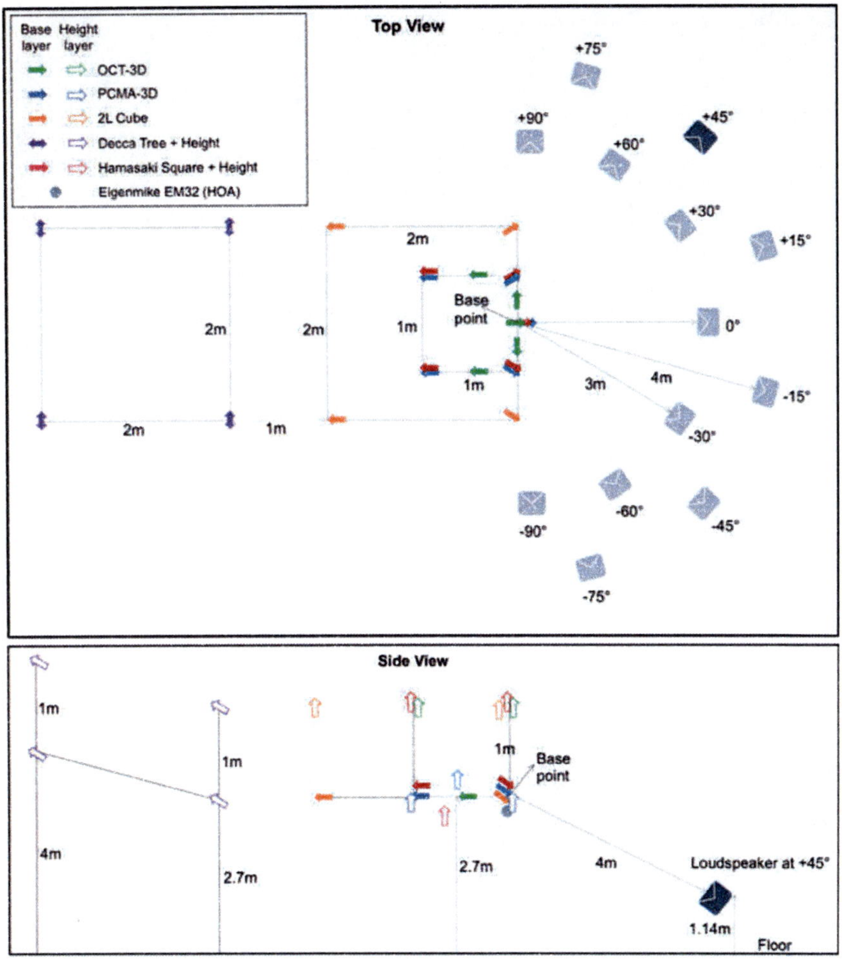

Fig. 10.19 Physical layout of the microphones and loudspeakers used for capturing the multichannel room impulse responses (MRIRs) in 3D-MARCo. For the objective measurements, presented in Fig. 10.22, the MRIRs for the source-speaker at + 45° were used. Recording location: St. Paul's church (graphic equivalent to Fig. 2 from Lee and Johnson 2021)

arrays main axis' as sound source: while the DRR is usually in favor of the diffuse sound component for all microphones that are facing away from the sound source, this is not necessarily the case for the 1st order and 4th order Eigenmike® signals. The fact that the signals of both decoded 'Left height' microphone channels (for Front and Rear; i.e. FLh and RLh) have positive values in the favor of direct sound components clearly shows that the various decodings of the Eigenmike EM32 (1st order vs. 4th order) result in a much lower directivity than for the other microphone techniques which used dedicated single microphone patterns. Hence it should be no

Fig. 10.20 3D-multichannel microphone setup with musicians as part of the 3D-MARCo experiment at St. Paul's church, University of Huddersfield (from Lee 2019)

Fig. 10.21 Detail view of 3D-multichannel microphone setup with musicians as part of the 3D-MARCo experiment (from Lee 2019)

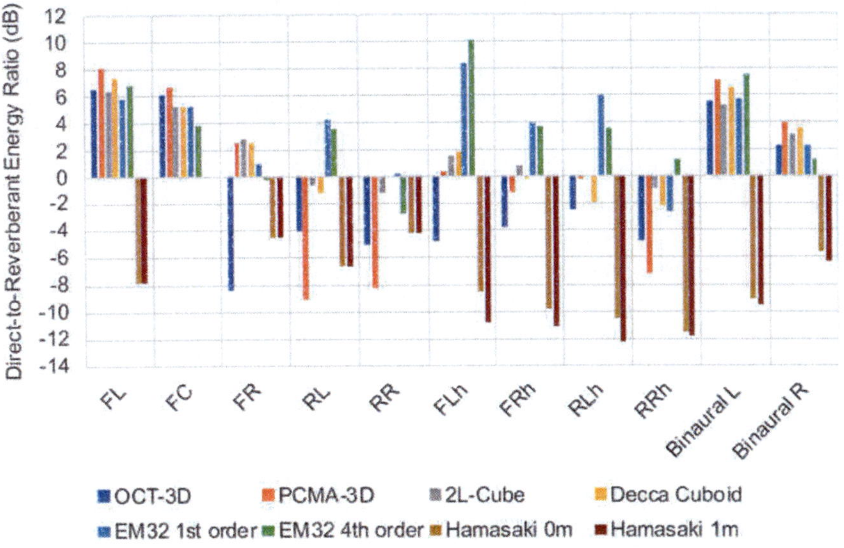

Fig. 10.22 Direct-to-reverberant ratio (DRR) for each microphone and ear-input signal (Fig. 9 from Lee and Johnson 2021)

surprise if localization resolution and—for sure—spatial impression proved worse for the EM32 1st and 4th order recordings—in relation to the other techniques—in subjective listening tests.

The overall significantly higher signal correlation in both EM32 Ambisonics recordings shows very clearly also in the results of the inter aural cross correlation measurements, which have been derived from a binaurally synthesized 9-channel 3D loudspeaker system and are displayed in Fig. 10.23. While the IACCs for the various 3D microphone arrays usually stay below a value of 0.4 for the base layer, as well as the height layer (with the exception of the base-layer signals the PCMA-3D and OCT-3D techniques), for the EM32 1st order and EM32 4th order these values are usually above 0.6, i.e. highly correlated. In broad terms: the Ambisonic techniques sound more 'narrow' (i.e. less wide ASW—Apparent Source Width) in comparison to the other techniques.

A very interesting and up-to-date comparison between several first-order Ambisonics, as well as Higher-Order Ambisonics commercially available microphones including the Sennheiser 'Ambeo,'Core Sound 'TetraMic'; SoundField 'MK V,' MHAcoustics 'Eigenmike®' (with 32 transducer elements) and Zoom 'H2n' has been made by Enda Bates and his colleagues about which they report in Bates et al. (2016, 2017). As it turns out, it seems dubious whether HOA is really able to provide overall superior sonic results in comparison to FOA, as the increase of the number of capsules also seems to have its trade-offs: In Bates et al. (2017) it is reported that "… As might be expected, the more elaborate designs of the Eigenmike® produced

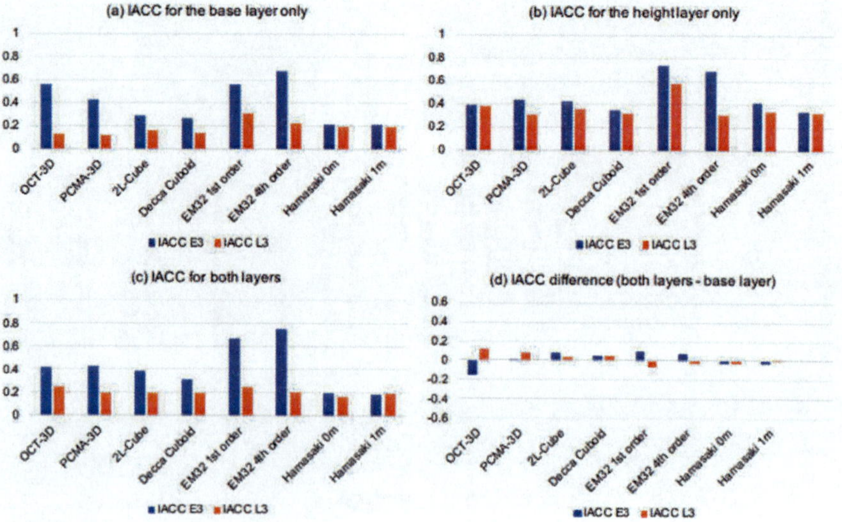

Fig. 10.23 Interaural cross-correlation coefficients (IACCs) for ear-input signals resulting from different microphone signals reproduced from a binaurally synthesized 9-channel 3D loudspeaker system (Fig. 8 from Lee and Johnson 2021)

the best results overall in terms of directional accuracy, but with decreased performance in terms of timbre. As with part 1 of this study, this again suggests a trade-off between timbre quality and directionality, particularly when the number of individual capsules in the microphone is significantly increased." Therefore, it is interesting to note that the FOA SoundField MKV microphone, despite being the oldest version of the Ambisonic microphones tested in Bates et al. (2016, 2017) still seems to be—'overall quality' wise—at par with the HOA 'Eigenmike'® with the latter being superior in terms of localization accuracy, but less favorable in terms of 'timbre' or sound-color …

The results from the objective measurements on 3D microphone arrays by Lee and Johnson (2021) also indicate that going to a higher order in Ambisonics does not necessary resolve all problems "…this might suggest that, in the current 9-channel loudspeaker reproduction, the well-known limitation of Ambisonic loudspeaker reproduction regarding phasiness during head movement would still exist even at higher order" (Lee and Johnson 2021).

Later-on other recordings were undertaken in the same venue, using real acoustic instruments in the form of a small ensemble (see Fig. 10.20 and 10.21), as well as single instruments. However, at the time (i.e. summer 2022) no formal subjective listener evaluation has taken place, but Prof. Lee has generously made the recordings freely available for research purposes, downloadable from this source: https://zenodo.org/record/3477602#.YgUh-6btxjs.

The main microphone arrays consisted of PCMA-3D (a layout based on proposals found in Lee and Gribben (2014), which is horizontally spaced and vertically coincident), OCT-3D (based on Theile and Wittek 2012), 2L-Cube (after Lindberg 2014), Decca Cuboid, First-order Ambisonics (FOA—Sennheiser Ambeo), Higher-order Ambisonics (HOA—Eigenmike EM32) and Hamasaki Square with height. In addition, ORTF, side/height, 'Voice of God' and floor channel microphones as well as a dummy head and spot microphones were included. The sound sources recorded were string quartet, piano trio, piano solo, organ, acappella group, various single sources and room impulse responses of a virtual ensemble with 13 source positions captured by all of the microphones.

10.7 Informal 3D Microphone Array Comparison (Gericke and Mielke)

A comparative 3D microphone array test in which some of the same systems were used as in the experiment documented in Lee and Johnson (2021) was realized by Harald Gericke and Olaf Mielke in February 2020, which they report about in Gericke and Mielke (2022). Among the microphone systems were 2L-Cube, OCT-3D, ORTF-3D, PCMA-3D and Sennheiser 'Ambeo' (i.e. Ambisonics 1st order).

The microphones were tested on the occasion of an orchestra recording at the opera house of Rostock in Germany, as well as with a chamber music (quintet) recording in a studio in Hannover.

Following is an English translation of the informal impressions which these two seasoned sound-engineers with several decades of experience in classical music recording have published (Gericke and Mielke 2022): "... The first impression when switching through the various microphone setups was that the 2L-Cube has the 'most beautiful' basic sound, a large room with pronounced depth and width and a very natural sound quality. Compared to this, the directional systems sound a bit nasal. Of course, they have significantly less low-frequency content and the room appears a bit cramped and sometimes a bit 'compressed'. As the omni-surround microphones (LS/RS) in the 2L-Cube are significantly louder than, for example, the rear-facing cardioids of the PCMA-3D—in the course of the listening tests—we decided to lower their level a little." [Rem.: in this context see also Sect. 1.10 on 'Distribution of direct- and diffuse-sound reproduction in multichannel loudspeaker systems', as well as Sect. 1.8 with reference to the findings by Lee (2011), Wallis and Lee (2016, 2017) according to which—in order to avoid unwanted upwards shifting of source image in 3D reproduction—a direct sound captured or reproduced from a height channel (i.e. vertical inter-channel crosstalk) should be at least 7 dB attenuated compared to the same sound captured or reproduced from the main channel] "... Localization appears somewhat more precise with the directional systems, individual instruments are more clearly defined, but the full orchestra has less depth and breadth and appears somewhat flat. The extent to which reflections from the hall ceiling have an effect, can unfortunately only be guessed at.

What is interesting, when comparing the various height microphones, is that the four omnis (independent of whether with or without EQ, which was set to reduce the bass, which was too strong) appear to create a large, homogeneous but somewhat 'central' sound plane above the listening position. On the other hand, the four super-cardioids of the OCT seem to leave a hole in the top center, while one believes to locate the position of the four height speakers quite accurately. In addition, the height signal of the super-cardioids significantly jumps towards the rear, compared to the four omnis. However, a further comparison also shows that with the omnis as height microphones, one has to expect comb filter effects and localization fuzziness. The super-cardioids capture significantly less direct signal and thus have significantly more indirect sound components or more spatial sound. They sound significantly brighter than the omnis (which also have a low-cut). We also have the impression that—with loud source signals—the super-cardioids generate a significantly higher loudness than the omnis, which means that there is a risk that the localization can shift upwards with volume. We did not observe this effect with the omnis. Compared to the omnis, the super-cardioids 'look further' into the distance. The super-cardioids of the PCMA-3D on the lower layer obviously capture significantly more direct signal, which is more colored as a result. Therefore we prefer height mics that are at physically at a higher level than the L-C-R-LS-RS setup. One should also experiment with the direction of the microphones. If the super-cardioids were not pointed directly to the ceiling, but more towards the corners of the room, they might capture fewer annoying reflections. [Rem.: in this respect see the proposal of the AB-BPT-3D 9.1 microphone setup, Sect. 9.6].

With the directional microphone setups, it is noticeable that one has to reckon with sonic distortions, probably due to captured reflections of the room, more so than with the omnis, which seem less susceptible to this problem. For example, in the super-cardioids of the upper layer, we perceive very strong reflections of the trumpets and the triangle in some loud parts, which results in that the location of the trumpets in the loud parts moves significantly upwards. The super-cardioids are also significantly brighter and more pointed during applause and cause that the applause moves up, compared to the omnis.

In the listening comparison of the various directional setups, the Ambeo microphone falls off sharply in our music recordings, both in terms of sound and spatial imaging. Apparently it is designed for other use cases.

By using eight super-cardioids, the ORTF-3D produces a very narrow, strongly nasal, distorted sound and offers only a small width and depth in spatial terms, but allows precise localization of individual sources. The OCT-3D has a more pronounced center localization than the ORTF-3D, also quite strongly colored by the outward-facing super-cardioids on the lower layer, but depicts a slightly larger area with a clear mono center than the ORTF-3D. The PCMA-3D takes a big step forward in terms of sound and spatial imaging. Probably due to the larger distances in the lower layer and the use of five cardioids, it sounds much more natural and also depicts a nicer, more natural room. However, none of the directional setups can match the natural sound and great spatial impression of the 2L cube.

When comparing these 'unprocessed' 3D setups, we had a bit of a trouble with the loudness comparison. Due to the significantly higher low-frequency content of the 2L-Cube compared to all directional setups, a balanced loudness also means that the directional systems also sound brighter than the omnis, which is why the 2L-Cube appears even darker in sound. The [Schoeps] MK-2H used on the lower layer also only have a slight treble boost, which does not compensate for this difference to the cardioids and super-cardioids" (from Gericke and Mielke 2022).

After the above informal evaluation of the sonic character of the various 3D microphone arrays, the authors arrive at the following conclusions: "... We found that none of the 3D setups made us completely happy in the unaltered state. For this reason, we tried to come a little closer to our ideal [with trying some combinations of the above]. For a more precise front localization, the level of the rear spheres in the 2L-Cube should definitely be lowered a bit. Alternatively, you can also work with cardioids or wide cardioids for LS/RS, which capture less of the direct signal from the stage and also sound a bit brighter than the omnis and therefore also ensure better front localization and are less thick at the bottom. Using the cardioid as a center mic makes the center appear a little less thick/dense and makes the base width appear a little wider. But, as a consequence, the trumpet in the orchestral recording was very sharp and direct in loud passages. As you can see, there is of course not the one-and-only true system for 3D miking either. There are many good-sounding setups possible, which everyone would certainly judge differently for themselves. After these listening comparisons, we are sure that we would always use omnis, at least for L and R—just like we do with stereo recordings. For us, they offer the best and most natural sound over the entire frequency spectrum and convey the most beautiful spatial impression. [Rem.: in this respect see the proposal of the AB-BPT-3D 9.1 microphone setup, Sect. 9.6] Conversely, this means that both an ORTF-3D and an OCT-3D are rather unsuitable for music productions, as—at least for our taste—they distort the sound too much through the exclusive use of (strongly) directional microphones. The PCMA-3D is certainly an interesting alternative to the 2L-Cube, although we would prefer to place the height microphones in an upper layer with some distance to the lower layer.

The clear winner in terms of handling is the ORTF-3D: eight microphones in such a small space in an easy-to-use setup are certainly extremely attractive for many applications (e.g. live sports). The Ambeo microphone could also be interesting for certain applications due to its ease of use—but in terms of sound, in our opinion, both are ruled out for recordings of classical music.

With the other microphone positions, we evaluated the advantages and disadvantages of the various directional characteristics differently over the course of our listening tests. In terms of sound, the omnis are great, especially as a single signal, which can be adjusted in terms of level and possibly with filters in terms of sound and taste. It is certainly also conceivable to use the ball attachments for the Schoeps omni capsules or comparable systems from other manufacturers, which achieve increased directivity in the range from 1 to 4 kHz. The use of directional mics for the height and surround channels, on the other hand, has the advantage of picking up less of the direct signal from stage (assuming the musicians aren't positioned in a circle around the main mic).

As a result, the sound impression is more spatial, but also thinner and sometimes significantly brighter. However, there is a greater risk of catching unwanted reflections or direct signals with the directional microphones, which are much less noticeable with omnidirectional microphones. When using nine or even eleven omnis (7.1.4) one has to consider that the low-frequency range appears much louder than with only one comparable stereo-AB or the directional microphones. Certainly the set-up of the musical ensemble and the hall also play an important role in the selection of the 3D setup. For recordings where the musicians are placed in a circle around the main mic, we would definitely use omnis in the bottom layer" (from Gericke and Mielke 2022).

10.8 3D Audio Spaced Mic Array Versus Near-Coincident Versus Coincident Array Comparison (Kamekawa and Marui)

Looking at the rather elaborate multi-channel multi-layer microphone setups in Figs. 10.18, 10.24 and 10.25 we notice that various microphone pattern characteristics are used, from omni via wide-cardioid (WC) to cardioid (C) and super-cardioid (SC), and even Figure-of-Eight in the case of the Hamaski-square.

In connection to the resulting signal correlations of microphone signals (as documented e.g. in Figs. 1.6–1.11 from Chap. 1 as well as Fig. 10.23), of course the question arises, whether there is an overall 'ideal' mic-pattern characteristic, which could be used in 3D-audio microphone system setups, or—at least—if it was possible to relate a particular 'overall sound characteristic' to microphone systems which use essentially only one specific type of mic pattern. In this respect an interesting experiment, including subjective listening test, applying repertory grid technique and MDS (Multi Dimensional Scaling) methodology, has been conducted at Tokyo University of the Arts. The experiment has been carried out by Kamekawa and Marui and deals with a comparison and subjective listener evaluation of 3D-audio microphone systems (Kamekawa and Marui 2020). For details of the experiment see Figs. 10.25, 10.26, 10.27, 10.28 and 10.29.

In their study, the researchers compared three microphone techniques for 22.2 multichannel sound, specifically a spaced microphone array (using 17 omni-microphones and 3 cardioids), a near-coincident microphone array using 24 shortgun-microphones, and a coincident microphone array (First Order Ambisonics FOA). First, the evaluation attributes were extracted by referring to the repertory grid technique. Through this the following ten attributes where selected for listener evaluation: 'rich', 'bright', 'hard', 'wide sound image' (width), 'near', 'clear sound image' (clear), 'listener envelopment' (LEV), 'wide space' (broad), 'more reverberation'

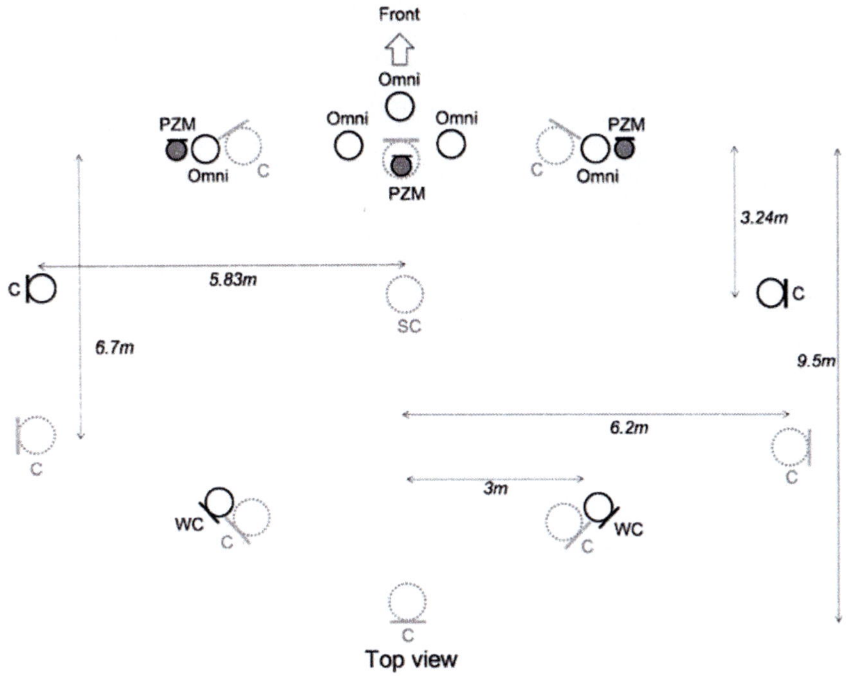

Fig. 10.24 Top view of Howie et al. s 9 + 10 + 3 microphone arrangement (Howie et al 2017) used for recording a large orchestra. The solid black and dotted grey circles represent the middle and upper layer microphones, respectively. The filled grey circles represent bottom layer microphones. This layout uses the Decca Tree as the main microphone array; middle layer height = 3 m, upper layer height = 5.5 m. (C = cardioid, SC = Super cardioid, WC = wide cardioid, PZM = pressure zone microphone) (partial reproduction of Fig. 9 from Lee 2021)

(rev), 'clear localization' (loc), 'the sense of being there' (presence), and 'preference' (pref). Using these attributes, participants had to compare the differences between these microphone techniques, including the difference in the listening position by means of two experiments. From the results it was observed that the difference, depending on the listening position, was the smallest in the spaced array. The FOA seemed to give the impression of sounding 'hard'. The near-coincident array appeared as 'rich' and 'wide', while the spaced array gave the impressions of 'clear' and 'presence'. Furthermore, 'presence' was evaluated from the viewpoints of clarity and richness of reverberation, with a negative correlation with the spectral centroid and a positive correlation with the reflection from lateral and vertical sides. [Rem.: the 'Spectral Centroid' is a measure used to characterize a frequency spectrum. It indicates where the 'Center of mass' of the spectrum is. Perceptually, it has a robust connection with the impression of 'brightness' of a sound (see Grey and Gordon 1978). It is calculated as the weighted mean of the frequencies present in the signal, determined using a Fourier transform, with their magnitudes as weights (see IRCAM 2003; Schubert et al. 2004)].

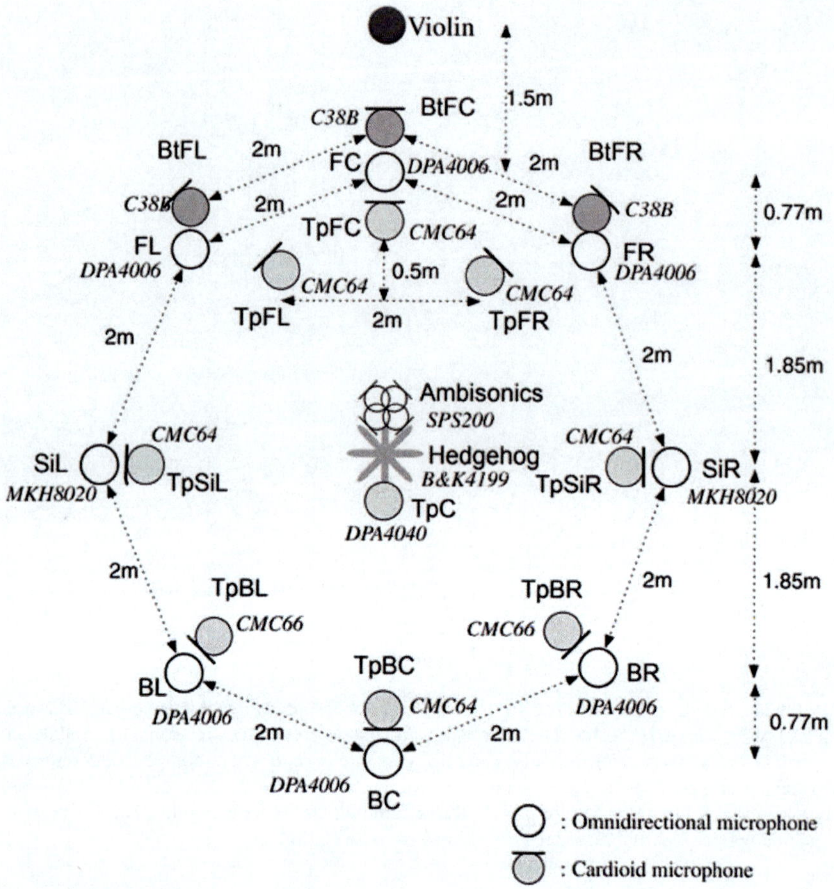

Fig. 10.25 Top view of microphone layout as used in the experiment of (Kamekawa and Marui 2020) Italics indicate the model names of microphone (graphic is upper part of Fig. 1 from Kamekawa and Marui 2020)

10.9 Attempt at a Qualitative Ranking of 3D-Audio Microphone Arrays, Conclusion and Outlook

Before undertaking the daring attempt to arrive at a qualitative ranking of 3D-audio microphone techniques, the author would like to state his conviction that—of course—there is certainly not one particular microphone technique, which could be considered 'best' for all possible recording situations. As most recording engineers will agree, the variables of sound source size and complex (i.e. frequency-dependent) radiation characteristics, room acoustics, external noise sources, the necessity to be able to rebalance individual sound sources in an ensemble, etc., all contribute to the fact that the appropriate microphone technique for capturing a sonic event needs

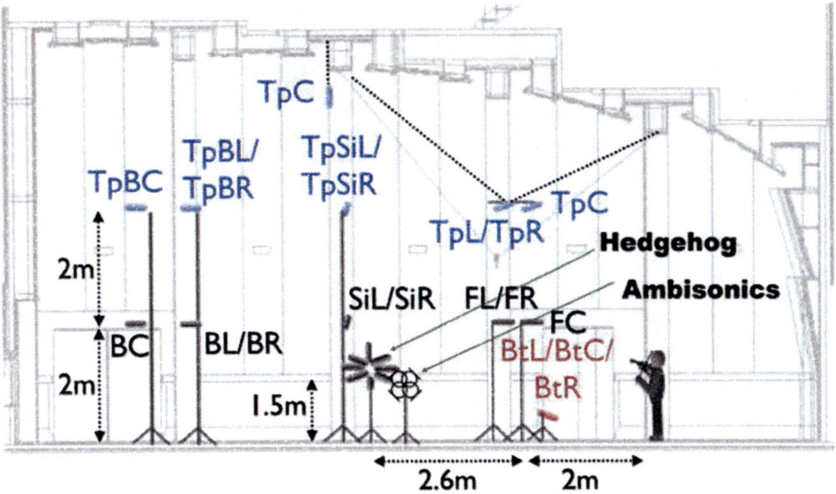

Fig. 10.26 Sectional view of microphone layout as used in Kamekawa and Marui (2020). Symbols such as FL (Front Left) correspond to the channels of the 22.2 multi-channel audio system. "Tp" indicates the top layer and "Bt" indicates the bottom channel (graphic is lower part of Fig. 1 from Kamekawa and Marui 2020)

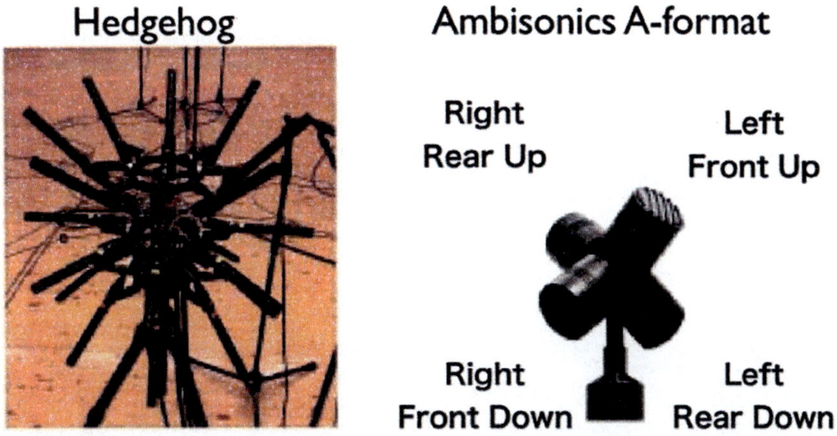

Fig. 10.27 The "Hedghog" (near coincident) and First-Order Ambisonic (coincident) microphone systems as used in the experiment of Kamekawa and Marui (2020) (graphic is Fig. 2 from Kamekawa and Marui 2020)

to be chosen very carefully on an individual case-by-case basis. Therefore, it can happen that a microphone technique does not render satisfactory results in one situation, but may have an advantage over other techniques under different conditions. In listening tests, sometimes the program selection itself (e.g. solo instrument instead

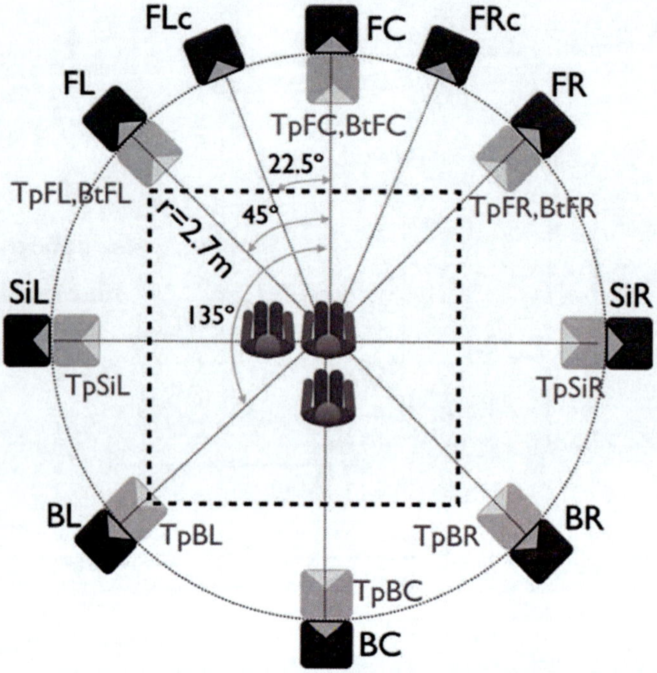

Fig. 10.28 Top view of the loudspeaker setup of listening part of the experiment by Kamekawa and Marui (2020). The participants moved to each of these three listening positions as shown in the above figure (graphic is upper part of Fig. 3 from Kamekawa and Marui 2020)

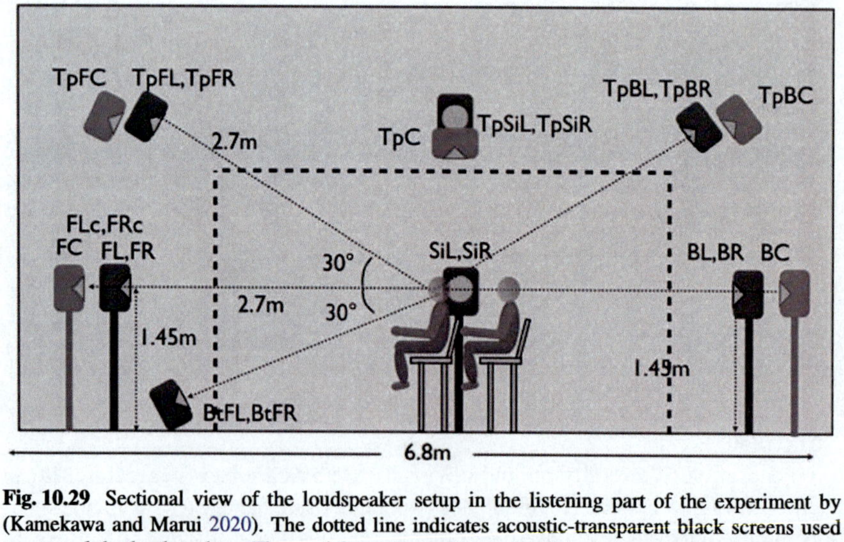

Fig. 10.29 Sectional view of the loudspeaker setup in the listening part of the experiment by (Kamekawa and Marui 2020). The dotted line indicates acoustic-transparent black screens used to conceal the loudspeakers. The participants moved to each of these three listening positions as shown in the figure above. (Graphic is lower part of Fig. 3 from Kamekawa and Marui 2020)

of small ensemble as sound source) may already be a determining factor, without even changing the room acoustics. Nevertheless—with knowledge gained through experience, experiments and well founded scientific understanding of underlying acoustic principles—it is possible to make 'informed decisions' which should be very helpful in choosing an appropriate technique for a particular event beforehand.

As—unfortunately—results from a subjective listening test of the very extensive 3D-audio microphone comparison (3D-Marco) as documented in Lee (2021) and Lee and Johnson (2021) have not been obtained yet, we have to try to 'stitch together' results from various other studies in the field:

The comparative recording by Luthar and Maltezos (2015) (Decca vs. Fukada/OCT-Tree) was an investigation in the effect of the polar pattern of the microphones on perceived spatial impression and cohesiveness of the sound image. For this recording, the authors' opinion was that the set of microphones producing the most natural image in the space were the 9 omnidirectional microphones. The acoustics of the concert hall were ideal, and the omnidirectional microphones were the best at recreating that space. However, with the cello as solo-instrument, it seemed favorable to use a cardioid pattern for the center microphone for better stabilization of the sound image and also to bring out more detail in the cello. On the other hand, using only cardioid patterns for the L, C, R microphones caused an auditory disconnect between the height channels and the horizontal sound stage. "The use of only cardioids had a tendency to flatten the frontal image and the sound was perceived as more aggressive and lacking any sense of 'air'. In general, the height channels helped provide a natural, but improved, spatial impression with a greater sense of envelopment. The radiation patterns of the instruments were reproduced more accurately with the addition of the height channels. For example, the bassoon in the 'woodwind' excerpt was more present and full with the height channels engaged. It was also easier to localize lower frequencies (from the double basses, for example)" (from Luthar and Maltezos 2015)

This revelation coincides very well with the results from the research by Kamekawa and Marui (2020), in which the authors compared three microphone techniques for 22.2 multichannel sound, namely a spaced microphone array (using 17 omni-microphones and 3 cardioids), a near-coincident microphone array using 24 shortgun-microphones, and a coincident microphone array (First Order Ambisonics FOA). Participants had to compare the differences between these microphone techniques, including the difference in the listening position by means of two experiments. It was observed that the difference, depending on the listening position, was the smallest in the spaced array. Also it was found that FOA gave the impression of 'hard'-sounding, the near-coincident array resulted as 'rich' and 'wide', while the spaced array gave the impressions 'clear' and 'presence'.

Part of the findings by Kamekawa and Marui (2020) correspond very well also with the results of a comparative test between 'Twins Square' and Double-MSZ by Ryaboy (2015): "... The [spaced] Twins Square technique demonstrated a more enveloping soundstage, wider ensemble width, and a more spacious environment compared to the Double-MSZ technique, which showed better localization in vertical and horizontal planes, stable imagery, and a more intimate perspective" (Ryaboy 2015).

In Howie et al (2017) 'Howie 3D-Tree' and 'Hamasaki 3D' were compared with 'Higher Order Ambisonics' (HOA). All three techniques were set up for an orchestral recording for 22.2 immersive audio. The 'Howie 3D-Tree' consisted only of omnidirectional microphones, with microphone orientation typically mirroring assigned playback channel orientation: for example, the TpFL microphone would have a horizontal orientation of around 60°, and a vertical orientation of approximately 45°. 'Hamasaki 3D' consists of five supercardioid microphones, placed at equal intervals across the sound stage and ambient sound was captured with an array of laterally oriented bi-directional microphones (see Fig. 10.7). The placement and spacing of the bi-directional microphones ensured minimal capture of direct and rear wall sound, and that the ambient sound field is decorrelated across the audible frequency spectrum. For the third technique, an Eigenmike(R) HOA microphone by mh acoustics, with 32 capsules, was chosen. One conclusion from the results of the study presented in Howie et al. (2017) was that "… the Eigenmike [was] performing poorly for Scene Depth, Environmental Envelopment and Sound Source although a convenient alternative to large spaced microphone arrays, may not yet be suited to professional 3D music recording." Also "… Frequently observed by both subjects and researchers were unpleasant and unnatural 'out of phase' sonic artefacts present in Technique 3 [i.e. HOA]. This may explain why amongst all attributes, Technique 3s mean rating was lowest for 'naturalness'." (from Howie et al. 2017) In conclusion both the 'Howie 3D-Tree' and the 'Hamasaki 3D' were observed to be of significantly higher quality than the 'HOA' (for details on attributes evaluated see Fig. 10.8).

Howie et al. in (2018) addresses another comparative microphone test (concerning the 3D audio recording of solo-piano), including both subjective listening tests as well as objective measures for multichannel audio (This was based on IACC measurements; a second set was derived from a binaural model designed to predict room acoustic attributes, which have proven to correlate with subjective assessments of 'reverberance', 'clarity', 'apparent source width' and 'listener envelopment'. The third set of features were designed to characterize signals' monaural frequency content. These features were the Spectral Centroid, Spectral Crest factor, Spectral Flatness, Spectral Kurtosis, Spectral Skew, Spectral Spread, and Spectral Variation.) Howie concludes this as in the following quote:

> The results of the subjective listening test showed the two spaced techniques as being equally highly rated for the subjective attributes 'naturalness of sound scene', 'naturalness of timbre', and 'sound source image size'. "Technique 3 (OCT-9) was rated lower than both spaced techniques, but higher than technique 4 (Double-MS + Z)." Listeners rated the coincident technique significantly lower than all other techniques under investigation for all perceptual attributes. Binaural recordings of the stimuli were analyzed using several different objective measures, some of which were found to be good predictors for the perceptual attributes 'envelopment' and 'sound source image size'. (from Howie et al. 2018).

If we try to make at a qualitative ranking, in regards to a microphone technique's ability to convey a good sense of spatial impression, based on the above test results in an order which uses mathematical operands we arrive at the following:

10.9.1 Qualitative Ranking

Largely Spaced 3D-Techniques (omnis, wide omnis) > Near Coincident 3D-Techniques (directional, card.) > Coincident 3D-Techniques (Double MS + Z $\geq$ HOA). [rem.: usually associated microphone patterns in round brackets ()]

While the wide spaced techniques usually have the advantage of a more convincing spatial impression (due to sufficient signal de-correlation between channels) which can result in a more enveloping soundstage, wider ensemble width (ASW), and a more spacious environment, the near coincident and coincident techniques at times achieve better localization in vertical and horizontal planes and higher image stability, possibly delivering a more intimate perspective at the expense of spatial impression.

This outcome as such is not a great surprise, as similar results have already been observed in connection with stereo, as well as surround recording techniques. (A detailed analysis on this can, e.g., be found in Chap. 9 of Pfanzagl-Cardone 2020).

As discussed above, the qualitative ranking as outlined above is also reflected in the outcome of the already above mentioned study by Kamekawa and Marui (2020), in which the spaced array was discovered to give the impressions 'clear' and 'presence' (with 'presence' being evaluated from the viewpoints of clarity and richness of reverberation, with a negative correlation with the spectral centroid and a positive correlation with the reflection from lateral and vertical sides), the near-coincident array resulted as 'rich' and 'wide' and the First—Order Ambisonics based system was judged to give the impression of 'hard'—sounding.

A more detailed analysis concerning the differentiation into Large-, Medium and Small-AB spaced microphones systems, taking the 'reverberation radius' (or 'critical distance') of the recording venue into account, is provided in Chap. 3, Sect. 3.3.5 of Pfanzagl-Cardone (2020).

The physical capsule separation in omni (pure pressure transducer) based microphones is of course of major importance for signal correlation [for further details see Fig. 1.2 and 1.24 in Chap. 1 of this book and Chap. 2 on "Correlation and Coherence" in microphone systems in Pfanzagl-Cardone (2020), see Link: *The Art and Science of Surround and Stereo Recording|SpringerLink* (https://link.springer.com/book/10.1007/978-3-7091-4891-4)]. It is known from previous research that a low signal correlation, especially at frequencies below 500 Hz (and even more so below 200 Hz) is a key factor for 'spatial impression' for the human listener (see—among others—Barron and Marshall 1981; Griesinger 1986; Hidaka et al. 1995, 1997; Beranek 2004).

10.9.2 Some Thoughts on Localization and Diffuse Sound

In Theile and Wittek (2011) we find the following statement: "… Diffuse sound (i.e. reverb or background noise) needs to be reproduced diffusely. This can be achieved using Auro-3D if appropriate signals are fed to the extra speakers. Diffuse signals must be sufficiently different on each speaker, that is, they need to be decorrelated over the entire frequency range. A sufficient degree of independence is necessary, in particular, in the low-frequency range as it is the basis of envelopment perception (for an example, see Griesinger 1998). However, increasing the number of channels that need to be independent makes recording more complex. It is a tough job to generate decorrelated signals using first-order microphones—for example, a coincident array such as a double-MS array or a Soundfield microphone allows for generating a maximum of four channels providing a sufficient degree of independence (Wittek et al. 2006). Therefore, the microphone array needs to be enlarged to ensure decorrelation."

The layout of a 3D microphone system will therefore have to aim for overall low signal correlation (over the entire audio bandwidth) for the diffuse sound components in a recording (which can either be achieved by sufficiently large omni-capsule spacings, or the use of appropriate opening angles with directional capsules, e.g. a Blumlein-pair of crossed figure-of-eight microphones) in order to achieve good spatial impression, while at the same time providing solid localization properties, including an appropriate representation of sound source width (ASW) based on the direct sound components.

As a good example for the struggle in achieving both aims we can take a look at the OCT surround microphone system (see Theile 2000, 2001). This system seems to be characterized by good localization properties, but the overall spatial impression is degraded due to high signal-correlation at low frequencies, as evident in Fig. 1.13 (see Chap. 1), which has also been noticed and commented by practicing sound engineers (see Appendix B, p. 395 in Pfanzagl-Cardone (2020), downloadable here as "backmatter" PDF: https://link.springer.com/content/pdf/bbm%3A978-3-7091-4891-4%2F1.pdf).

Further evidence of the relatively high cross-correlation value found with the OCT-technique, due to the rather small capsule distance of only 40 cm between the backward facing cardioid microphones and the front microphones, appears in the measurements of Lee and Johnson (2021), as displayed in Fig. 1.11 (see OCT-3D 'IACCs for the base layer only', E3 early segment in blue).

10.9.3 Combined Microphone Systems

As pointed out in Chapt. 3, Sect. 3.4 of Pfanzagl-Cardone (2020) it is often the combination of more than one 'main-microphone' system (e.g. the combination of large-AB omnis with a Blumlein-Pair or BPT-microphone system) which can provide sonic results that are superior to any of the single systems involved. In case of the large-AB/Blumlein combination the overall system profits of the high sense of spatial impression which large-AB is able to provide and the very high localization accuracy of the Blumlein (or BPT) system.

Arguably, the very elaborate and extensive 3D-microphone arrays by the likes of Howie (see Howie et al. 2016, 2017, 2018) and Hamasaki (see Hamasaki and van Baelen 2015) may also be seen as a combination of more than one microphone system.

But, in reality, of course it is not only the frequency dependent inter-channel correlation coefficients—as caused by physical capsule spacing and applied microphone patterns—which determine the final sonic impression. The adequate positioning (and levelling in the final mix) of spot-microphones, as is common practice for sound engineers, plays a crucial role as well. In this respect it is interesting to note—as already pointed out in Chap. 1 of this book—that sound engineers involved in 3D audio recordings now tend to use more microphones with narrower directivity patterns (like super- or hyper-cardioid) than previously, when working only in stereo or surround. This is likely due to the need of more precise localization of the single sound sources (i.e. instruments) being picked up by the spot microphones, caused by the higher overall amount of diffuse sound in the 3D-microphone main array signals.

10.9.4 Relative Volume Levels for Height and Bottom Layers

Regarding this, an interesting research about the relation between relative levels of the height layer and bottom layer in comparison to the middle layer of a 3D loudspeaker replay system was published in Eaton and Lee (2022). Based on and downmixed from an original $9 + 10 + 3$ (i.e. NHK 22.2) recording of classical orchestral music "... it was found that the preferred levels of the upper and bottom layers were around 4 and 6 dB lower than that of the middle layer, on average." (cited from Eaton and Lee 2022). Assuming that the upper (height) and bottom layers would mainly contain diffuse sound (in addition to first ceiling and floor reflections, at least for the front channels of both layers), while the middle layer contains mainly direct sound, this result already gives an indication of appropriate balancing of direct versus diffuse sound.

In this context it is interesting to mention the approach adhered to by Morten Lindberg with his '2L'-technique, for which he maintains correct gain staging (with unity gain for all microphones involved) from the recording until mastering in order to ensure authentic reproduction of direct as well as diffuse sound components (For more details see Chap. 9, Sect. 9.13).

In 2-channel stereo direct sound, as well as diffuse sound need to be reproduced through the same loudspeakers. Multichannel sound—with a 5.1 speaker layout, for example—enables the recording engineer to largely split the reproduction of direct sound from the reproduction of diffuse sound by putting the microphone signals intended for direct sound pickup in the front speakers, while the microphones for diffuse sound pickup are being routed to the rear speakers. For 3D audio—which uses height speakers in addition to the horizontal layer of surround speakers—this task is more refined and the capturing of the diffuse sound-field with an appropriate microphone technique is of even higher importance.

Further studies on the effects of diffuse-field correlation would certainly be useful for determining the required minimum spacing and angles of microphone pairs. In

Riekehof-Böhmer et al. (2010) coincident, equivalent, and delay-based main stereo microphone techniques have been examined in respect to their suitability for eliminating diffuse-field correlation. For this, the term DFI (Diffuse Field Image) predictor has been introduced and the outcome was that only six of the examined microphone arrays were fulfilling the requirement of sufficient diffuse-field decorrelation: the Blumlein pair array, consisting of two coincident figure-eight microphones, angled at $\pm$ 45° in relation to the sound source), two equivalent cardioid arrays, and three delayed omnidirectional arrays where the microphones were placed more than 35 cm apart (see Chap. 1, Fig. 1.1).

Further research on diffuse-field correlation (DFC) in various microphones techniques can be found in Wittek (2012) (see Chap. 1, Fig. 1.2).

10.9.5 Introducing an Artificial Head as 'Human Reference'

The results of an empirical study regarding frequency dependent signal correlation (FCC) concerning not only stereo, but also surround microphone systems have first been published in (Pfanzagl-Cardone and Höldrich 2008) as well as (Pfanzagl-Cardone 2011). A more in-depth analysis can also be found in Chaps. 2, 7 and 8 of Pfanzagl-Cardone (2020): In order to make the FCC (Frequency dependent Cross-correlation Coefficient) more meaningful, an artificial human head has been introduced during the recording- (see Fig. 10.30) and re-recording process (see Fig. 10.31), and—for simulation—a software plug-in, using the HRTFs of the KEMAR dummy head) to serve as a 'human reference' when measuring the FIACC (Frequency-dependent Inter-Aural Cross-correlation Coefficient)—see Chap. 8 in Pfanzagl-Cardone (2020) for more details, as well as Pfanzagl-Cardone and Höldrich (2008).

By measuring signal cross-correlation between channels of the various microphone techniques, it has been shown that every stereo and surround microphone technique can be characterized by such a 'sonic fingerprint'. See Fig. 10.32 for an analysis of signal-correlation between adjacent channels of the OCT-surround microphone system—[or the binaural equivalent FIACC 'Frequency dependent Inter Aural Cross-correlation Coefficient' of the final result of replaying the recordings via such microphone arrays over a standard 2-channel stereo or 5.1 surround loudspeaker setup; see Fig. 10.33 for such measurements as carried out for OCT, DECCA, KFM and ABPC: more details can be found in Chapt. 7 of Pfanzagl-Cardone (2020)]. The specific pattern of signal correlation/decorrelation in the various frequency ranges is quite meaningful and can serve as a first indicator for the sonic features of the respective microphone technique under test.

In this context it would also be interesting to conduct more research on the defined BQIrep ('Binaural Quality Index for reproduced music') (see Chap. 1 of this book for more details), which intends to serve as a measure for the quality of reproduced music. In the example of the Salzburg Festival 'Grand Hall' it was shown that it is possible to re-create a sound-field, very similar to that of a best-seat position in

Fig. 10.30 Neumann KU81i artificial human head set up in the sixth row at the Salzburg Festival hall for a binaural recording (Fig. 6.3 from Pfanzagl-Cardone 2020)

the concert-hall, by using an appropriate microphone technique that takes sufficient signal-decorrelation into account (see Chap. 11 in Pfanzagl-Cardone 2020 as well as Pfanzagl-Cardone 2012). In Pfanzagl-Cardone and Höldrich (2008) and (Pfanzagl-Cardone 2020, Chap. 6, p. 207 ff), it was shown that listener preference is primarily related to the perceived 'naturalness', which in turn is highly correlated to sound color. Also, preference is highly correlated to spaciousness, as well as sound color. (For a listing of the exact values of these attribute correlations please see Table 1.1 in Chap. 1 of this book) In combination with the results of the acoustic measurements and by means of correlation analysis of the results of subjective listening test (see Pfanzagl-Cardone 2011 as well as Pfanzagl-Cardone 2012) it has become clear that 'listener preference' actually goes toward those surround microphone systems, which were also the best in approximating the binaural signal correlation characteristics (FIACC) of the original sound field in the concert hall (for details see Pfanzagl-Cardone 2012) by trying to optimize a physical re-creation of the sound field of the concert hall also in the listening room. (Rem.: this was essentially achieved by combining the

Fig. 10.31 Re-recording of the surround microphone signals by means of a Neumann KU81i artificial human head in the sweet spot of the control room of the IEM—Institute of Electronic Music and Acoustics at the KUG University of Music and Performing Arts, Graz, Austria (rear-channel loudspeakers outside of the picture; rem: clothes placed on and around the mixing desk to avoid unwanted reflections) (Fig. 6.6 from Pfanzagl-Cardone 2020)

high spatial impression provided by a large-AB main mic technique with the highly accurate localization properties of a center-fill system in the form of an ORTF-Triple for the front microphone arrangement, as used in the AB-PC technique (documented in Pfanzagl-Cardone 2002; Pfanzagl-Cardone and Höldrich 2008; Pfanzagl-Cardone 2020). A similar approach can also be found in the combination of large AB with a Blumlein-Pfanzagl-Triple (BPT) as center-fill (see Pfanzagl-Cardone and Höldrich 2008).

The outcome of the above mentioned subjective listening test statistics-data correlation analysis, which can be found in Chap. 1, actually finds its parallels in other studies (Berg and Rumsey 2001a, b; Gabrielsson and Lindström 1985). Among others, the correlation found between the attributes 'preference' and 'envelopment' in Berg and Rumsey (2001a, b) coincides well with values for the correlation between 'preference' and 'spaciousness' of the study the results of which have been published in Pfanzagl-Cardone and Höldrich (2008), as well as Pfanzagl-Cardone (2012). Also, the high correlation between 'spaciousness' and 'naturalness' that was found matches well with the results of previous research by Toole (1985).

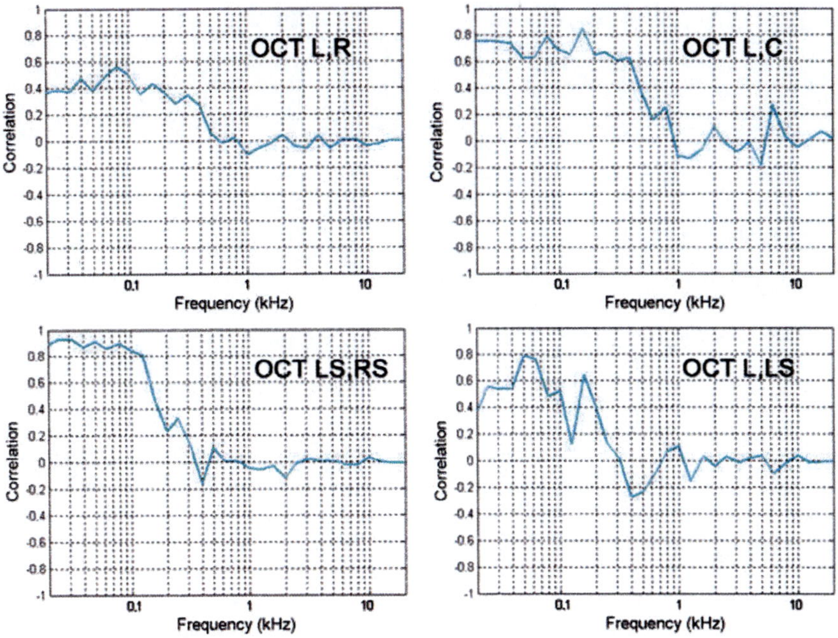

Fig. 10.32 Paired channel correlation over frequency (i.e. FCC) for OCT surround microphone system signals (2048-point DFT grouped to 31 frequency bands, 1/3rd octave; center frequencies according to ISO; orchestral music, 60 s, concert hall); capsule spacing for L/C and C/R is 31 cm, capsule spacing for L/LS, R/RS is 41 cm, capsule spacing for LS/RS about 80 cm (Fig. 7.13 from Pfanzagl-Cardone 2020)

10.9.6 *BQI_{rep}—Binaural Quality Index of Reproduced Music*

The BQI_{rep}, being mainly related to the FIACC of certain frequency bands is therefore also a strong indicator for the quality of spatial impression and hence of major interest for the creation of convincing 3D audio recordings (see Chap. 1, Sect. 1.5 for more details) (Rem.: the original 'Binaural Quality Index' BQI, defined by Keet in 1968, has been identified as "one of the most effective indicators of the acoustic quality of concert halls" by Beranek in 2004).

A similar approach to the research documented in Pfanzagl-Cardone and Höldrich (2008) as well as Pfanzagl-Cardone (2011), relating the outcome of the commonly found subjective listening tests with the results of acoustic signal analysis -based on binaural dummy head signals- can be found in Howie et al. (2018) as well, including inter-aural cross-correlation coefficient (IACC) and spectral centroid measurements, among others.

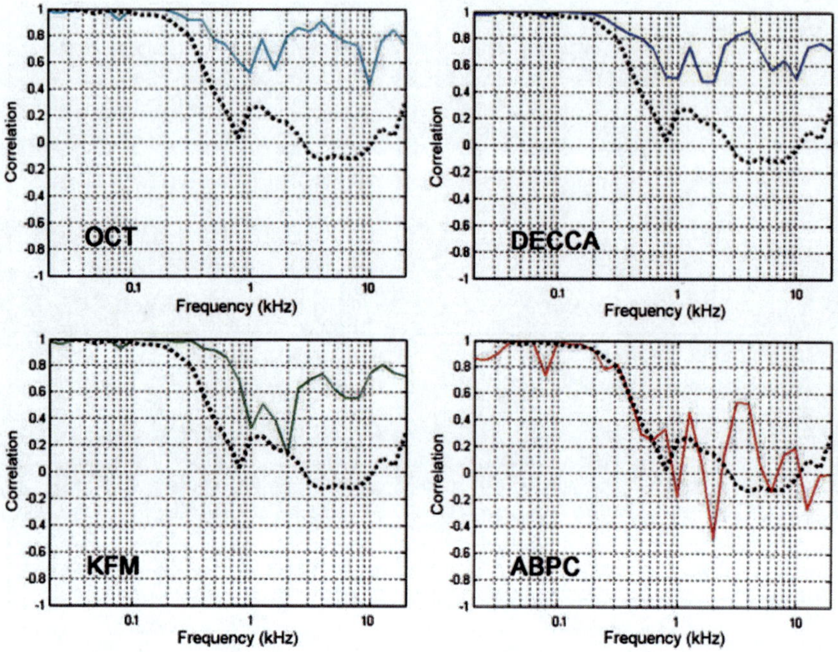

Fig. 10.33 FIACC of various surround-microphone techniques (OCT, DECCA, KFM, ABPC) as re-recorded through a Neumann KU81i artificial head (solid line) compared with the original FIACC, recorded through the same artificial head in a "best seat" position (see Fig. 10.30) in the concert hall (dotted line); (orchestral sample of 60 s duration, soft to loud dynamics) (from Fig. 7.4 in Pfanzagl-Cardone 2020)

10.9.7 FCC (Frequency-Dependent Cross-Correlation) and FIACC (Frequency Dependent Inter Aural Cross-Correlation Coefficient) in 3D Audio Recordings

To make measurements (or calculations) of the FCC's of all neighboring channels (like it has been done for the signals of the OCT-surround microphone system in Fig. 10.32) for multichannel 3D-audio recordings would be a very labor intensive task. Also, in 3D audio each loudspeaker of course contributes less to the 'overall sonic picture' which results at the ears and brain of the listener, as multi-channel loudspeaker layouts are employing at least 9 full-range speakers (with Auro/Atmos 9.1, for example; see front part of studio in Fig. 10.34) or even many more in the case of a Hamasaki 22.2 layout or larger scale cinematic Dolby Atmos /DTS:X/Auro-3D systems.

Therefore it may be a more efficient approach to measure, record or simulate (see Howie et al. 2018) the resulting FIACC at the ear entrances of a dummy head, also in the case of 3D audio recordings (as it has already been done for various

Fig. 10.34 Front part of the 9.1 3D-audio monitoring loudspeaker setup at the Salzburg Festival hall control room, made up of Dynaudio BM6A with BM10S subwoofer (in addition: a pair of Bowers and Wilkins 801 (Series 80) floor standing reference loudspeaker for stereo mixing; Yamaha digital mixing console and the book-author)

surround microphone systems; in Fig. 10.33 one can see that the signals of the ABPC microphone system deliver the best match in terms of frequency dependent cross-correlation coefficient to the original dummy head recording from the concert hall. For more details see Pfanzagl-Cardone (2020), Chap. 7).

A short series of educational videoclips concerning microphone technique analysis (mainly for 2-channel stereo microphone techniques, but also including 3-channel techniques like DECCA and BPT), based on FCC measurements using the "2BCmultiCORR" frequency dependent signal cross-correlation plug-in can be found on the author's Youtube channel "futuresonic100", when searching for 'mic tech analysis' or 'Nevaton BPT'.

10.9.8 Conclusion and Outlook

Especially with 3D-Audio—for which the number of channels interacting with each other has been augmented significantly in relation to 2D surround sound—it does make very much sense to make use of the above mentioned 'human reference' (binaural dummy head recording) in the form of measuring the FIACC (or BQI_{rep}, derived from it) to arrive at a qualitative evaluation of a respective microphone technique: no matter how many reproduction loudspeaker channels are involved as part of the 3D replay setup—finally the sound-vibrations have to enter the two ear canals of the human head, so the resulting FIACC measured between the ear-drums is what counts in the end …

References

Barbour JL (2003) Elevation perception: phantom images in the vertical hemisphere. In: Proceedings to the 24th audio engineering society international; conference: multichannel audio. The New Reality

Barrett N (2012) The perception, evaluation and creative application of higher order Ambisonics in contemporary music practice. Ircam Composer Res Rep 2012:1–6

Barron M, Marshall AH (1981) Spatial impression due to early lateral reflections in concert halls: the derivation of a physical measure. J Sound Vibr 77:211–232

Bates E, Gorzel M, Ferguson L, O'Dwyer H, Boland FM (2016) Comparing Ambisonics microphones: part 1. Paper presented at the audio engineering society conference on sound field control, Gilford

Bates E, Doonery S, Gorzel M, O'Dwyer H, Ferguson L, Boland FM (2017) Comparing Ambisonics microphones—part 2. Paper 9730 presented at the 142nd audio engineering society convention, Berlin

Beranek L (2004) Concert halls and opera houses: music, acoustics and architecture, 2nd edn. Springer, New York

Berg J, Rumsey F (2001a) Verification and correlation of attributes used for describing the spatial quality of reproduced sound. Paper presented at the audio engineering society 19th international conference

Berg J, Rumsey F (2001b) Optimization and subjective assessment of surround sound microphone arrays. Paper 5368 presented at the 110th audio engineering society convention, Amsterdam

Bowles D (2015) A microphone array for recording music in surround-sound with height channels. Paper 9430 presented at the 139th audio engineering society convention, New York

Camerer F, Sodl C (2001) Classical music in radio and TV—a multichannel challenge. The IRT/ORF surround listening test. http://www.hauptmikrofon.de/stereo-surround/orf-surround-techniques. Accessed 31 May 2019

Chapman M et al (2009) A standard for interchange of Ambisonic signal sets. In: Ambisonics symposium 2000, Graz, Austria

Choisel S, Wickelmaier F (2006) Relating auditory attributes of multichannel sound to preference and to physical parameters. Paper presented at the 120th audio engineering society convention, Paris

Daniel J (2009) Evolving Views on HOA: From Technological to Pragmatic Concerns. In: Ambisonics Symposium 2009, Graz, Austria, June 2009, pp 1–18

de Keet WV (1968) The influence of early lateral reflections on spatial impression. In: 6th international congress on acoustics, Tokyo

Eaton C, Lee H (2022) Subjective evaluations of three-dimensional, surround and stereo loudspeaker reproductions using classical music recordings. Acoust Sci Tech 43(2):149–161

Fukada A (2001) A challenge in multichannel music recording. Paper presented at the19th international conference of the audio engineering society, Schloss Elmau, Germany

Gabrielsson A, Lindström B (1985) Perceived sound quality of high-fidelity loudspeakers. J Audio Eng Soc 33(1/2):33

Geluso P (2012) Capturing height: the addition of Z microphones to stereo and surround microphone arrays. Paper 8595 presented at the 132nd audio engineering society convention

George S et al (2010) Development and validation of an unintrusive model for predicting the sensation of envelopment arising from surround sound recordings. J Audio Eng Soc 58(12):1013–1031

Gericke H, Mielke O (2022) 3D-Haupmikrofon-setups im Hörvergleich. Magazin Des Verbands Der Deutschen Tonmeister 3:35–39

Gerzon M (1973) Periphony: with-height sound reproduction. J Audio Eng Soc 21(1)

Grey JM, Gordon JW (1978) Perceptual effects of spectral modifications on musical timbres. J Acoust Soc Am 63(5):1493–1500

Griesinger D (1986) Spaciousness and localization in listening rooms and their effects on the recording technique. J Audio Eng Soc 34(4):255–268

Griesinger D (1997) Spatial impression and envelopment in small rooms. Paper 4638 presented at the 103rd audio engineering society convention

Griesinger D (1998) General overview of spatial impression, envelopment, localization and externalization. In: Proceedings to the audio engineering society 15th international conference on small room acoustics, Denmark, pp 136–149

Hamasaki K et al (2004) Advanced multichannel audio systems with Superior Impressions of Presence and Reality. Paper presented at the 116th Audio Eng Soc Convention, Berlin, May 2004

Hamasaki K, Hiyama K (2003) Reproducing spatial impression with multichannel audio. Paper presented at the audio engineering society 24th international conference on multichannel audio. The New Reality, Banff, Canada

Hamasaki K, Van Baelen W (2015) Natural sound recording of an orchestra with three-dimensional sound. Paper presented at the 138th audio engineering society convention, Warsaw

Heller A, Benjamin EM (2014) The Ambisonic decoder toolbox: extensions for partial-coverage loudspeaker arrays. In: Linux audio conference 2014, Karlsruhe

Heller A, Lee R, Benjamin EM (2008) Is my decoder Ambisonic? Paper presented at the 125th convention of the audio engineering society, San Francisco, USA

Hidaka T, Beranek L, Okano T (1995) Interaural cross-correlation, lateral fraction, and low- and high-frequency sound levels as measures of acoustical quality in concert halls. J Acoust Soc Am 98(2)

Hidaka T, Beranek L, Okano T (1997) Some considerations of interaural cross correlation and lateral fraction as measures of spaciousness in concert halls. In: Ando Y, Noson D (eds) Music and concert hall acoustics. Academic Press, London

Howie W, King R (2015) Exploratory microphone techniques for three-dimensional classical music recording. Convention E-Brief presented at the 138th audio engineering society convention, Warsaw, Poland

Howie W, King R, Martin D (2016) A three-dimensional orchestral music recording technique, optimized for 22.2 multichannel sound. Paper 9612 presented at the 141st audio engineering society convention, Los Angeles

Howie W, King R, Martin D, Grond F (2017) Subjective evaluation of orchestral music recording techniques for three-dimensional audio. Paper 9797 presented at the 142nd audio engineering society convention, Berlin

Howie W, Martin D, Benson DH, Kelly J, King R (2018) Subjective and objective evaluation of 9ch three-dimensional acoustic music recording techniques. Paper presented at the audio engineering society international conference on spatial reproduction. Aesthetics and Science, Tokyo

Ikeda M et al (2016) New recording application for software defined media. Paper presented at 141st audio engineering society convention, Los Angeles

IRCAM (2003) A large set of audio features for sound description. Section 6.1.1, Technical Report IRCAM

ITU Recommendation ITU-R BS.775-3 (2012) Multichannel stereophonic sound system with and without accompanying picture. International telecommunications union 08-2012

ITU-R BS.2051-0 (2014) Advanced sound system for programme production

Kamekawa T, Marui A (2020) Evaluation of recording techniques for three-dimensional audio recordings: comparison of listening impressions based on difference between listening positions and three recording techniques. J Acoust Sci Tech 41:1

Kearney G (2017) SADIE binaural database. Available at: https://www.york.ac.uk/sadie-project/binaural.html, 2016, https://doi.org/10.15124/89aac2f4-2caa-4c01-8dcb-6771635e1a1c, online. Accessed 19 June 2017

King R et al (2016) A survey of suggested techniques for height channel capture in multi-channel recording. Paper presented at the 140th audio engineering society convention, Paris

kolor.com (2017) GoPro VR Player. http://www.kolor.com/gopro-vr-player. Accessed 8 Aug 2017

Kronlachner M (2014) Plug-in suite for mastering the production and playback in surround sound and ambisonics. Paper presented at the 136th audio engineering society convention (student design competition), Berlin

Lee H (2011) The relationship between interchannel time and level differences in vertical sound localization and masking. Paper 8556 presented at the 131st audio engineering society convention, New York

Lee H (2012) Subjective evaluations of perspective control microphone array (PCMA). Paper presented at the 132nd audio engineering society convention

Lee H (2017) Capturing and Rendering 360° VR audio using cardioid microphones. http://eprints.hud.ac.uk/29582/1/Audio.Eng.Soc.com. Accessed 9 June 2017

Lee H (2019) 3D microphone array comparison (3D-MARCo). Resolution magazine V18.7 (December)

Lee H (2021) Multichannel 3D microphone arrays: a review. J Audio Eng Soc 69(1/2):5–26. https://doi.org/10.17743/jaes.2020.0069

Lee H, Gribben C (2013) On the optimum microphone array configuration for height channels. e-Brief 93 presented at audio engineering society convention

Lee H, Gribben C (2014) Effect of vertical microphone layer spacing for a 3D microphone array. J Audio Eng Soc 62(12):870–884

Lee H, Johnson D (2021) 3D microphone array recording comparison (3D-MARCo): objective measurements. https://doi.org/10.5281/zenodo.4018009

Lindberg M (2014) 3D recording with the '2L-cube'. http://www.2l.no/artikler/2L-VDT.pdf. Accessed 10 Feb 2022

Luthar M, Maltezos E (2015) Polar pattern comparisons for the left, center, and right channels in a 3-D microphone array. Paper presented at the 139th convention of the audio engineering society, New York

Martin B et al (2016a) Microphone arrays for vertical imaging and three-dimensional capture of acoustic instrument. Paper presented at conference of the audio engineering society on sound field control, Guilford

Martin B et al (2016b) Subjective graphical representation of microphone arrays for vertical imaging and three-dimensional capture of acoustic instruments, part I. Paper presented at the 141th audio engineering society convention, Los Angeles

Mason R, Rumsey F (2000) An assessment of the spatial performance of virtual home theatre algorithms by subjective and objective methods. Paper 5137 presented at the 108th audio engineering society convention, Paris, May

Mason R, Rumsey F (2002) A comparison of objective measurements for predicting selected subjective spatial attributes. Paper 5591 presented at the 112th audio engineering society convention, Munich

Olabe I (2014) Técnicas de grabación de música clásica. Evolución histórica y propuesta de nuevo modelo de grabación. Dissertation, Universitat de les Illes Balears. http://hdl.handle.net/10803/362938. Accessed 12 Oct 2015

Peeters G et al (2011) The timbre toolbox: extracting audio descriptors from musical signals. J Acoust Soc Am 130(5):2902–2916

Pfanzagl-Cardone E (2002) In the light of 5.1 surround: why AB-PC is superior for symphony-orchestra recording. Paper 5565 presented at the 112th audio engineering society convention, Munich

Pfanzagl-Cardone E (2011) Signal-correlation and spatial impression with stereo- and 5.1 surround-recordings. Dissertation, University of Music and Performing Arts, Graz, Austria. https://iem.kug.ac.at/fileadmin/media/iem/altdaten/projekte/dsp/pfanzagl/pfanzagl_diss.pdf. Accessed Oct 2018

Pfanzagl-Cardone E (2012) 'Naturalness' and related aspects in the perception of reproduced music. In: Proceedings to the 27. Tonmeistertagung des VTD, Köln

Pfanzagl-Cardone E (2020) The art and science of surround and stereo recording. Springer-Verlag GmbH Austria.https://doi.org/10.1007/978-3-7091-4891-4

Pfanzagl-Cardone E, Höldrich R (2008) Frequency-dependent signal-correlation in surround- and stereo-microphone systems and the Blumlein-Pfanzagl-Triple (BPT). Paper 7476 presented at the 124th audio engineering society convention, Amsterdam

Power PJ (2015) Future spatial audio: subjective evaluation of 3D surround systems. Dissertation, University of Salford, UK

Power P et al (2014) Investigation into the impact of 3D surround systems on envelopment. Paper presented at the 137th audio engineering society convention, Los Angeles

Riaz H, Stiles M, Armstrong C, Chadwick A, Lee H, Kearney G (2017) Multichannel microphone array recording for popular music production in virtual reality. e-brief presented at the 134rd audio engineering society convention, New York

Riekehof-Böhmer H, Wittek H, Mores R (2010) Voraussage der wahrgenommenen räumlichen Breite einer beliebigen stereofonen Mikrofonanordnung. In: Proceedings to the 26. Tonmeistertagung des VDT, pp 481–492

Rossing TD (2007) Acoustics in halls for speech and music. Springer Handbook of Acoustics. Springer New York

Rumsey F (2002) Spatial quality evaluation for reproduced sound: terminology, meaning, and a scene-based paradigm. J Audio Eng Soc 50(9):651–666

Ryaboy A (2015) Exploring 3D: a subjective evaluation of surround microphone arrays catered for Auro-3D reproduction system. Paper 9431 presented at the 139th convention of the audio engineering society, New York

Schoeps Mikrofone (2017a) ORTF-3D outdoor set. http://www.schoeps.de/en/products/ortf-3D-outdoor-set. Accessed 11 June 2017a

Schoeps Mikrofone (2017b) IRT cross. http://www.schoeps.de/en/products/irt-cross-set. Accessed 12 June 2017b

Schubert E, Wolfe J, Tarnopolsky A (2004) Spectral centroid and timbre in complex, multiple instrumental textures. In: Proceedings to the 8th international conference on music perception and cognition, North Western University, Illinois

Sennheiser (2017) Sennheiser AMBEO VR microphone. http://en-uk.sennheiser.com/microphone-3d-audio-ambeo-vr-mic. Accessed 12 June 2017

spook.fm (2017) SpookSyncVR. http://www.spook.fm/spooksyncvr/. Accessed 20 Aug 2017

Stiles M (2018) Recording spatial audio. Resolution 17(2):49–51

Theile G (2000) Mikrofon- und Mischungskonzepte für 5.1 Mehrkanal-Musikaufnahmen. In: Proceedings to the 21. Tonmeistertagung des VDT, Hannover, p 348

Theile G (2001) Multichannel natural music recording based on psychoacoustic principles. In: Proceedings to the 19th international conference of the audio engineering society, pp 201–229

Theile G, Wittek H (2011) Principles in surround recordings with height. Paper 8403 presented at the 130th audio engineering society convention, London

Theile G, Wittek H (2012) 3D audio natural recording. In: Proceedings to the 27. Tonmeistertagung des VDT, Cologne, pp 731

Toole FE (1985) Subjective measurements of loudspeaker quality and listener performance. J Audio Eng Soc 33(1/2):2–32

Wallis R, Lee H (2016) Vertical stereophonic localization in the presence of interchannel crosstalk: the analysis of frequency-dependent localization thresholds. J Audio Eng Soc 64(10):762–770

Wallis R and Lee H (2017) The reduction of vertical interchannel crosstalk: the analysis of localisation thresholds for natural sound sources. Appl Sci 7(278)

Williams M (1991) Microphone arrays for natural multiphony. Paper 3157 presented at the 91st audio engineering society convention, New York

Williams M (2003) Multichannel sound recording practice using microphone arrays. Paper presented at the audio engineering society 24th international conference on multichannel audio. The New Reality

Williams M (2012) Microphone array design for localization with elevation cues. Paper presented at the 132nd audio engineering society convention, Budapest

Williams M (2013) The psychoacoustic testing of the 3D multiformat microphone array design, and the basic isosceles triangle structure of the array and the loudspeaker reproduction configuration. Paper 8839 presented at the 134th audio engineering society convention, Rome

Williams M (2016) Microphone array design applied to complete hemispherical sound reproduction—from Integral 3D to Comfort 3D. Paper presented at the 140th audio engineering society convention, Paris

Wittek H (2002) Image assistant V2.0. http://www.hauptmikrofon.de. Accessed 24 June 2008

Wittek H (2012) Mikrofontechniken für Atmoaufnahmen in 2.0 und 5.1 und deren Eigenschaften. In: Proceedings to the 27. Tonmeistertagung des VDT, Cologne, p 804

Wittek H, Theile G (2017) Development and application of a stereophonic multichannel recording technique for 3D Audio and VR. Paper presented at the 143rd audio engineering society convention, New York

Wittek H, Haut C, Keinath D (2006) Doppel-MS—eine surround-Aufnahmetechnik unter der Lupe. In: Proceedings to the 24. Tonmeistertagung des VDT, Leipzig

Woszczyk W (1990) Acoustic pressure equalizers. Pro Audio Forum pp 1–24

Zacharov N et al (2016) Next generation audio system assessment using the multiple stimulus ideal profile method. Paper presented at the 8th international conference on quality of multimedia experience, Lisbon, Portugal

Zotter F, Frank M (2012) All-round Ambisonic panning and decoding. J Audio Eng Soc

Index

E. Pfanzagl-Cardone, *The Art and Science of 3D Audio Recording*,
https://doi.org/10.1007/978-3-031-23046-2

T

U

V